EVOLUTIONARY CLASSIFICATION AND ENGLISH-BASED NOMENCLATURE IN CRETACEOUS PLANKTIC FORAMINIFERA

EARTH SCIENCES IN THE 21ST CENTURY

Additional books in this series can be found on Nova's website
under the Series tab.

Additional e-books in this series can be found on Nova's website
under the e-book tab.

EVOLUTIONARY CLASSIFICATION AND ENGLISH-BASED NOMENCLATURE IN CRETACEOUS PLANKTIC FORAMINIFERA

M. DAN GEORGESCU

AND

CHARLES M. HENDERSON

EDITORS

nova publishers

New York

Library of Congress Cataloging-in-Publication Data

Evolutionary classification and English-based nomenclature in Cretaceous planktic foraminifera / editors, M. Dan Georgescu and Charles M. Henderson (University of Calgary, Department of Geosciences, Calgary, Alberta, Canada).
    pages cm. -- (Earth sciences in the 21st century)
  Includes index.
  ISBN 978-1-63321-959-5 (hardcover)
  1. Foraminifera, Fossil--Classification. 2. Foraminifera, Fossil--Nomenclature. 3. Protista, Fossil--Classification. 4. Protista, Fossil--Nomenclature. 5. Plankton, Fossil. 6. Paleobiology. I. Georgescu, M. Dan., editor. II. Henderson, Charles Murray, 1956- editor.
  QE772.E96 2014
  561'.994--dc23
                            2014034857

Published by Nova Science Publishers, Inc. † New York

CONTENTS

PREFACE

The year 1551 is paramount in paleontology because the work in three books *De Re Metallica, Hoc Est, de Origine, Varietate, & Natura Corporum Metallicorum, Lapidum, Gemmarum, atq ; aliarum, que ex fodinis, eruuntur, rerum, ad Medicine usum deservientium* by Christophorus Encelius was published in Frankfurt. This work includes the first unequivocal scientific illustrations of fossils: one bivalve and one gastropod. Surprisingly instead of considering them among the molluscs, Encelius included them in the Chelonitid category, which was named and briefly described by Pliny the Elder as resembling a tortoise. The answer to why Encelius assigned two fossil mollusc shells to the Chelonitid group can be only found through a thorough documentation of the period literature. Five years before Encelius a work of extraordinary influence was published: *De Natura Fossilium* by Georg Bauer in which the author described the fossils based solely on their resemblance with other objects. Now it appears logical why Encelius assigned the two illustrated shells to Chelonitids in the light of Georg Bauer's new vision of recognizing fossils: the hard body parts of a tortoise, a bivalve and a gastropod shell are convex on one side, concave on the other and hollow at the interior. Besides the merit of being probably the earliest classification of fossils, Christophorus Encelius gave us an early example of anomalies that can occur when only form is taken into account into classification.

In 1859, with the publication of *The Origin of Species* by Charles Robert Darwin, a shift occurred at paradigm level in understanding changes in the living world on Earth. One of the most important components in the argumentation for the Theory of Evolution was prefigured in 1855 by Alfred Russel Wallace in an article titled *On the Law which has regulated the Introduction of New Species* where the hypothesis of the Theory of Evolution was enunciated. This component is represented by those data collected from the fossil record in which the shift in paradigm is clearly visible: species are not unchangeable units, but evolving ones and connected through ancestor-descendant relationships. Despite this breakthrough, fossil classification did not advance much in concept and practice in over one hundred fifty years and much of this stagnation is due to the development of the Rules and then the Codes of nomenclature in the nineteenth and twentieth century respectively. Nomenclatural stability and classification practicality were invoked often by many of the advocates of the thorough use of these codes and affected the very foundation of paleontological and biological classification such that experimentation was drastically reduced and frequently denied.

Foraminiferal classification was Aristotelian in nature since its beginnings and the developments in foraminiferal evolution followed a similar path. Change came only late and was generated by the extensive use of the scanning electron microscope on specimens collected from throughout the stratigraphic ranges of various species. It became possible to understand that evolution can be observed with relative ease at the level of test microstructure such as ornamentation, wall ultrastructure and porosity-related features. The evolutionary classification that emerged from these advances in understanding the fossil record is based upon lineages rather than species, and a new nomenclature system in English is herein developed for it.

It is a privilege for us to acknowledge the development of a publication environment that encourages and supports innovation in science by Nova Science Publishers. Such an environment is an indispensable condition for the advancement of science, and the results in the foraminiferal articles published during the last three years by Nova Science Publishers demonstrate this well.

Calgary, July 7, 2014

PART 1: LINEAGE REVISIONS IN EVOLUTIONARY CLASSIFICATION

In: Evolutionary Classification ...   ISBN: 978-1-63321-959-5
Editors: M. Dan Georgescu and C. M. Henderson   © 2015 Nova Science Publishers, Inc.

Chapter 1

NEW DATA AND INSIGHTS ON THE POLYPHYLETIC ORIGIN OF THE CRETACEOUS PLANKTIC FORAMINIFERA

M. Dan Georgescu[*], *Randall M. Burke and Caterina J. Heikkinen*

Department of Geosciences, University of Calgary, Calgary, Alberta, Canada

ABSTRACT

New data from the Cretaceous help in understanding the polyphyletic nature of the planktic foraminifera. The new species *Praeplanctonia atlantis* Burke and Georgescu from the lower upper Albian sediments of the eastern North Atlantic Ocean (offshore Morocco) presents intermediate morphological features between the pleurostomellid benthics and *P. globifera* Georgescu 2009, which is the earliest known species of *Praeplanctonia*. Additional data on the benthic foraminifera with biserial, biserial-triserial and trochospiral chamber arrangement seemingly suggest multiple originations of foraminifera with planktic habitat during the Cretaceous. A new taxon, *Globallomorphina globosa* Heikkinen and Georgescu - new genus and species, from the late Santonian of the New Jersey coastal plain may indicate such a transition from the benthic allomorphinid group to the planktic gubkinellids. This study shows that evolutionary and typological classifications are not compatible and they can be used as tools that function as the degree of knowledge on a certain fossil group.

Keywords: Foraminifera, benthics, planktics, evolution, new taxa

INTRODUCTION

The polyphyletic nature of the planktic foraminifera is one of the major topics in the scientific literature of the last two decades, which was initiated with the genetic studies on the large and small units (LSU and SSU respectively) of the ribosomal ribonucleic acid (rRNA)

---
[*] Corresponding author: dgeorge@ucalgary.ca.

gene (Pawlowski et al. 1994, 1996; Darling 1996, 1997; Wade et al. 1996); based on the data provided by the SSU rRNA it was shown that some planktic foraminiferal groups possibly originated from benthic ancestors, although the benthic groups that could have evolved a planktic mode of life are not known with precision (Darling et al. 1997, p. 261). Of particular interest are those data that demonstrated the occurrence of both planktic and benthic stages in the same modern foraminiferal species (Banner et al. 1985; Smart and Thomas 2007; Darling et al. 2009).

The polyphyletic nature of the planktic foraminifera was prefigured by Ujiié and others (2008) based on those data from the SSU of rRNA that the triserial planktic *Gallitellia vivans* (Cushman 1934) plots close to the rotaliid genus *Stainforthia* Hofker 1956, which has a triserial-biserial gross test architecture. This study shows according to its authors that *G. vivans* evolved independently from the rotaliid group but the morphological relationship between the two taxa was considered only "plausible" (Ujiié and others, 2008, p. 337) indicating therefore a considerable uncertainty in the documentation of their alleged *Stainforthia-Gallitellia* lineage. Examination of the morphology of *Stainforthia fusiformis* (Williamson 1858) and *G. vivans* indicates that such an ancestor-descendant relationship cannot be inferred; the longitudinally elongate chambers and dense pores with uneven distribution (Gooday and Alve 2001) are features that occur in the former species and they contrast with the globular-chambered *G. vivans*, which have rare pores. These features suggest that *G. vivans* is a species with primitive morphological features while *S. fusiformis* has a more evolved morphology (Figure 1); if there would have been an evolutionary relationship between the two taxa then *G. vivans* should have been the ancestor and *S. fusiformis* the descendant, which is the contrary of what was suggested by Ujiié and others (2008). However, the evolutionary transition from the tests with triserial-biserial architecture to triserial tests as pattern in the iterative occurrence of the triserial planktics in the fossil record was mentioned but in the absence of any mention of such a specific lineage from the fossil record the study by Ujiié and others (2008) it is merely an intuitive one, without a solid basis for interpretation.

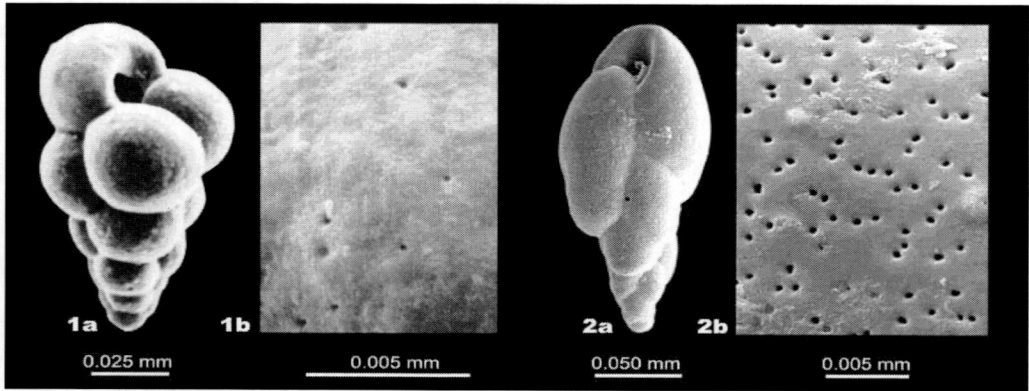

Figure 1. Two specimens of *Gallitellia vivans* (Cushman 1934) and *Stainforthia fusiformis* (Williamson 1858) showing the general test architecture (1, 3) and wall ultrastructure and porosity (2, 4). 1-2 Specimen of *G. vivans* illustrated by Loeblich and Tappan (1994, pl. 179, figs 4 and fragment of 5 respectively). 3-4 Specimen of *S. fusiformis* illustrated by Gooday and Alve (2001, pl. 3, figs B and fragment of J respectively).

It is evident that the datasets for such studies come from genetics, biology, ecology and oceanography, and for this reason they have small chance to impact significantly interpretations based on the fossil record. Genetic and biologic data are not available in the fossil species, and the paleoecological and paleoceanographical interpretations based on the fossil taxa are at much lower level of resolution when compared to those based on the living ones. Two attempts to combine modern data and those from the fossil record proved limited results (Leckie 2009; BouDagher-Fadel 2012).

Planktic foraminifera were traditionally considered a monophyletic group that evolved in the late early Jurassic (Toarcian) (Hart et al 2002, 2003, Hudson and others 2009). It was considered that the evolution in the planktic foraminiferal group happened between subgroups of planktic foraminifera; for example it was generally accepted that the serial planktic foraminifera evolved from the planispiral ones due to the occurrence of one early planispiral coil in the early portion of a number of heterohelicid tests. The possibility that other groups of foraminifera might have been planktics was also taken in consideration by other authors. For example Bettenstaedt and Spiegler (1982) considered that their new species *Pleurostomella gracilis* of the lower upper Albian sediments adopted a planktic mode of life, which is documented by the "… increased size of the aperture, elongation of the test, and the appearance of slim forms …" (Bettenstaedt and Spiegler 1982, p. 446). Another example of a genus assigned to the heterohelicid group based on the chamber architecture is *Elhasaella* described by Hamam (1976) from the Maastrichtian of Jordan, and reported only one time after by Hamam and Hayes (1977); its valid status in the heterohelicid group was not recognized in any subsequent revision of the Cretaceous planktic foraminifera although the aperture morphology, wall structure, ornamentation and porosity indicate close resemblances some globular hedbergellid and globigerinelloidid species. The taxonomic status of both *P. gracilis* and *Elhasaella* require additional high resolution study.

The polyphyletic nature of the Cretaceous planktic foraminifera was demonstrated by Georgescu (2009), who showed, based on data from quasi-complete occurrences, that the serial planktics evolved from *Praeplanctonia* Georgescu 2009 in the late Albian, and its descendants *Archaeoguembelitria* Georgescu 2009 (triserial throughout) and *Protoheterohelix* Georgescu and Huber 2009 (biserial throughout) independently achieved a planktic mode of life; *Praeplanctonia* origins were recognized among the representatives of the pleurostomellid group. This study was based on the high resolution observations of the test ultrastructure, ornamentation, porosity, and high detail morphological features of the test, which were collected from throughout the stratigraphic ranges of the species analyzed. In a continuation of this new stream in the search for the origins of the Late Cretaceous planktics Georgescu et al. (2011) considered that their new species *Neobulimina jerseyensis* with the chamber surface ornamented with pore mounds is most likely the ancestor of the triserial *Guembelitria cretacea* Cushman 1933. These data demonstrated the multiple origins of the Mesozoic planktic foraminifera, and is also consistent with the iterative evolution pattern known in this group (Fleisher and Steineck 1978).

New data in our search for the evolutionary processes that led to the evolution of the serial planktic foraminifera in the late Albian come to clarify the early evolution of *Praeplanctonia*. Although Georgescu (2009) mentioned that this genus/directional lineage evolved from *Pleurostomella*, no details of this process were given due to the relatively narrow stratigraphic interval investigated, and consequently the lack of information from the lower part of the upper Albian. Those data collected from the lower upper Albian sediments of the Deep Sea Drilling Project Site 370 (offshore Morocco) suggest that the *Pleurostomella-Praeplanctonia* evolution was a long and complex process, which requires further investigation.

A New Species of Praeplanctonia from the Late Albian (Early Cretaceous)

A new species of the benthic foraminiferal genus *Praeplanctonia* Georgescu 2009 is herein recognized and formally described from the lower upper Albian sediments from DSDP Site 370 situated in the North Atlantic Ocean, offshore Morocco. The section sampled encompasses the Valanginian-lower Cenomanian stratigraphic interval; 101 samples were collected from throughout this stratigraphic interval (Figure 2). The biostratigraphic framework at this site was given by Georgescu (2012a) and is followed herein. Evolutionary classification units are after Georgescu (2010) at lineage level, and Georgescu (2012b) at species level.

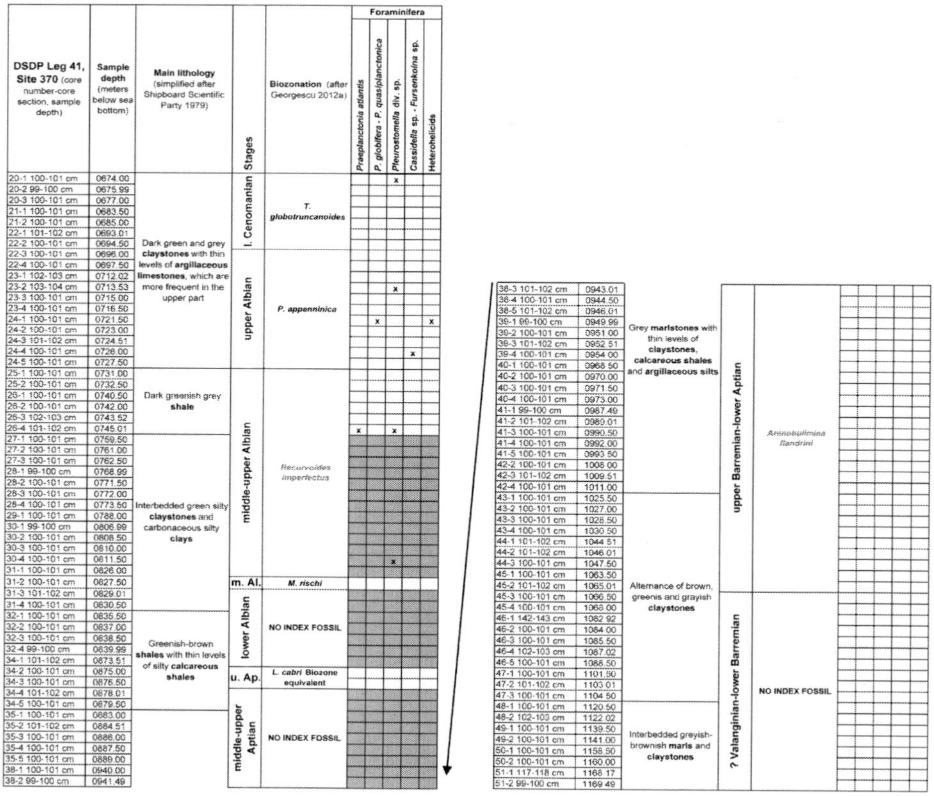

Figure 2. Main lithological types, samples collected, biostratigraphic framework and occurrences of benthic and planktic foraminifera in the Valanginian-lower Cenomanian sediments at DSDP Site 370. Stratigraphic intervals for which the sedimentation below lysocline is inferred from the micropaleontological assemblage composition are shown on grey background. Note the lysocline fluctuations in middle Aptian-middle Albian stratigraphic interval.

Directional Lineage: *Praeplanctonia* Georgescu 2009

Species included. Initiating species (IS): *P. atlantis* - new, First Descendant Species (FDS): *P. globifera* Georgescu 2009, Second Descendant Species (SDS): *P. quasiplanctonica* Georgescu 2009.

Age. Late Albian.

Geographic distribution. Western North Atlantic Ocean (Blake Plateau), Eastern North Atlantic Ocean (offshore Spanish Sahara and offshore Morocco) and Caribbean region (Trinidad).

IS: *Praeplanctonia atlantis* Burke and Georgescu - new species
(Figure 3:1-4)

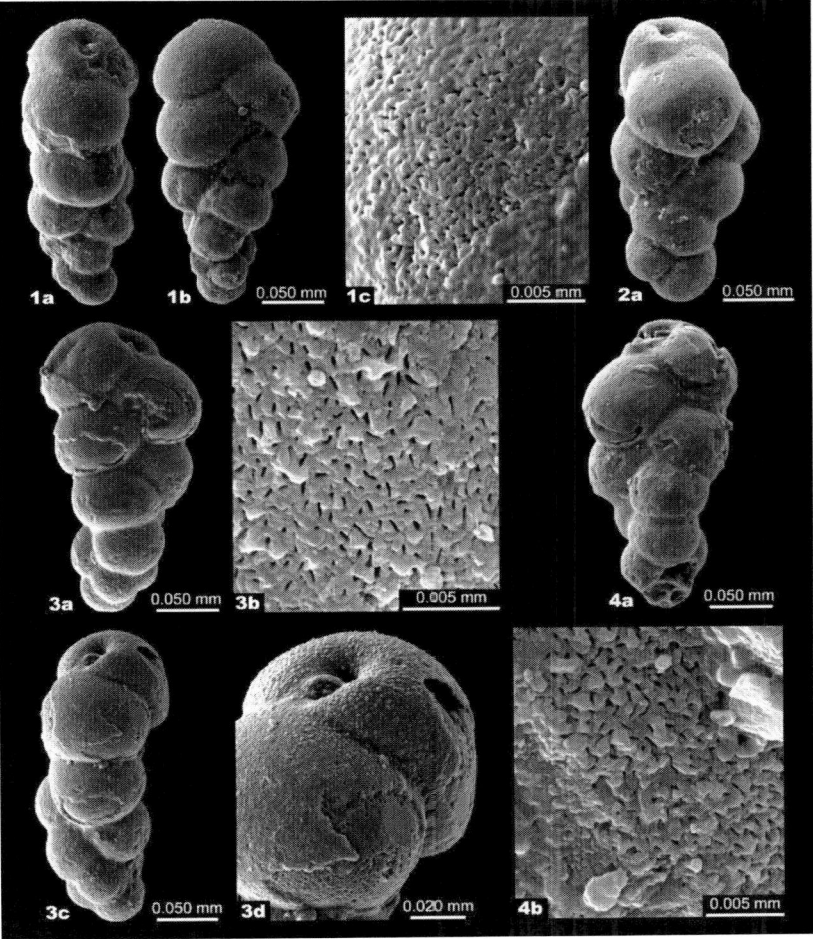

Figure 3. Specimens of *Praeplanctonia atlantis* Burke and Georgescu - new species from the lower upper Albian sediments of DSDP Site 370 (offshore Morocco). All specimens are collected from Sample 26-4, 101-102 cm. 1 Holotype (WKB 010148), 2-4 paratypes.

Holotype. Specimen WKB 010148 (Figure 3:1).

Holotype dimensions. Length: L=0.2498 mm; width: W=0.1213 mm; thickness: T=0.0980 mm.

Paratypes. Five specimens, WKB 010149-010153.

Dimensions. L=0.1852-0.2573 mm; W=0.1155-0.1388 mm; T=0.088-0.1044 mm. Ranges based on the average measurements of the holotype and paratypes.

Material. Eleven specimens: one holotype and ten paratypes.

Type locality. DSDP Site 370, Eastern North Atlantic Ocean (offshore Morocco), geographical coordinates: 32° 50.2' N and 10° 46.6 W.

Type level. Lower upper Albian dark greenish grey shale, Sample 26-4, 101-102 cm.

Derivation. The name is derived from the Atlantic Ocean, where the type locality is situated.

Diagnosis. *Praeplanctonia* with large proloculus and deeply incised sutures.

Description. Test is twisted along the growth axis; the large-sized proloculus (0.0396-0.0483 mm in diameter) is followed by a triserial stage that represents one fourth to one half of the test length, and the biserial adult stage. Chambers are globular to subglobular and overlap the previous one or two at variable rates; chambers increase slowly in size. Sutures are distinct and depressed, deeply incised and easily recognizable between all the chambers of the test. Aperture is elliptical in shape and subterminal in position; it is partly bordered by an asymmetrical imperforate rim, and provided with a toothplate. A supplementary suture occurs on the chamber anterior side at least in the adult stage; the supplementary suture stretches from the aperture to the suture between the last-formed two chambers. Chamber surface is smooth, without ornamentation elements. Test wall is calcitic, hyaline, simple and perforate; pores are circular to elongate, with a diameter or maximum dimension of 0.0002-0.0007 mm.

Remarks. *Praeplanctonia atlantis* differs from the other species of the lineage mainly by the larger proloculus (0.0396-0.0483 mm rather than 0.0280-0.0440 mm in *P. globifera* and 0.0110-0.0140 mm in *P. quasiplanctonica*).

Age. Early late Albian.

Geographic distribution. Eastern North Atlantic Ocean (offshore Morocco).

PRAEPLANCTONIA: A DEEP OCEANIC DIRECTIONAL LINEAGE AT THE BOUNDARY BETWEEN PLANKTICS AND BENTHICS

Although a fully benthic foraminifer, *Praeplanctonia* offers an excellent record in the transition of the benthic foraminifera towards the planktic habitat. Notably the earliest heterohelicid species, namely *Protoheterohelix washitensis* (Tappan 1940) is a direct descendant of the FDS of the DL *Praeplanctonia*: *P. globifera*; this descent is of particular importance as it helps in defining the direction of the benthic-planktic transition among the species of *Praeplanctonia*, namely from *P. atlantis* (IS) to *P. quasiplanctonica* (SDS). The evolution within the *Praeplanctonia* lineage shows that most of the changes in the test architecture happened in the chamber overlapping, and the species can be recognized mainly by the suture incision, which is deeper in the IS and shallower in the FDS and SDS, and the proloculus size that decreases in size from the IS to the SDS. Among the conservative features, namely those that do not present apparent changes throughout the lineage evolution are the main aperture in subterminal position and the occurrence of one supplementary suture that stretches from the aperture to the suture formed between the last-formed chambers; notably, the supplementary suture was considered the effect of the wall "wrapping" happened during the *Pleurostomella-Praeplanctonia* evolution (Georgescu 2009). The occurrence of mixed circular and elongate pores, which were reported in the FDS and SDS, is confirmed in the IS; the functional-morphological significance of the elongate pores is still unknown.

The evolution of two lineages, which was initiated from the FDS of *Praeplanctonia*, led to the development of the planktic mode of life in the taxa with serial chamber arrangement: *Archaeoguembelitria* Georgescu 2009 and *Protoheterohelix* Georgescu and Huber 2009. The data from the fossil record indicates that this evolutionary process happened through major chamber rearrangements, position of the aperture, nature of the periapertural structures, and test wall porosity. The two planktic lineages that evolved from triserial-biserial *Praeplanctonia* record the complete disappearance of the biserial stage in *Archaeoguembelitria* and the triserial stage in *Protoheterohelix*; divergent evolution is the process through which the two lineages initiated. All the other features that evolved in the transition from the benthic to planktic mode of life present parallel evolution: aperture change from subterminal towards the base of the last-formed chamber, development of periapertural structures consisting of rims, and disappearance of the elongate pores (Georgescu 2009).

The result of these studies on the *Praeplanctonia* lineage led to the evolutionary continuum between the benthic and planktic lineages. As a result, recognizing the habitat of a foraminiferal species or specimen recorded in a certain region can be done with higher accuracy. However, the process of taxon identification requires certain care for the high detail test morphological features, and only the general test appearance (e.g., triserial, biserial, biserial-triserial, etc) does not any longer determine a correct identification at least in the case of the Cretaceous heterohelicids. Based on the high resolution observations acquired with the aid of the SEM and more accurate taxon definitions it became possible to avoid such errors in taxa identification. For example, the specimen illustrated as *Heterohelix* sp. by Caron and Petrizzo (in Gale and others 2011, fig 44: M-N) from the upper Albian (*P. ticinensis* Biozone) sediments of the Hautes-Alpes (France) can be reassigned accurately to *P. globifera* despite the insufficient illustration in which the edge view was not given.

Praeplanctonia gives additional information in recognizing the evolution of the planktic mode of life from the smaller benthic foraminifera and its reason. All three species of this lineage were deep water species as suggested by their paleobiogeographical distribution and paleoenvironmental reconstruction at each site; the lineage appears restricted to the Eastern and Western North Atlantic Ocean and Caribbean region (Trinidad). The evolution from the benthic to planktic habitat is well documented in the late Albian *P. ticinensis* Biozone, which corresponds to the beginning of the black shale accumulation in the early late Albian (*P. ticinensis* Biozone) that preceded the oceanic anoxic event in the *P. appenninica* Biozone in the terminal Albian (OAE 1d). Therefore, it appears reasonable to consider that the achievement of the planktic habitat in the foraminiferal group, which resulted in the evolution of the heterohelicids, is the result of the initiation of the OAE 1d event.

BEFORE *PRAEPLANCTONIA*: THE NEW DATASET

The origins of *Praeplanctonia* require further study. Although it is evident from this study and that by Georgescu (2009) that this directional lineage evolved from *Pleurostomella*, little is known about the evolution process through which *Praeplanctonia* evolved. The new material collected from DSDP Site 370 yielded interesting tests, which indicate that the process of evolution from *Pleurostomella* to *Praeplanctonia* was extremely complex, and most likely long.

Probably the most interesting tests collected from the upper Albian sediments at DSDP Site 370 are the pleurostomellids in which the early stage exhibits a loose initial coil, resembling that in the early stage of *P. atlantis*. Such specimens with bandyellid appearance are considered with intermediate morphology between *Pleurostomella obtusa* Berthelin 1880 and *P. atlantis* (Figure 4). The aperture of these specimens is symmetrical, similar to that known in the former species. Therefore these specimens demonstrate that the evolution from *P. obtusa* to *P. atlantis* happened initially at the level of chamber arrangement, and the loss of the aperture symmetry and development of the supplementary suture are rather late evolutionary characters in this lineage. This observation further confirms the idea that the morphological changes in the aperture and periapertural structures are of paramount importance in the foraminiferal group during the transition from the benthic to planktic habitat.

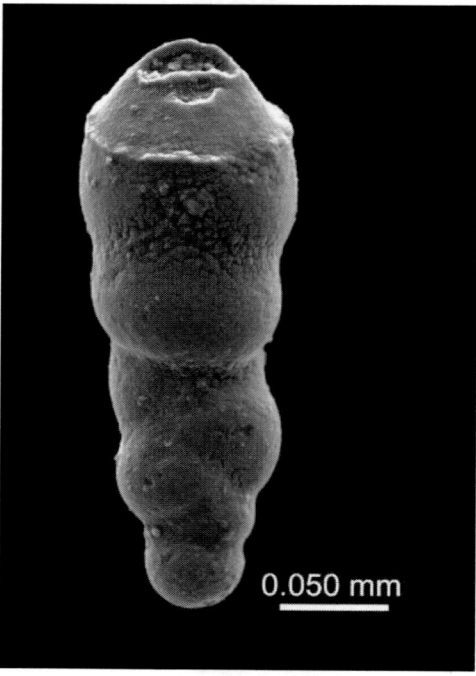

Figure 4. Specimen with bandyellid appearance collected from DSDP Site 370, Sample 26-4, 101-102 cm. The precise systematic position of such specimens requires further studies.

An interesting specimen of *Pleurostomella gracilis* Bettenstaedt and Spiegler (1982, pl. 7.3-4: 8) from the lower Upper Albian of northwestern Germany indicates clearly through its laterally migrated apertural teeth the possibility that more than one species of *Pleurostomella* could have attempted to evolve the planktic habitat. Moreover, the specimen illustrated by Bettenstaedt and Spiegler (1982) was collected from a stratigraphic level situated below and in the proximity of the evolutionary occurrence of *Praeplanctonia*.

RECURRING *PRAEPLANCTONIA ATLANTIS* TEST ARCHITECTURE DURING ALBIAN-CENOMANIAN

Praeplanctonia atlantis presents a general test architecture that is known at three stratigraphic levels during the Albian-Coniacian (Figure 5). The overall appearance of this species can be described as a combination of the following features: triserial-biserial chamber arrangement, subglobular chambers, deeply incised sutures, and aperture bordered by a curved and narrow lip. These taxa are presented below in stratigraphical order.

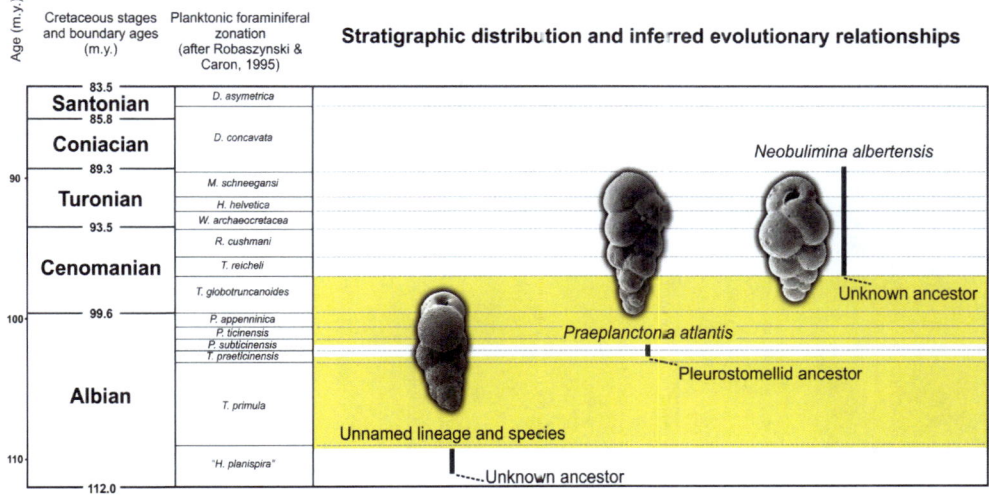

Figure 5. The three stratigraphic intervals with recurring *Praeplanctonia atlantis* test architecture occurrences during the Albian-Coniacian. The gaps marked with yellow background are indicative for the iterative evolution of this test architecture.

The earliest occurrence of this test architecture is known from the lower Albian sediments of the Blake Plateau (western North Atlantic Ocean), Ocean Drilling Program Hole 1049B. Holbourn and Kuhnt (2001, fig. 8:7) reported specimens they assigned to *Neobulimina minima* Tappan 1940 from Core 12; this species was described from the Grayson Formation of Texas by Tappan (1940, p. 117, pl. 19, figs 5-6). Tappan (1940) considered a late Albian age for this lithostratigraphic unit, but this was later reviewed and assigned a latest Albian-early Cenomanian age (Pessagno 1967; Michael 1972). This taxon is re-examined (Figure 6) and its identification as *N. minima* is questioned, especially due to the deeper sutures and higher rate of chamber size increase in the adult stage. Test wall in the Blake Plateau specimens is simple in the adult stage but larger specimens present incipiently reticulate wall over the earlier portion of the test due to the addition of successive layers of calcite. Pore shape is circular in the adult stage and often elliptical in the earlier portion of the test; pores are 0.0009-0.0015 mm in diameter or maximum dimension, but larger pores occur occasionally around the aperture. The aperture is situated at the base of the last-formed chamber and is bordered by one curved asymmetrical rim. The systematic position of these specimens is unknown and requires examination of additional material.

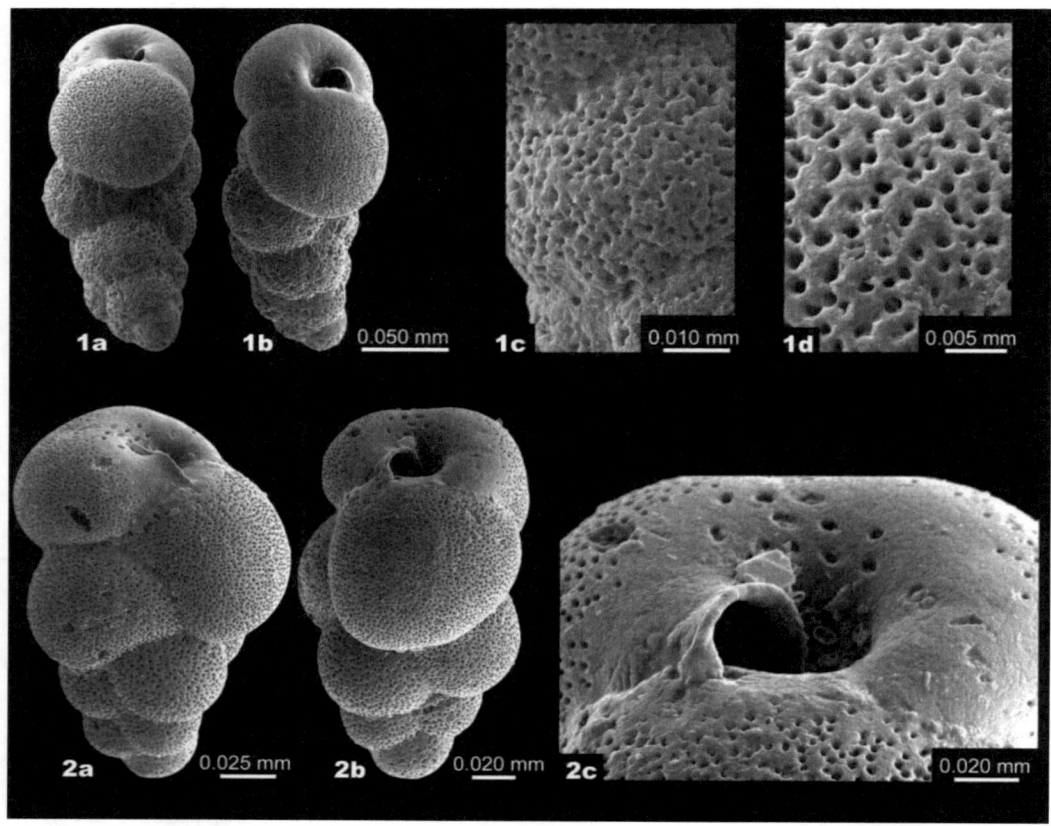

Figure 6. Specimens with the test architecture of *Praeplanctonia atlantis* from the lower Albian sediments from ODP Hole 1049B, Sample 12-3, 12-15 cm. This taxon precise systematic position is not precisely known. Note the large-sized pores with circular or elliptical shape and incipiently reticulate test wall (1c, 1d).

The second occurrence of this test architecture occurs in the late Albian, and it is known only in the new species *P. atlantis*. The aperture is situated in subterminal position, and is not adjacent to the base of the last-formed chamber; one supplementary suture, which is oblique to the test axis of growth stretches from the aperture base to the suture between the last-formed two chambers. Test wall is simple; pores are circular to elongate, with a diameter or maximum dimension of 0.0002-0.0007 mm. The similarities in the gross test architecture makes that practically only the wall ultrastructure and pore size can be used to discriminate between *P. atlantis* of the late Albian and the tests of the early Albian.

The highest occurrence of this test architecture is known from the middle Cenomanian-Turonian of Canada (Alberta and Manitoba) (Stelck and Wall 1954; McNeil and Caldwell 1981), and United States of America (Alaska, Colorado, Kansas, Montana, South Dakota, Wyoming) (Tappan 1962; Fox 1954; Eicher 1965; Eicher and Worstell 1970); such tests are assigned to *Neobulimina albertensis* (Stelck and Wall 1954) (Figure 7). The test wall in *N.*

albertensis is simple, with small-sized pores, at sizes comparable to those known in *P. atlantis*; the fine pores confer a smooth appearance to the test wall, which is evident in well-preserved specimens. The aperture is situated at the base of the last-formed chamber and is bordered by an asymmetrical and curved rim. The major difference between *N. albertensis* and *P. atlantis* is in the position of the aperture, which is situated at the base of the last-formed chamber in the former and subterminal in position in the latter; in addition the supplementary suture occurs only in the latter.

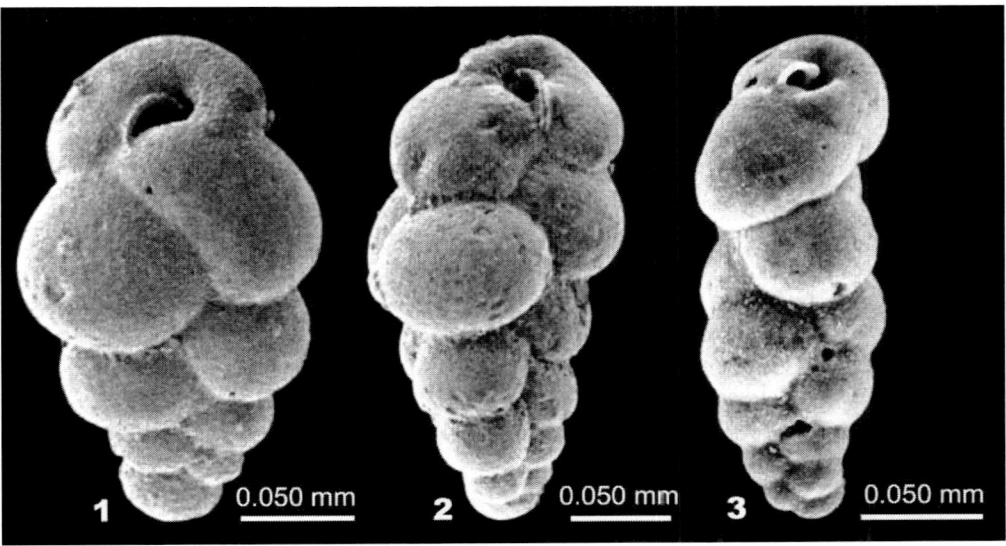

Figure 7. Three specimens of *Neobulimina albertensis* (Stelck and Wall 1954) from the upper part of the Greenhorn Formation of Wyoming; specimens originally illustrated by Eicher and Worstell (1970, pl. 4, figs 2, 3 and 4 respectively). Note the *Praeplanctonia atlantis* test architecture and curved periapertural lip.

The distribution of the Albian-Turonian species that have the general test appearance resembling that in *P. atlantis* show that there are significant gaps between them and their stratigraphic ranges that do not overlap. Morphological differences between these three taxa indicate that there are no evolutionary relationships between the early Albian tests, which present incipiently reticulate test wall over the earlier portion of the test, and the late Albian-Turonian ones, in which the test wall is simple. The evolution in the foraminiferal group happens from the simple to incipiently reticulate, therefore a direct ancestor-descendant relationship between the two groups cannot be taken into consideration. The existence of a gap spanning the latest Albian-early Cenomanian, namely between the extinction of *P. atlantis* and evolutionary occurrence of *N. albertensis*, apparently indicates that there is no evolutionary relationship between them, and the two species iteratively evolved from another benthic group with globular chambers. A direct evolutionary relationship between the two species can be considered in case future studies will reveal occurrences of specimens with morphological features intermediate between *P. atlantis* and *N. albertensis*, and in this case *Praeplanctonia* should be considered the ancestor of *Neobulimina* Cushman and Wickenden 1928.

Iterative Evolution in the Aptian-Maastrichtian Triserial Planktics

Triserial morphology in the Cretaceous (Aptian-Maastrichtian) planktics evolved three times (Georgescu 2009), and in all three cases the shift from benthic to planktic habitat was associated with the development of pore mound ornamentation. The three lineages are *Koutsoukosia* Georgescu 2009, *Archaeoguembelitria* Georgescu 2009, and *Guembelitria* Cushman 1933. Notably the recognition of the three lineages cannot be made without taking in consideration the iterative evolution pattern among planktic foraminifera.

Koutsoukosia sergipensis Koutsoukos 1994 occurs in upper Albian-lower upper Albian sediments (Koutsoukos 1994). Its ornamentation consists according to the original description and figuration of a mixture of pore mounds and pustules, but many of the pustules appear to be the result of diagenesis. In the original publication the "pore mounds" should have been mentioned rather than *small-sized pustules* as unintentionally occurred (Georgescu 2009, p. 112-113) and this is herein corrected; therefore, the ornamentation of *Koutsoukosia* should be considered as formed by a mixture of pore mounds and pustules. The blunt pustules on the surface of some specimens of the topotype material indicate the possibility that *K. sergipensis* may be multi-taxic. The upper Albian genus *Pseudoguembelitria* Huber and Leckie 2011, which is also ornamented with pore mounds, presents a distinctly coiled early stage. The evolutionary relationships between *Koutsoukosia* and *Pseudoguembelitria* require further study of better preserved material.

Archaeoguembelitria evolved from *Praeplanctonia globifera* in the late Albian through divergent evolution (Georgescu 2009, p. 137-140). Ornamentation evolution in this directional lineage shows that the IS has smooth chamber surface whereas the FDS is ornamented with pore mounds. The evolution of a planktic habitat was considered by Georgescu (2009) who documented by the pore mounded ornamentation. *Archaeoguembelitria* crossed the Cenomanian/Turonian boundary and became extinct towards the top of the Turonian. No triserial tests with pore mounds ornamentation are known from the uppermost Turonian-lower Santonian.

Triserial tests with pore mounds evolved again in the late Santonian: *Guembelitria*; late Santonian occurrences are known from Wyoming (Frerichs 1979) and Tunisia (Georgescu and others 2011). The diversification in the *Guembelitria* lineage occurred in the Maastrichtian (Arz and others 2010; Georgescu and others 2011). *Guembelitria cretacea* is the most frequent species of this branched lineage in the fossil record and is known from the Santonian to the top of the Maastrichtian and is also one of the two Cretaceous planktic foraminiferal species that survived the Cretaceous/Paleogene boundary event. The origins of *G. cretacea* are not precisely known. Georgescu (2009) considered that the species *G. turrita* Kroon and Nederbragt 1990, which is ornamented with small-sized pustules over the early portion of the test. A different evolution path was suggested when Georgescu and others (2011, p. 126) described *Neobulimina jerseyensis* from the Late Cretaceous (late Campanian) of the New Jersey coastal plain; this species shows the typical triserial-biserial neobuliminid architecture and is ornamented with pore pounds. However, the *N. jerseyensis-G. cretacea* ancestor-descendant relationship requires additional study because *N. jerseyensis* is known only from late Campanian, which is situated circa 11 m.y. after the evolutionary occurrence of its alleged descendant *G. cretacea* (Georgescu and other 2011). Additional high resolution studies on the Late Cretaceous neobuliminids are necessary in order to understand their evolution and evolutionary relationships with the triserial planktics.

THE ORIGINS OF THE PLANKTICS WITH ALTERNATE CHAMBER ADDITION

Planktic foraminifera with alternate chamber addition are also referred to as heterohelicids; in the typological classification they are traditionally included in the family Heterohelicidae Cushman 1927. The group evolved in the late Albian (*P. ticinensis* Biozone) from *Praeplanctonia globifera* through divergent evolution; the two species that directly evolved from *P. globifera* are the triserial throughout *Archaeoguembelitria cenomana* (Keller 1935) and biserial throughout *Protoheterohelix washitensis* (Tappan 1940). The earlier heterohelicids were included in the genus *Protoheterohelix* by Georgescu and Huber (2009); *Protoheterohelix* accommodates a directional lineage in evolutionary classification, which consists of *P. washitensis* as IS and *P. obscura* Georgescu and Huber 2009 as FDS. The earliest species (*P. washitensis*) has asymmetrical test and periapertural structures; its direct descendant (*P. obscura*) presents symmetrical tests in edge view, but the periapertural structures are still asymmetrical. The complete symmetry of the test and periapertural structures in the heterohelicid group was achieved in the middle Cenomanian (*T. reicheli* Biozone) with *Planoheterohelix moremani* (Cushman 1938).

The test asymmetry in the heterohelicids of the upper Albian-lower Cenomanian is irrefutable evidence that this group of planktic foraminifera evolved from asymmetrical benthic ancestors. Georgescu (2009, pl. 1, figs 7-10) illustrated specimens of *P. washitensis* with the aperture bordered by one curved rim, which closely resembles the periapertural structures of many small-sized benthic species; however, the well-developed biserial chamber arrangement indicates that such specimens clearly belong to the heterohelicid group. Another feature that indicates the evolutionary relationship between the small-sized benthics of the praeplanctonid group and the heterohelicid planktics is the smooth chamber surface in both groups. Georgescu and Huber (2009) demonstrated that the costate ornamentation in the heterohelicid group was achieved in the late Cenomanian, and the oldest species with ornamented chamber surface are *Globoheterohelix paraglobulosa* Georgescu and Huber 2009 and *Planoheterohelix postmoremani* Georgescu and Huber 2009.

A different point of view was proposed by BouDagher-Fadel (2012, p. 113) who considered that the planktic foraminifera with biserial chamber arrangement evolved from the benthic genus *Brizalina* Costa 1856 without providing an ancestral species belonging to this genus; the arguments taken in consideration are the microperforate test wall, imperforate longitudinal costae and the basal loop aperture of *Brizalina*. This idea cannot be considered valid mainly because it does not fulfil the basic requirement of an evolution reconstruction that the ancestor is older than the descendant; the evolutionary occurrence of *Brizalina* is in the Late Cretaceous (Campanian) (Loeblich and Tappan 1987), circa 20 m.y. after the evolutionary occurrence of the heterohelicid group from the Early Cretaceous (late Albian). Moreover, *Brizalina* presents features (e.g., compressed tests, angular or subangular periphery, etc) that appear the result of a long process of evolution from globular-chambered ancestors, and such morphological features cannot be considered ancestral to those of the earliest heterohelicids, which are primitive (e.g., globular to subglobular chambers, broader periphery, etc).

BISERIAL OR TRISERIAL-BISERIAL PLANKTICS BEFORE HETEROHELICIDS?

The recognition of the firm first occurrence of the heterohelicid group in the late Albian (P. ticinensis Biozone) creates the opportunity for a better evaluation of the reports of the representatives of this group from older sediments, and the possibility of achievement of the planktic habitat in foraminifera with biserial chamber arrangement before the heterohelicid group.

The oldest alleged heterohelicid tests were described from the Oxfordian (Late Jurassic) of Poland by Fuchs (1973, p. 463-464), and were assigned to the new genus and species *Eoheterohelix prima* (Figure 8: 1-2). This taxon was not re-illustrated with the aid of the SEM or ESEM ever since its publication. The examination of the type figures in the original publication indicates that they represent very low trochospiral tests with rapid chamber size increase in the last whorl; therefore they are not assignable to the heterohelicid group. Two heterohelicid species were described by Fuchs (1971, p. 38) from the middle Barremian (Early Cretaceous) of Austria: *Heterohelix hohenemsensis* and *H. trochospiralis* (Figure 8: 3-5 respectively). The two species present general test features that are common among the Santonian-Maastrichtian (Late Cretaceous) representatives of the heterohelicid group. Occurrence of an early planispiral coil in *H. trochospiralis* is evident in the original figure of the holotype; it was proven that chamber coil in the early portion of the test in the heterohelicid group occurred in a small number of taxa from the Santonian-Maastrichtian stratigraphic interval (Brown 1969; Georgescu and Abramovich 2008, 2009), and represent a late development in the heterohelicid evolution. Therefore the two species described from the middle Barremian of Austria should be carefully considered, and even the possibility of contamination from younger sediments should not be *a priori* ruled out.

A possibly new group of foraminifera with triserial-biserial chamber arrangement and large symmetrical aperture is of particular interest in the context of the study of polyphyletic nature of the Cretaceous planktics (Figure 9). These tests form a minor part of the foraminiferal assemblage in the DSDP Site 370, Sample 26-4, 101-102 cm, therefore they occur at a stratigraphic level situated below the heterohelicid evolutionary occurrence. Examination of the text ultrastructure and porosity showed that these two features have close resemblances with the earliest species of the heterohelicid group of late Albian-middle Cenomanian age; pore size is identical in the two groups: 0.0002-0.0007 mm. No evolutionary relationship between the early late Albian triserial-biserial foraminifera can be inferred due to the periapertural structure asymmetry in *Protoheterohelix*. Additional well-preserved material is necessary to assess the systematic position and habitat of the newly reported foraminifera.

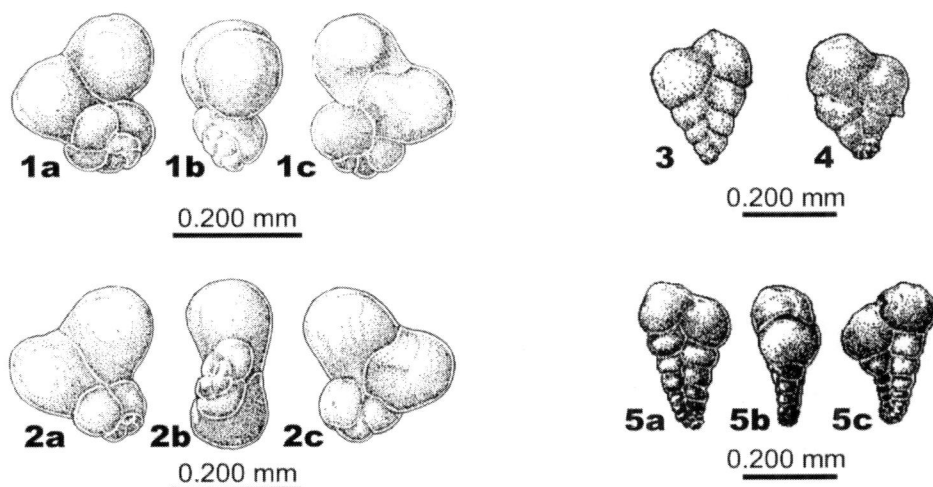

Figure 8. Specimens Albian assigned to the heterohelicid group from older sediments than upper. 1-2 Specimens of *Eoheterohelix prima* figured by Fuchs (1973, pl. 3, fig. 4 and pl. 4, fig 3 respectively) from the Oxfordian from Poland. 3-4 Specimens of *Heterohelix hohenemsensis* illustrated by Fuchs (1971, pl. 10, fig. 2 and pl. 10, fig. 19 respectively) from the middle Barremian of Austria. 5 Specimen of *H. trochospiralis* illustrated by Fuchs (1971, pl. 10, fig. 22) from the middle Barremian of Austria.

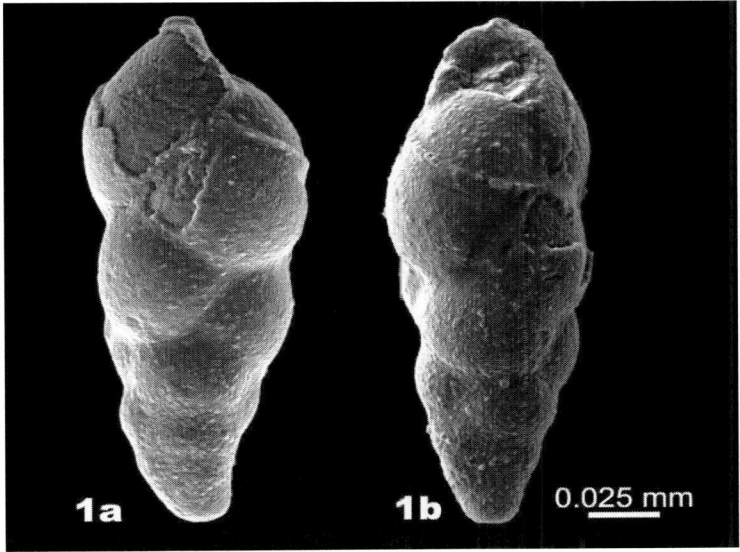

Figure 9. Specimen with triserial-biserial chamber arrangement and large-sized symmetrical aperture from the lower upper Albian sediments of the DSDP Site 370, Sample 26-4, 101-102 cm. The test ultrastructure and pore size show close similarities with those known in the earliest heterohelicids of the late Albian-middle Cenomanian. The precise systematic position of such specimens is unknown.

POSSIBLE ALLOMORPHINID ANCESTORS IN THE SANTONIAN

Our data on the Cretaceous foraminifera show unequivocally that the evolution of the planktic species from benthic ancestors happened among the taxa with globular and subglobular chambers; this is evident in the divergent evolution from *Praeplanctonia* to *Archaeoguembelitria* and *Protoheterohelix* during the late Albian. A less explored strategy is the development of globular chambers from compressed ones in parallel with the evolution of test features, which are frequent occurrences in the planktic group (e.g., periapertural lip, similar porosity pattern, etc). Such strategy would impact significantly the efforts towards discovering the benthic groups that evolved a planktic way of life. A new foraminifer from the Santonian of the ODP Leg 174AX of the New Jersey coastal plain comes apparently to confirm the possibility of evolution of the planktic habitat by other groups.

The sediments accumulated during the Santonian transgression of the Cheesequake-Merchantville sedimentary cycle yielded the tests of a new foraminiferal taxon. Foraminiferal assemblages consist of both planktic and benthic taxa, and their diversity and abundance are controlled mostly by the relative sea-level changes (Georgescu 2006). The new foraminiferal taxon with general allomorphinid appearance is recorded in one sample situated in the proximity of the late Santonian transgression peak. The lack of data on the stratigraphic distribution and high-detail morphological features is acute for the Cretaceous benthic foraminifera; therefore the new genus and species *Globallomorphina globosa* is described in typological classification. This shows that typological and evolutionary classifications are not incompatible, and typology can be used until the development of the evolutionary classification. The typological classification units follow those of Loeblich and Tappan (1987).

Suborder **ROTALIINA** Delage and Hérouard 1896
Superfamily **CHILOSTOMELLACEA** Brady 1881
Family **CHILOSTOMELLIDAE** Brady 1881
Subfamily **CHILOSTOMELLINAE** Brady 1881

Genus *Globallomorphina* Heikkinen and Georgescu - new

Type species. *Globallomorphina globosa* Heikkinen and Georgescu - new species.

Diagnosis. Tests with allomorphinid early stage and adult stage consisting of subglobular to globular chambers.

Description. Test with early trochospiral stage with allomorphinid appearance and adult stage consisting of three globular or subglobular chambers. Sutures are distinct and depressed. Periphery is broadly rounded and simple. Aperture is a low arch in umbilical position and is bordered by a thin lip. Chamber surface is ornamented with low-relief irregular areas or more rarely dome-like pustules. Test wall is calcitic, hyaline and sparsely perforate.

Remarks. *Globallomorphina* differs from *Allomorphina* Reuss in Cžjžek 1849 mainly by the globular chambers in the adult stage and sparsely perforate test wall; the genus appears with intermediary morphological features between *Allomorphina* and *Gubkinella* Suleymanov 1955.

Derivation. Latin prefix *Globo-* (=globular) is added to the pre-existing foraminiferal name *Allomorphina*.

Species included. *Globallomorphina globosa* Heikkinen and Georgescu - new species and *G. conica* (Cushman and Todd 1949).

Age. Late Santonian, Paleocene.

Geographic distribution. USA (New Jersey), Caribbean region (Trinidad), Europe (Czech Republic).

Globallomorphina globosa Heikkinen and Georgescu - new species
(Figures 10:1-8, 11:1-6)

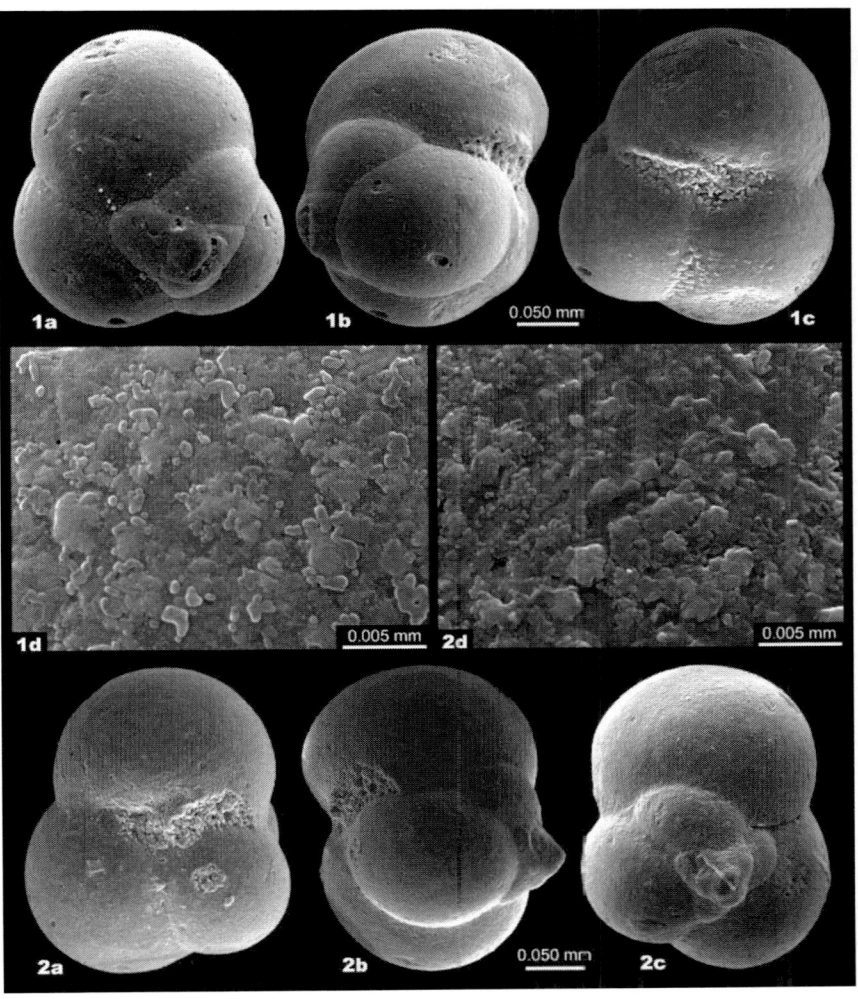

Figure 10. Specimens of *Globallomorphina globosa* Heikkinen and Georgescu - new genus and species from the upper Santonian sediments of ODP Leg 174AX, Sample 505.35-505.38 m. 1 holotype (WKB 010154), 2 paratype. Note the sparsely perforate test wall (1d and 2d).

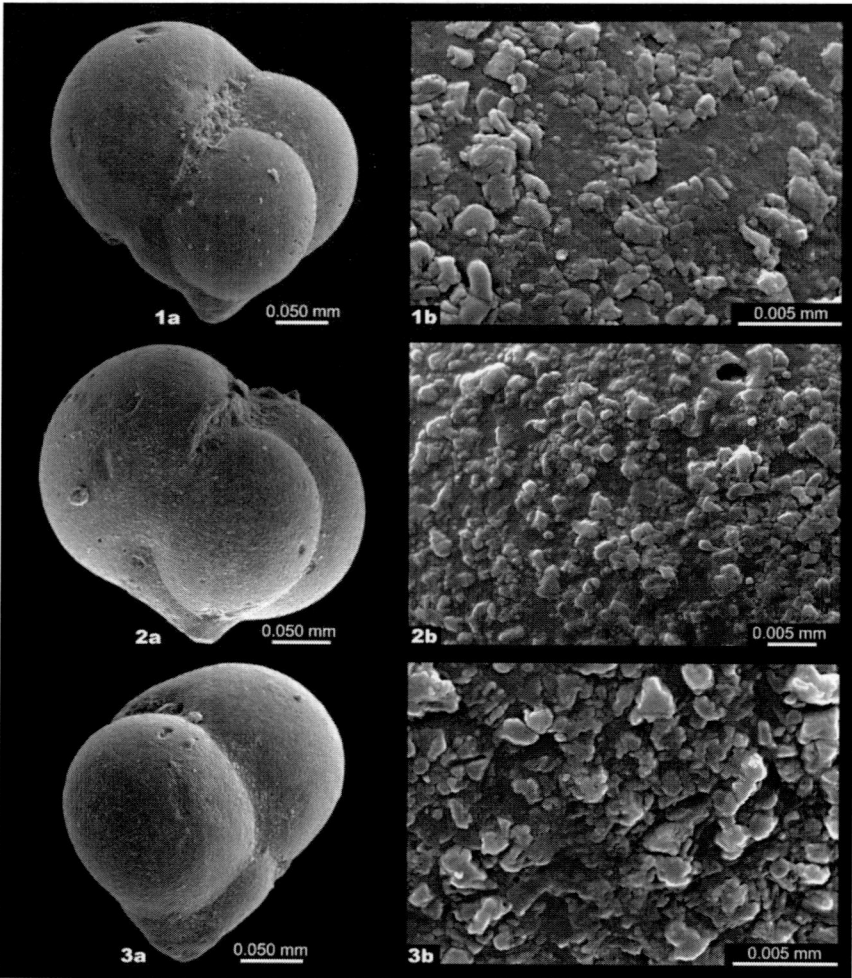

Figure 11. Paratypes of *Globallomorphina globosa* Heikkinen and Georgescu - new genus and species from the upper Santonian sediments of ODP Leg 174AX, Sample 505.35-505.38 m. Note the sparsely perforate test wall and ornamentation consisting of low relief irregular areas. The pores that penetrate through the ornamentation areas demonstrate that these in-relief structures are not the result of diagenesis.

Holotype. Specimen WKB 010154 (Figure 10:1).

Holotype dimensions. Maximum test diameter: D_{max}=0.2856 mm; minimum test diameter: D_{min}=0.2335 mm; height: H=0.2390 mm.

Paratypes. Twelve specimens, WKB 010155-010166.

Dimensions. D_{max}=0.1980-0.2958 mm; D_{min}=0.1688-0.2382 mm; H=0.2365-0.3111 mm. Ranges based on the average measurements of the holotype and paratypes.

Material. Over 300 specimens.

Type locality. ODP 174AX New Jersey coastal plain (Bass River Site), geographical coordinates: 39° 36' 42" N and 74° 26' 12" W.

Type level. Upper Santonian Merchantville Formation glauconitic clays, Sample 505.35-505.38 m.

Derivation. Name derived from the Latin adjective *globosus, -a* (=globular, spherical).

Diagnosis. *Globallomorphina* with deeply incised sutures and slightly axially elongate chambers.

Description. Test is trochospiral, with early allomorphind stage. The adult stage consists of two to three globular chambers. Chambers present gradual size increase and variable overlapping rate. Sutures are distinct and depressed, straight or curved. Periphery is broadly rounded and simple, without peripheral structures. Aperture is a low arch at the base of the last-formed chamber, and is situated in umbilical position; one thin lip borders the aperture. Chamber surface is ornamented with irregular regions and rarely with small-sized dome-like pustules. Test wall is calcitic, hyaline, simple and perforate; pores are circular, and with the average diameter of 0.0001 mm or smaller.

Remarks. *Globallomorphina* globosa differs from *G. conica* mainly by the deeply incised sutures, and more lobate test in the views from either spiral or umbilical sides. There is no overlapping between the stratigraphic ranges of the two species included in *Globallomorphina*.

Age. Late Santonian.

Geographic distribution. USA (New Jersey).

CONCLUSION

Polyphyletic origin of the planktic foraminifera is a topic that requires extensive and continuous study of the fossil record and modern taxa. Initiated through the results based on the genetic characteristics of the modern species, the study of the modern representatives of the foraminiferal group proved its limitations in less than one decade. This happened when Ujiié and others (2008) defined one triserial planktic origination from biserial-triserial ancestors based on the resemblances in the SSU of rRNA, but the features invoked by these authors indicate that the alleged descendant taxon presents primitive features whereas the ancestor is more evolved. These data from the fossil record showed that the polyphyletic ancestry of the planktic foraminifera as shown by Georgescu (2009), demonstrated that two the planktic lineages *Archaeoguembelitria* and *Protoheterohelix* evolved in the late Albian from the benthic *Praeplanctonia*.

New data from lower upper Albian sediments of the DSDP Site 370 (offshore Morocco) show the occurrence of a new species of *Praeplanctonia*, which presents intermediate features between the pleurostomellid stock and *Praeplanctonia globifera*, *P. atlantis*. Based on these new data, *Praeplanctonia* is redefined as a directional lineage in evolutionary classification, with *P. atlantis* its initiating species.

High resolution studies on the test ultrastructure and porosity of some taxa of the Cretaceous (Albian-Santonian) indicate the possibility to recognize additional originations of planktic foraminifera. Among them is a new taxon formalized as *Globallomorphina globosa* new genus and new species from the Santonian of New Jersey, which appears transitional between the allomorphinid and gubkinellid foraminifera. Evolution of globular chambers from axially compressed ones is a rare evolutionary process among the foraminifera and its causes require careful further study; in addition *G. globosa* evolved one thin periapertural lip and broadly rounded and simple periphery, which are features that occur in the primitive planktic foraminifera. This case clearly shows the necessity to expand our study on the foraminiferal groups that were traditionally considered benthic in habitat in order to recognize other originations of planktic representatives of this group.

ACKNOWLEDGMENTS

The authors thank the DSDP and ODP Headquarters for providing the samples from which the fossil material used in this study was collected. Dr M. Schoel (Microscopy and Imaging Facility, University of Calgary) is thanked for the professional and enthusiastic support during the SEM operations.

REFERENCES

Arz, J.A., Arenillas, I., Nañez, C., 2010. Morphostatistical analysis of Maastrichtian populations of *Guembelitria* from El Kef, Tunisia. *Journal of Foraminiferal Research*, 40, 148-164.

Banner, F.T., Pereira, C.P.G., Desai, D., 1985. "Tretomphaloid" float chambers in the Discorbidae and Cymbaloporidae. *Journal of Foraminiferal Research*, 15, 159-174.

Berthelin, G., 1880. Mémoire sur les foraminifères fossils de l'étage Albien de Montcley (Doubs). *Mémoires de la Societé Géologique de France*, 1(5), 1-84.

Bettenstaedt, F., Spiegler, D., 1982. *Pleurostomella* (Foram.) in der Unterkreide Nordwestdeutschlands. *Geologisches Jahrbuch*, A63, 445-479.

BouDagher-Fadel, M.K., 2012. Biostratigraphic and geological significance of planktonic foraminifera. *Elsevier Developments in Palaeontology and Stratigraphy*, 22, 1-289.

Brown, N.K., Jr., 1969. Heterohelicidae Cushman, 1927, amended, a Cretaceous planktonic foraminiferal family. In: *Proceedings of the First International Conference on Planktonic Microfossils, Geneva 1967* (P. Brönnimann, P. and H.H. Renz, Eds). Leiden: E.J. Brill, 2, 21-67.Costa, O.G., 1856. Paleontologia del regno di Napoli, Parte II. *Atti Academia Pontanitana di Napoli*, 7: 113-378.

Cushman, J.A., 1927. An outline of a re-classification of the foraminifera. *Contributions from the Cushman Laboratory for Foraminiferal Research*, 3, 1-105.

Cushman, J.A., 1933. Some new foraminiferal genera. *Contributions from the Cushman Laboratory for Foraminiferal Research*, 9, 32-38.

Cushman, J.A., 1934. A recent *Guembelitria* (?) from the Pacific. *Contributions from the Cushman Laboratory for Foraminiferal Research*, 10, 105.

Cushman, J.A., 1938. Cretaceous species of *Gümbelina* and related genera. *Contributions from the Cushman Laboratory for Foraminiferal Research*, 14, 2-28.

Cushman, J.A., Todd, R., 1949. Species of the genera *Allomorphina* and *Quadrimorphina*. *Contributions from the Cushman Laboratory for Foraminiferal Research*, 25, 59-72.

Cushman, J.A., Wickenden, R.T.D., 1928. A new foraminiferal genus from the Upper Cretaceous. *Contributions from the Cushman Laboratory for Foraminiferal Research*, 4, 12-13.

Cžjžek, J., 1849. Über zwei neue Arten von Foraminiferen aus den Tegel von Baden und Möllersdorf. *Mitteilungen von Freunden der Naturwissenschaften*, 5(6), 50-51.

Darling, K.F., Kroon, D., Wade, C.M., Brown, A.J.L., 1996. Molecular phylogeny of the planktic foraminifera. *Journal of Foraminiferal Research*, 26, 324-330.

Darling, K.F., Thomas, E., Kasemann, S.A., Seears, H.A., Smart, C.W., Wade, C.M., 2009. Surviving mass extinction by bridging the benthic/planktic divide. *Proceedings of the National Academy of Sciences*, 106, 12629-12633.

Darling, K.F., Wade, C.M., Kroon, D., Brown, A.J.L., 1997. Planktic foraminiferal molecular evolution and their phylogenetic origins from benthic taxa. *Marine Micropaleontology*, 30, 251-266.

Delage, Y., Hérouard, E., 1896. *Traite de zoologie concrete. Tome I: La cellule et les protozoaires*. Paris: Schleicher Frères, 584 p.

Eicher, D.L., 1965. Foraminifera and biostratigraphy of the Graneros Shale. *Journal of Paleontology*, 39, 875-909.

Eicher, D.L., Worstell, P., 1970. Cenomanian and Turonian foraminifera from the Great Plains, United States. *Micropaleontology*, 16, 269-324.

Fox, S.K. Jr., 1954. Cretaceous foraminifera from the Greenhorn, Carlile and Cody Formations, South Dakota, Wyoming. *United States Geological Survey Professional Paper*, 254E, 97-124.

Frerichs, W.E., 1979. Planktonic foraminifera from the Sage Breaks Shale, Centennial Valley, Wyoming. *Journal of Foraminiferal Research*, 9, 159-184.

Fuchs, W., 1971. Eine alpine Foraminiferenfauna des tieferen Mittel-Barrême aus den Drugsbergschichten vom Ranzenberg bei Hohenems in Vorarlberg (Österreich). *Abhandlungen der Geologischen Bundesanstaldt*, 27, 1-49.

Fuchs, W., 1973. Über Ursprung und Phylogenie des Trias-"Globigerinen" und die Bedeutung dieses Foramenkreises für das echte Plankton. *Verhandlungen der Geologischen Bundesanstaldt*, 110, 135-176.

Gale, A.S., Bown, P., Caron, M., Crampton, J., Crowhurst, S.J., Kennedy, W.J., Petrizzo, M.R., Wray, D.S., 2011. The uppermost Middle and Upper Albian succession at the Col de Palluel, Hautes-Alpes, France: An integrated study (ammonites, inoceramid bivalves, planktonic foraminifera, nannofossil, geochemistry, stable oxygen and carbon isotopes, cyclostratigraphy). *Cretaceous Research*, 32, 59-130.

Georgescu, M.D., 2006.Santonian-Campanian planktonic foraminifera in the New Jersey Coastal Plain and their distribution related to the relative sea-level changes. *Canadian Journal of Earth Sciences*, 43, 101-120.

Georgescu, M.D., 2009. On the origins of Superfamily Heterohelicacea Cushman, 1927 and the polyphyletic nature of plantic foraminifera. *Revista Española de Micropaleontología*, 41, 107-144.

Georgescu, M.D., 2010. Origin, taxonomic revision and evolutionary classification of the late Coniacian-early Campanian (Late Cretaceous) planktic foraminifera with multichamber growth in the adult stage. *Revista Española de Micropaleontología*, 42, 59-118.

Georgescu, M.D., 2012a. Morphology, taxonomy, stratigraphic distribution and evolutionary classification of the schackoinid planktic foraminifera (late Albian-Maastrichtian, Cretaceous). In: *Deep-Sea Marine Biology, Geology, and Human Impact* (Bailey, D.R. and S.E. Howard, Eds). New York: Nova Publishers, 1-52.

Georgescu, M.D., 2012b. Iterative evolution, taxonomic revision and evolutionary classification of the praeglobotruncanid planktic foraminifera, Cretaceous (late Albian-Santonian). *Revista Española de Micropaleontología*, 43, 173-207.

Georgescu, M.D., Abramovich, S., 2008. Taxonomic revision and phylogenetic classification of the Late Cretaceous (Upper Santonian-Maastrichtian) serial planktonic foraminifera (Family Heterohelicidae Cushman, 1927) with peripheral test wall flexure. *Revista Española de Micropaleontología*, 40, 97-114.

Georgescu, M.D., Abramovich, S., 2009. A new Late Cretaceous (Maastrichtian) serial planktonic foraminifera (Family Heterohelicidae) with early planispiral coil and revision of *Spiroplecta* Ehrenberg, 1844. *Geobios*, 42, 687-698.

Georgescu, M.D., Arz, J.A., Macauley, R.V., Kukulski, R.B., Arenillas, I., Pérez-Rodriguez, I., 2011. Late Cretaceous (Santonian-Maastrichtian) serial foraminifera with pore mounds or pore mound-based ornamentation structures. *Revista Española de Micropaleontología*, 43, 109-139.

Georgescu, M.D., Huber. B.T., 2009. Early evolution of the Cretaceous serial planktic foraminifera (late Albian-Cenomanian). *Journal of Foraminiferal Research*, 39: 335-360.

Gooday, A.J., Alve, E., 2001. Morphological and ecological parallels between sublittoral and abyssal foraminiferal species in the NE Atlantic: a comparison of *Stainforthia fusiformis* and *Stainforthia* sp. *Progress in Oceanography*, 50, 261-283.

Hamam, K.A., 1976. *Elhasaella*, a new planktic foraminifer from the Maastrichtian of Jordan. *Revista Española de Micropaleontología*, 8, 453-458.

Hamam, K.A., Haynes, J.R., 1977. Upper Cretaceous-lower Tertiary biostratigraphy and planktic foraminifera of Abu El Awafi succession, Jordan. *Revista Española de Micropaleontología*, 9, 49-68.

Hart, M.B., Oxford, M.J., Hudson, W., 2002. The early evolution and paleobiogeography of Mesozoic planktonic foraminifera. In: *Paleobiogeography and Biodivesity Change: the Ordovician and Mesozoic-Cenozoic radiations* (Crame, J.A. and Owen, A.W., Eds). *Geological Society of London, Special Publications*, 194, 115-125.

Hart, M.B., Hylton, M.D., Oxford, M.J., Price, G.D., Hudson, W., Smart, C.W., 2003. The search for the origins of planktic Foraminifera. *Journal of the Geological Society, London*, 160, 341-343.

Hofker, J., 1956. Tertiary foraminifera of coastal Ecuador, Part II. Additional notes on the Eocene species. *Journal of Paleontology*, 30, 891-958.

Holbourn, A., Kuhnt, W., 2001. No extinctions during the Oceanic Anoxic Event 1b: the Aptian-Albian benthic foraminiferal record of ODP Leg 171. In: *Western North Atlantic Paleogene and Cretaceous Palaeoceanography* (Kroon, D., Norris, R.D. and Klaus A., Eds). *Geological Society of London, Special Publications*, 183, 73-92.

Huber, B.T., Leckie, R.M., 2011. Planktic foraminiferal species turnover across deep-sea Aptian/Albian boundary sections. *Journal of Foraminiferal Research*, 41, 53-95.

Hudson, W., Hart, M.B., Smart, C.W., 2009. Paleobiogeography of early planktonic foraminifera. *Bulletin de la Société Géologique de France*, 180, 27-38.

Keller, B. M., 1935. Microfauna of the Upper Cretaceous in the Dnjepr-Donets valley and some other adjacent regions. *Byiulletin Moskovskovo Ovacestva Prirodii (Geologii)*, 43(13), 522-558 [in Russian].

Koutsoukos, E.A.M., 1994. Early stratigraphic record and phylogeny of the planktic genus *Guembelitria* Cushman, 1933. *Journal of Foraminiferal Research*, 24, 288-295.

Kroon, D., Nederbragt, A.J., 1990. Ecology and paleoecology of triserial planktic foraminifera. *Marine Micropaleontology*, 16, 25-38.

Leckie, M.R., 2009. Seeking a better life in the plankton. *Proceedings of the National Academy of Sciences*, 106, 14183-14184.

Loeblich, A.R. Jr., Tappan, H., 1987. *Foraminiferal Genera and Their Classification*. New York: Van Nostrand Reinhold Company, 970 p.

Loeblich, A.R. Jr., Tappan, H., 1994. Foraminifera from the Sahul Shelf and Timor Sea. *Cushman Foundation for Foraminiferal Research, Special Publication*, 31, 1-661.

McNeil, D.H., Caldwell, W.G.E., 1981. Cretaceous Rocks and Their Foraminifera in the Manitoba Escarpment. *The Geological Association of Canada Special Paper*, 21, 1-439.

Michael, F.Y., 1972. Planktonic foraminifera from the Comanchean Series (Cretaceous) of Texas. *Journal of Foraminiferal Research*, 2, 200-220.

Pawlowski, J., Bolivar, I., Guiard-Maffia, J., Gouy, M., 1994. Phylogenetic position of foraminifera inferred from LSU rRNA gene sequences. *Molecular Biology and Evolution*, 11, 929-938.

Pawlowski, J., Bolivar, I., Fahrni, J., De Vargas, C., Gouy, M, Zaninetti, L., 1996. Early origin of foraminifera suggested by SSU rRNA gene sequences. *Molecular Biology and Evolution*, 13, 445-450.

Pessagno, E.A. Jr., 1967. Upper Cretaceous planktonic foraminifera from the western Gulf coastal plain. *Palaeontographica Americana*, 5(37), 243-445.

Shipboard Scientific Party, 1978. Site 370: Deep Basin off Morocco. In: *Initial Reports of the Deep Sea Drilling Project, Volume 41* (Lancelot, Y., Seibold, E. and others, Eds). Washington, D. C.: United States Government Printing Office, 421-491.

Smart, C.W., Thomas, E., 2007. Emendation of the genus *Streptochilus* Brönnimann and Resig 1971 (Foraminifera) and new species from the lower Miocene of the Atlantic and Indian Oceans. *Micropaleontology*, 53, 73-103.

Stelck, C.R., Wall, J.H., 1954. Kaskapau Foraminifera from Peace River area of western Canada. *Research Council of Alberta, Reports*, 68, 1-38.

Suleymanov, I.S., 1955. A new genus, *Gubkinella*, and two new species of the family Heterohelicidae from the upper Senonian of the southeastern Kyzyl-Kumy. *Doklady Akademya Nauk SSSR*, 102, 623-624. [in Russian]

Tappan, H., 1940. Foraminifera from the Grayson Formation of northern Texas. *Journal of Paleontology*, 14, 93-126.

Tappan, H., 1962. Foraminifera from the Arctic Slope of Alaska. Part 3, Cretaceous Foraminifera. *United States Geological Survey Professional Paper*, 236C, 91-206.

Ujiié, Y., Kimoto, K., Pawlowski, J., 2008. Molecular evidence for an independent origin of modern triserial planktonic foraminifera from benthic ancestors. *Marine Micropaleontology*, 69, 334-340.

Wade, C.M., Darling, K.F., Kroon, D., Brown, A.J.L., 1996. Early evolutionary origin of the planktic foraminifera inferred from small subunit rRNA sequence comparisons. *Journal of Molecular Evolution*, 43, 672-677.

Williamson, W.C., 1858. *On the Recent foraminifera of Great Britain*. London: Ray Society, 105 p.

In: Evolutionary Classification ...
Editors: M. Dan Georgescu and C. M. Henderson

ISBN: 978-1-63321-959-5
© 2015 Nova Science Publishers, Inc.

Chapter 2

REINSTATEMENT OF THE CRETACEOUS PLANKTIC FORAMINIFER *BRONNIMANNELLA* MONTANARO GALLITELLI 1956 AS DIRECTIONAL LINEAGE IN EVOLUTIONARY CLASSIFICATION

M. Dan Georgescu[*]

Department of Geosciences, University of Calgary,
Calgary, Alberta, Canada

ABSTRACT

Genus *Bronnimannella* Montanaro Gallitelli 1956 is revised as directional lineage in evolutionary classification. It includes the initiating species *Ventilabrella plummerae* Sandidge 1932 and first descendant species *Gümbelina nuttalli* Voorwijk 1937. *Bronnimannella* presents the test ornamented with longitudinal leptocostae and is the first lineage that evolved chamber transversal elongation; it evolved in the Santonian and became extinct in the late Maastrichtian.

Keywords: Foraminifera, planktic, Santonian, Campanian, Maastrichtian, evolutionary classification

INTRODUCTION

Chamber transversal elongation was the first feature recognized by Rzehak (1891) in the group of Cretaceous planktic foraminifera with chambers alternately added with respect to the test growth axis following the pioneering studies on the representatives of the group by Ehrenberg (1838, 1841, 1844, 1854). The genus *Pseudotextularia*, which was proposed by this author for the tests having this feature, was widely but not unanimously accepted in the

---

[*] Corresponding author: dgeorge@ucalgary.ca.

next decades; the alternative taxonomic solution was to include the tests with transversally elongate chambers within the genus *Gümbelina* Egger 1899 based on the occurrence of biserially arranged chambers throughout the ontogeny.

Montanaro Gallitelli (1955, 1956) was the first to realize the importance of chamber transversal elongation in the typological classification and above the species level. This author considered *Pseudotextularia* valid and in addition described the genus *Bronnimannella* having as type species *Gümbelina plummerae* Loetterle 1937 for the test presenting incipient chamber transversal elongation. This taxonomic framework was not followed further and one year later Montanaro Gallitelli (1957) considered *Bronnimannella* one junior synonym of *Pseudotextularia*. This synonymy was validated in all the taxonomic revisions of the group that followed (Loeblich and Tappan 1964, 1987; Pessagno 1967; Brown 1969; Masters 1977; Weiss 1983; Nederbragt 1989, 1991). Moreover, it was accepted in all the regional reports dedicated to the heterohelicid group (Darmoian 1975; Abdel-Kireem 1986; Georgescu 1995).

Nederbragt (1991) recognized for the first time ancestor-descendant relationships between the species with chamber transversal elongation, which were all included within the genus *Pseudotextularia*. It is evident that the evolution within the lineage recognized by Nederbragt (1991) involves the increase in ornamentation prominence and subsequently the development of incipient chamber proliferation.

The material on which this study is based was collected from the following Deep Sea Drilling Project (DSDP)/Ocean Drilling Program (ODP) sites and holes: DSDP Site 305 (Shatsky Rise, Central Pacific Ocean), DSDP Site 463 (western Mid-Pacific Mountains, Central Pacific Ocean), ODP Hole 761B (Wombat Plateau, East Indian Ocean), ODP Holes 762C and 763B (Exmouth Plateau, East Indian Ocean) and Holes 1050C and 1052E (Blake Nose, North Atlantic Ocean). Well-preserved specimens from spot samples from the DSDP Site 384 (*J*-Anomaly Ridge, North Atlantic Ocean) and Eureka well 67-128 (Gulf of Mexico, Caribbean region) were studied in the van Morkhoven Collection at the National Museum of Natural History, Smithsonian Institution, Washington, D.C.

As a result of recent studies based on the high-resolution observations with the aid of the Scanning Electron Microscope (SEM) it became possible to reassess the taxonomic role of the chamber elongation in the group systematic. One of the outcomes of this high-resolution taxonomic study was the recognition of three distinct lineages that resulted in the evolution of chamber transversal elongation; all these lineages evolved in the Santonian-Maastrichtian stratigraphic interval. Test ornamentation proves an accurate discriminator between the three lineages: one lineage evolved in the Santonian and it consists of tests with leptocostate ornamentation (*Bronnimannella*-emended herein), whereas the two lineages that evolved in the late Campanian present pycnocostate ornamentation (*Pseudotextularia* and *Racemiguembelina*). The pycnocostate lineages are presented in a different article of this volume.

SYSTEMATIC DESCRIPTIONS

The units in evolutionary classification are after Georgescu (2010, 2011, 2013). Species kind abbreviations: IS-initiating species, FDS-first descendant species.

Directional Lineage: *Bronnimannella* Montanaro Gallitelli 1956 - Emended

Bronnibrownia Montanaro Gallitelli 1955, p. 215 (invalid name).
Bronnimannella Montanaro Gallitelli 1956, p. 35.

Species included. IS: *B. plummerae* (Sandidge 1932) and FDS: *B. nuttalli* (Voorwijk 1937).

Diagnosis. Santonian-Maastrichtian directional lineage ornamented with longitudinal leptocostae that evolved chamber transversal elongation.

Description. Test with subglobular chambers with variable overlapping and gradual size increase. Chambers are alternately added with respect to the test growth axis throughout the ontogenetic development; one last-formed biaperturate chamber occurs occasionally in the FDS. Sutures are distinct and depressed between all the chambers of the test. Chambers present a distinct transversal elongation, which is an incipient stage in the IS and well-developed in the FDS. Periphery is broadly rounded and simple, without peripheral structures throughout. Aperture is an arch at the base of the last-formed chamber; it is low to medium high in the IS and low in the FDS. The aperture is bordered by two narrow metaflanges in the IS and the periapertural structures are strongly reduced and even absent in the case of the chambers with well-developed transversal elongation in the FDS; one imperforate lip borders the aperture between the two metaflanges. Chamber surface is ornamented with longitudinal leptocostae; a wide pustulose periapertural area consisting of dome-like pustules occurs in the anterior part of the chambers. Test wall is calcitic, hyaline, simple and perforate; pores are circular and are situated in the space between the leptocostae.

Remarks. *Bronnimannella* Montanaro Gallitelli 1956 was considered a junior synonym of *Pseudotextularia* 1891 shortly after its description (Montanaro Gallitelli 1957). It is herein redefined as a directional lineage in the evolutionary classification. The directional lineage *Bronnimannella* is the only leptocostate lineage that evolved chamber transversal elongation.

Age. Santonian-Maastrichtian.

Geographic distribution. Cosmopolitan.

IS: *Bronnimannella Plummerae* (Sandidge 1932) (Figure 1:1-10)

Ventilabrella plummerae SANDIDGE 1932, p. 195, pl. 19, figs 5-6.
Gümbelina plummerae LOETTERLE 1937, p. 33, pl. 5, figs 1-2.
Gümbelina plummerae Loetterle. Cushman 1938, p. 15, pl. 3, figs 3-5. Cushman and Deaderick 1942, p. 62, pl. 15, figs 2-4. Cushman 1944a, p. 10, pl. 2, fig. 18. Cushman 1944b, p. 90, pl. 14, fig. 3. Cushman 1946, p. 104, pl. 45, figs 1-3. Cushman 1948, pl. 24, fig. 2. Bolin 1952, p. 39, pl. 2, fig. 18. Hamilton 1953, p. 234, pl. 30, fig. 11.
Pseudotextularia elegans (Rzehak). Montanaro Gallitelli 1957, pl. 33, fig. 6. Graham and Clark 1961, p. 111, pl. 5, fig. 5. Frerichs and Dring 1981, p. 68, pl. 1, figs 15-16. Neagu 1987, p. 291, pl. 14, figs 31-32.
Pseudotextularia cushmani BROWN 1969, p. 55, pl. 2, figs 2-3, pl. 3, fig. 4.
Pseudotextularia plummerae (Loetterle). Brown 1969, p. 56, pl. 4, figs 6-7. Darmoian 1975, pl. 3, figs 18-19. McNeil and Caldwell 1981, p. 243, pl. 19, figs 5-6.
Pseudotextularia browni MASTERS 1976, p. 321, pl. 1, fig. 12. (only)

Heterohelix plummerae (Loetterle). Pandey 1980, p. 62, pl. 2, figs 19-22.

Pseudotextularia cushmani Brown. Abdel-Kireem 1986, p. 224, pl. 2, figs 8-9. Howe and others 2003, pl. 8, figs 1-2.

Pseudotextularia elongata Seiglie. Thompson 1991, p. 38, pl. 1, figs 10-11.

Pseudotextularia nuttalli (Voorwijk). Petrizzo 2000, fig. 11: 9. Howe and others 2007, fig. 7: A-B. Lamolda and others 2007, fig. 4: O. Farouk and Faris 2012, fig. 9: 1-2.

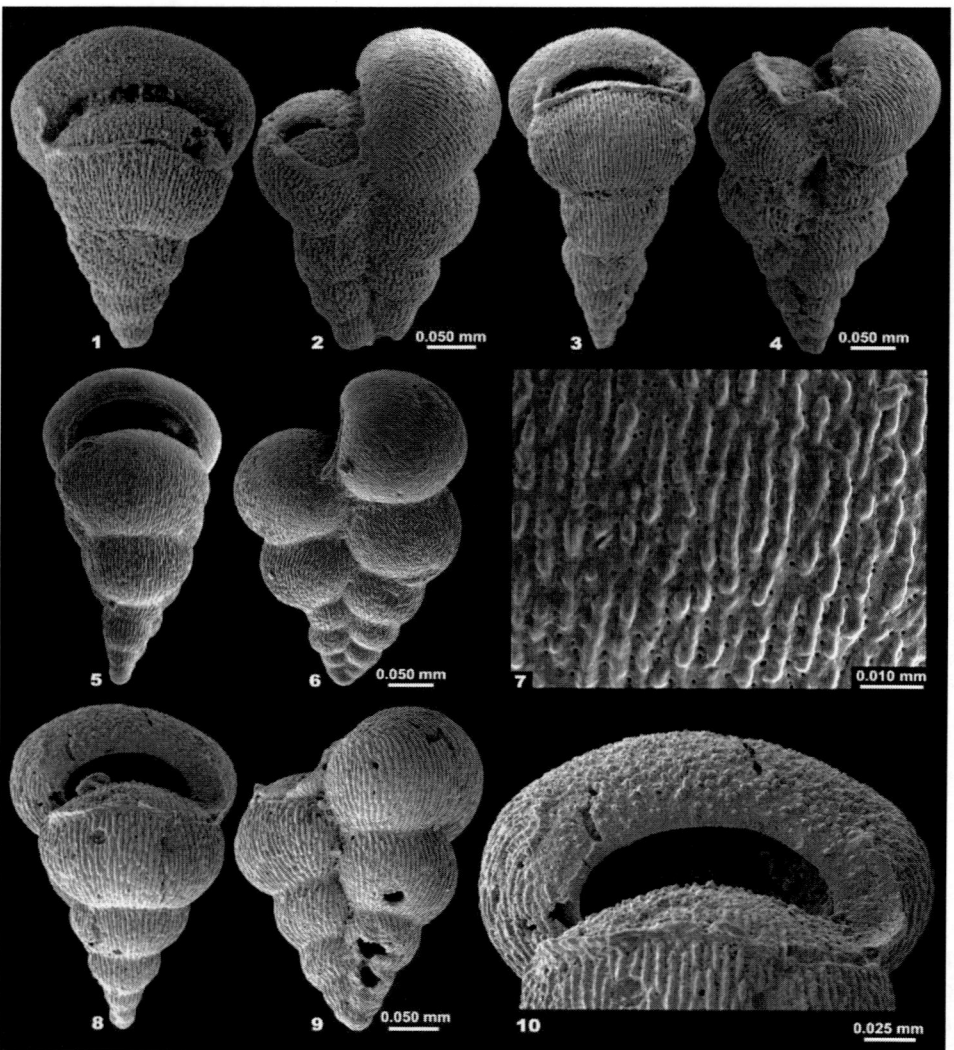

Figure 1. Specimens of *Bronnimannella plummerae* (Sandidge 1932) from the Yucatan Outer Shelf, Caribbean Region (1-4), New Jersey coastal plain, USA (5-7) and Western Australia (8-10). 1-2 Hypotype, specimen from Sample 10-95-15-4, 99.5-100.5 cm (upper Santonian, *D. asymetrica* Biozone). 3-4 Hypotype, specimen from Sample 10-95-15-4, 99.5-100.5 cm (upper Santonian, *D. asymetrica* Biozone). 5-7 Hypotype, specimen from Sample 174AX, 495.30-495.33 cm (upper Santonian, *G. arca* Biozone). 8-10 Hypotype, specimen from Toolonga Calcilutite (Santonian, Western Australia).

Diagnosis. *Bronnimannella* with incipient chamber transversal elongation.

Description. Test consists of the proloculus followed by 13-16 subglobular chambers that overlap at various rate and present a gradual size increase; chambers are alternately added with respect to the test growth axis throughout the ontogenetic development. Sutures are distinct, depressed, straight to slightly curved. Tests are symmetrical in edge view, with the chambers increasing rapidly in size resulting in a distinct transversal elongation, which is more apparent in the last-formed chambers. Periphery is broadly rounded and simple, without peripheral structures. Aperture is a medium to low arch situated at the base of the last-formed chamber. The aperture is bordered by two symmetrically developed narrow metaflanges, one on each side of the test; a narrow imperforate lip borders the aperture between the two metaflanges. Chamber surface is ornamented with longitudinal continuous leptocostae, but the continuity is less developed in the case of large specimens; the leptocostae present a thickness of 0.0033-0.0049 mm. A wide pustulose periapertural area consisting of dome-like pustules occurs in the anterior portion especially of the last-formed chambers. Test wall is calcitic, hyaline, simple and perforate. Pores are circular in shape and have diameters 0.0005-0.0020 mm; pores are situated in the spaces between the leptocostae.

Remarks. *Bronnimannella plummerae* differs from its ancestor *Mihaia reussi* (Cushman 1938) by the chamber transversal elongation, wider pustulose periapertural area consisting of dome-like pustules and larger pores (0.0005-0.0020 mm rather than 0.0006-0.0012 mm).

Age. Santonian-Campanian.

Geographic distribution. Cosmopolitan.

FDS: *Bronnimannella nuttalli* (Voorwijk 1937) (Figure 2:1-10)

Gümbelina acervulinoides EGGER 1899, pl. 14, figs 14-16 (only).

Gümbelina nuttalli VOORWIJK 1937, p. 5, pl. 2, figs 1-9.

Gümbelina plummerae Loetterle. Cole 1938, pl. 3, fig. 9. Cushman 1949, p. 7, pl. 3, figs 21-22.

Gümbelina elegans (Rzehak). Cita 1948, p. 125, pl. 2, fig. 6.

Pseudotextularia elegans (Rzehak). Noth 1951, p. 61, pl. 7, figs 15-17. Seiglie 1958, p. 55, pl. 1, figs 1, 3. Cati 1964, p. 261, pl. 42, fig. 10. Lehmann 1966, p. 316, pl. 2, fig. 10. Pessagno 1967, p. 268, pl. 75, figs 12-17, pl. 85, figs 10-11, pl. 88, figs 14-16, pl. 89, figs 10-11, Hanzliková 1969, p. 43 figs 11, 13 (only). Hanzliková 1972, p. 95, pl. 24, figs 11-12 (only). Smith and Pessagno 1973, p. 30, pl. 9, figs 5-15, pl. 10, figs 4-6. Wright and Apthorpe 1976, pl. 1, fig. 3 (only). Schreiber 1979, p. 41, pl. 1, figs 1-13, pl. 2, figs 1-4, pl. 3, figs 14-17. Butt 1981, pl. 18, fig. B. Petters 1983, p. 45, pl. 1, fig. 13. Weiss 1983, p. 61, pl. 8, figs 5-7. Gawor-Biedowa 1992, p. 74, pl. 11, figs 10-11. Abramovich and others 2002, pl. 1, fig. 7. Abramovich and others 2003, pl. 4, fig. 2.

Ventilabrella carseyae Plummer. Sellier de Civrieux 1952, p. 271, pl. 6, figs 17-18.

Gümbelina plummerae Loetterle. Said and Kenawy, 1956, p. 139, pl. 3, fig. 33.

Gümbelina nuttalli Voorwijk. Sacal and Debourle 1957 p. 12, pl. 3, fig. 2.

Pseudotextularia browni MASTERS 1976, p. 321, pl. 1, figs 10-11. (only)

Pseudotextularia intermedia de Klasz. Linares-Rodríguez 1977, pl. 48, fig. 5.

Pseudotextularia browni Masters. Masters 1977, p. 380, pl. 5, fig. 3-4.

Pseudotextularia nuttalli (Voorwijk). Nederbragt 1989, p. 204, pl. 8, figs 2-3, text-fig. 9. Malmgren 1991, pl. 1, fig. 10. Nederbragt 1991, p. 364, pl 10, figs 4, 6. Mancini and others

1996, fig. 5: 10. Zapeda 1998, p. 138, fig. 11: 6. Fondecave-Wallez and others 1999, pl. 2, figs 4-6. Campbell and others 2004, fig. 13: Y-Z. Ohmert 2011, pl. 3, figs 16-17.

Pseudotextularia cushmani Brown. Thompson 1991, p. 37, pl. 1, figs 8-9.

Figure 2. Specimens of *Bronnimannella nuttalli* (Voorwijk 1937) from the Gulf of Mexico, Caribbean Region (1-4), Wombat Plateau, East Indian Ocean (5-6) and J-Anomaly Ridge, North Atlantic Ocean (7-10). 1-2 Hypotype, specimen from the well Eureka 67-128 (upper Campanian, *R. calcarata* Biozone). 3-4 Hypotype, specimen from the well Eureka 67-128 (upper Campanian, *R. calcarata* Biozone). 5-6 Hypotype, specimen from Sample 122-761B-25-4, 75-76 cm (upper Campanian, *R. calcarata* Biozone). 7-8 Hypotype, specimen from Sample 43-384-14-2, 50-52 cm (upper Maastrichtian, *A. mayaroensis* Biozone). 9-10 Hypotype, specimen from Sample 43-384-14-2, 50-52 cm (upper Maastrichtian, *A. mayaroensis* Biozone).

Diagnosis. *Bronnimannella* with the last-formed chambers with well-developed transversal chamber elongation.

Description. Test consists of the proloculus followed by 15-18 subglobular chambers, which are subglobular in shape that overlap at various rates and present a gradual size increase. Chambers are alternately added with respect to the test growth axis throughout ontogeny. Occasionally a biaperturate last-formed chamber occurs in the anterior portion of the test. Sutures are distinct and depressed, straight to slightly curved. Chambers increase rapidly in thickness resulting in a well-developed transversal elongation, which is more apparent in the last-formed chambers. Periphery is broadly rounded and simple, without peripheral structures. Aperture is a low arch situated at the base of the last-formed chamber. The aperture is bordered by two symmetrically developed narrow metaflanges, which are strongly reduced or absent on the last-formed chambers that present well-developed transversal elongation; one imperforate lip borders the aperture between the two metaflanges. Chamber surface is ornamented with longitudinal leptocostae that present a thickness of 0.0032-0.0053 mm; a wide pustulose periapertural area consisting of dome-like pustules occurs in the anterior portion of the chambers. Test wall is calcitic, hyaline, simple and perforate; pores are simple, circular with a diameter of 0.0008-0.0025 mm and are situated in the spaces between the leptocostae.

Remarks. *Bronnimannella nuttalli* differs from its ancestor *B. plummerae* mainly by the well-developed rather than incipient chamber transversal elongation; in addition the pores present a slight increase in size: 0.0008-0.0025 mm rather than 0.0005-0.0020 mm.

Age. Campanian-Maastrichtian.

Geographic distribution. Cosmopolitan.

CONCLUSION

The taxonomic revision of the Cretaceous planktic foraminifera with the chambers alternately added with respect to the test growth axis that evolved for the first time in the group history transversally elongate chambers shows that they form a directional lineage. This directional lineage is named *Bronnimannella* after a genus described by Montanaro Gallitelli (1956), which is herein revised in evolutionary classification. *Bronnimannella* consists of two species: initiating species - *B. plummerae* (Sandidge 1932) and first descendant species - *B. nuttalli* (Voorwijk 1937). There is a gradual development in the chamber transversal elongation in this lineage; this feature is an incipient stage of development in the initiating species and well-developed in the first descendant species.

Bronnimannella plummerae evolved in the Santonian from *Mihaia reussi* (Cushman 1938) as indicated by the similarities in the gross test architecture, periapertural structures and test ornamentation. The evolution of chamber transversal elongation is gradual and tests in which this feature is well-developed occur for the first time in the proximity of the lower/middle Campanian boundary. The directional lineage *Bronnimannella* became extinct at the Cretaceous/Paleogene boundary.

REFERENCES

Abdel-Kireem, M.R., 1986. Planktonic foraminifera and stratigraphy of the Tanjero Formation (Maastrichtian), northeastern Iraq. *Micropaleontology*, 32, 215-231.

Abramovich, S., Keller, G., Adatte, T., Stinnesbeck, W., Hottinger, L., Stueben, D., Berner, Z., Ramanivosoa, B., Randiriamanantenasoa, A., 2002. Age and paleoenvironment of the Maastrichtian to Paleocene of the Mahajanga Basin, Madagascar: a multidisciplinary approach. *Marine Micropaleontology*, 47, 17-70.

Abramovich, S., Keller, G., Stüben, D., Berner, Z., 2003. Characterization of late Campanian and Maastrichtian planktonic foraminiferal depth habitats and vital activities based on stable isotopes. *Palaeogeography, Palaeoclimatology, Palaeoecology*, 202, 1-29.

Bolin, E.J., 1952. Microfossils of the Niobrara Formation of southeastern South Dakota. *University of South Dakota Report of Investigations*, 70, 1-74.

Brown, N.K., Jr., 1969. Heterohelicidae Cushman, 1927, amended, a Cretaceous planktonic foraminiferal family. In: *Proceedings of the First International Conference on Planktonic Microfossils, Geneva 1967* (P. Brönnimann, P. and H.H. Renz, Eds). Leiden: E.J. Brill, 2, 21-67.

Butt, A., 1981. Depositional environments of the Upper Cretaceous rocks in the northern part of the Eastern Alps. *Cushman Foundation for Foraminiferal research Special Publication*, 20, 1-121.

Campbell, R. J., Howe, R. W., Rexilius, J. P., 2004. Middle Campanian-lowermost Maastrichtian nannofossil and foraminiferal biostratigraphy of the northwestern Australian margin. *Cretaceous Research*, 25, 827-864.

Cati, F., 1964. Una microfauna campaniana dei Monti berici (Vicenza). *Giornale di Geologia*, 32, 199-271.

Cita, M.B., 1948. Ricerche stratigrafiche e micropaleontologiche sul Cretarcico e sull'Eocene di Tignale (Lago di Garda). II. Paleontologia. *Rivista Italiana di Paleontologia e Stratgrafia*, 54, 117-143.

Cole, W.S., 1938. Stratigraphy and micropaleontology of two deep wells in Florida. *Bulletin of the Florida State Geological Survey*, 16, 1-48.

Cushman, J.A., 1938. Cretaceous species of *Gümbelina* and related genera. *Contributions from the Cushman Laboratory for Foraminiferal Research*, 14, 2-28.

Cushman, J.A., 1944a. The foraminiferal fauna of the type locality of the Pecan Gap Chalk. *Contributions from the Cushman Laboratory for Foraminiferal Research*, 20, 1-16.

Cushman, J.A., 1944b. Foraminifera op the lower part of the Mooreville Chalk of the Selma Group of Mississippi. *Contributions from the Cushman Laboratory for Foraminiferal Research*, 20, 83-96.

Cushman, J.A., 1946. Upper Cretaceous foraminifera of the Gulf coastal region of the United States and adjacent areas. *United States Geological Survey Professional Paper*, 206, 1-241.

Cushman, J.A., 1948. Foraminifera from the Hammond well. *Cretaceous and Tertiary Subsurface Geology. The Stratigraphy, Paleontology, and Sedimentology of Three Deep Test Wells on the Eastern Shore of Maryland. Baltimore, Maryland*, 213-343.

Cushman, J.A., 1949. The foraminiferal fauna of the Upper Cretaceous Arkadelphia Marl in Arkansas. *United States Geological Survey Professional Paper*, 221A, 1-17.

Cushman, J.A., Deaderick, W.H., 1942. Cretaceous Foraminifera from the Brownstown Marl of Arkansas. *Contributions from the Cushman Laboratory for Foraminiferal Research*, 18, 50-66.

Darmoian, S.A., 1975. Planktonic foraminifera from the Upper Cretaceous of southeastern Iraq: Biostratigraphy and systematics of the Heterohelicidae. *Micropaleontology*, 21, 185-214.

Egger, J.G., 1899. Foraminiferen und Ostrakoden aus den Kreidemergeln der Oberbayerischen Alpen. *Abhandlungen der Mathematisch-Physikalischen Klasse der Königlich Bayerischen Akademie der Wissenschaften*, 21, 3-230. [published in 1902]

Ehrenberg, C.G., 1838. Über die Bildung der Kreidefelsen und des Kreidemergels durch unsichtbare Organismen. *Abhandlungen der Königlichen Akademie der Wissenschaften zu Berlin*, 1838, 59-147. [published in 1839]

Ehrenberg, C.G., 1841. Verbreitung und Einflufs des mikroscopischen Lebens in Süd- und Nord- Amerika. *Abhandlungen der Königlichen Akademie der Wissenschaften zu Berlin*, 1841, 291-445. [published in 1843]

Ehrenberg, C.G., 1844. Eine Mittbeilung über 2 neue Lager von Gebirgsmassen aus Infusorien als Meeres-Absatz in Nord-Amerika und eine Vergleichung derselben mit den organischen Kreide-Gebilden in Europa und Afrika. *Bericht über die zur Bekanntmachung geeigneten Verhandlungen der Königlich Preußischen Akademie der Wissenschaften zu Berlin*, 1844, 57-98.

Ehrenberg, C.G., 1854. *Mikrogeologie*. Leipzig: L. Voss, 374 p.

Farouk, S., Faris, M., 2012. Late Cretaceous calcareous nannofossil and planktonic foraminiferal bioevents of the shallow-marine carbonate platform in the Mitla Pass, west central Sinai, Egypt. *Cretaceous Research*, 33, 50-65.

Fondecave-Wellez, M.J., Eichène, P., Peybernès, B., Gourinard, Y., 1999. Les foraminifères planctoniques de la série campanienne continue d'Hendaye-Baie de Loya (Pyrénées-Atlantiques, France). *Géologie Méditerranéenne*, 26, 47-57.

Frerichs, W.E., Dring, N.B., 1981. Planktonic Foraminifera from the Smoky Hill Shale of West Central Kansas. *Journal of Foraminiferal Research*, 11, 47-69.

Gawor-Biedowa, E., 1992. Campanian and Maastrichtian foraminifera from the Lublin Upland, Eastern Poland. *Palaeontologica Polonica*, 52, 1-187.

Georgescu, M.D., 1995. Upper Cretaceous Heterohelicidae in the Romanian Western Black Sea offshore. *Revista Española de Micropaleontología*, 27, 91-106.

Georgescu, M.D., 2010. Origin, taxonomic revision and evolutionary classification of the late Coniacian-early Campanian (Late Cretaceous) planktic foraminifera with multichamber growth in the adult stage. *Revista Española de Micropaleontología*, 42, 59-118.

Georgescu, M.D., 2011. Iterative evolution, taxonomic revision and evolutionary classification of the praeglobotruncanid planktic foraminifera, Cretaceous (late Albian-Santonian). *Revista Española de Micropaleontología*, 43. 173-207. [published in 2012]

Georgescu, M.D., 2013. Revised evolutionary systematics of the Cretaceous planktic foraminifera described by C.G. Ehrenberg. *Micropaleontology*, 59, 1-49.

Graham, J.J., Clark, D.K., 1961. New evidence for the age of the "G-1 Zone" in the Upper Cretaceous of California. *Contributions from the Cushman Foundation for Foraminiferal Research*, 12, 107-114.

Hamilton, E.L., 1953. Upper Cretaceous, Tertiary, and Recent Planktonic Foraminifera from Mid-Pacific flat-topped seamounts. *Journal of Paleontology*, 27, 204-237.

Hanzlikova, E., 1969. The foraminifera of the Frýdek Formation (Senonian). *Sborník Geologických Věd, Paleontologie*, 11, 7-79.

Hanzliková, E., 1972. Carpathian Upper Cretaceous Foraminiferida of Moravia (Turonian-Maastrichtian). *Rozpravy Ústředního Ústavu Geologického*, 39, 1-160.

Howe, R.W., Campbell, R.J., Rexilius, J.P., 2003. Integrated uppermost Campanian-Maastrichtian calcareous nannofossil and foraminiferal biostratigraphic zonation of the northwestern margin of Australia. *Journal of Micropaleontology*, 22, 29-62.

Howe, R.W., Sikora, P.J., Gale, A.S., Bergen, J.A., 2007. Calcareous nannofossils and planktonic foraminiferal biostratigraphy of proposed stratotypes for the Coniacian/Santonian boundary: Olazagutía, northern Spain; Seaford Head, southern England; and Ten Mile Creek, Texas, USA. *Cretaceous Research*, 28, 61-92.

Lamolda, M.A., Peryt, D., Ion, J., 2007. Planktonic foraminiferal bioevents in the Coniacian/Santonian boundary interval of Olazagutia, Navarra Province, Spain. *Cretaceous Research*, 28, 18-29.

Lehmann, R., 1966. Description des Globotruncanidés et Hétérohelicidés d'une faune maestrichtienne du Prérif (Maroc). *Eclogae Geologicae Helvetiae*, 59, 309-317.

Linares-Rodríguez, D., 1977. *Foraminiferos planctonicos del Cretacico superior de las Cordilleras Beticas (sector central)*. Universidad de Málaga, Departamento de Geología, 410 p.

Loeblich, A.R. Jr., Tappan, H., 1964. Sarcodina Chiefly "Thecamoebians" and Foraminifera. In: *Treatise on Invertebrate Paleontology. Part C* (R.C. Moore, Ed.). The Geological Society of America and The University of Kansas Press, 900 p.

Loeblich, A.R. Jr., Tappan, H., 1987. *Foraminiferal Genera and Their Classification*. New York: Van Nostrand Reinhold Company, 970 p.

Loetterle, G.J., 1937. The micropaleontology of the Niobrara Formation in Kansas, Nebraska, and South Dakota. *Nebraska Geological Survey Bulletin*, 12, 1-73.

Malmgren, B.A., 1991. Biogeographic patterns in terminal Cretaceous planktonic foraminifera from Tethyan and warm transitional waters. *Marine Micropaleontology*, 18, 73-99.

Mancini, E.A., Puckett, T.M., Tew, B.H., 1996. Integrated biostratigraphic and sequence stratigraphic framework for Upper Cretaceous strata of the eastern Gulf Coastal Plain, USA. *Cretaceous Research*, 17, 645-669.

Masters, B.A., 1976. Planktic foraminifera from the Upper Cretaceous Selma Group, Alabama. *Journal of Paleontology*, 50, 318-330.

Masters, B.A., 1977. Mesozoic planktonic foraminifera. A world-wide review and analysis. In: *Oceanic Micropaleontology* (A.T.S. Ramsay, Ed.). London-New York-San Francisco: Academic Press, 1, 301-731.

McNeil, D.H., Caldwell, W.G.E., 1981. Cretaceous Rocks and Their Foraminifera in the Manitoba Escarpment. *The Geological Association of Canada Special Paper*, 21, 1-439.

Montanaro Gallitelli, E., 1955. Una revision della famiglia Heterohelicidae Cushman. *Atti e Memorie dell'Accademia di Scienze e Lettere Modena*, 13, 213-223.

Montanaro Gallitelli, E., 1956. *Bronnimannella*, *Tappanina* and *Trachelinella*, three new foraminiferal genera from the Upper Cretaceous. *Contributions from the Cushman Foundation for Foraminiferal Research*, 7, 35-39.

Montanaro Gallitelli, E., 1957. A revision of the foraminiferal family Heterohelicidae. In: *Studies in foraminifera* (A.R. Jr. Loeblich, Ed.). Washington, D.C.: *United States National Museum History Bulletin*, 215, 133-154.

Neagu, T., 1987. White Chalk foraminiferal fauna in southern Dobrogea (Romania). 1. Planktonic Foraminifera. *Revista Española de Micropaleontología*, 19, 281-314.

Nedebragt, A.J., 1989. Maastrichtian Heterohelicidae (planktic foraminifera) from the West North Atlantic. *Journal of Micropaleontology*, 8, 183-206.

Nedebragt, A.J., 1991. Late Cretaceous biostratigraphy and development of Heterohelicidae (planktic foraminifera). *Micropaleontology*, 37, 329-372.

Noth, R., 1951. Foraminiferen aus Unter- und Oberkreide des Österreichischen anteils an Flysch, Helvetikum und Vorlandvorkommen. *Jahrbuch der Geologischen Bundesanstaldt, Sonderband*, 3, 1-91.

Ohmert, W., 2011. Radiolarien-Faunen und Stratigraphie der Plattenau-Formation (Campanium bis Maastrichtium) im Helvetikum von Bad Tölz. *Zittelliana*, 51, 37-95.

Pandey, J., 1980. Cretaceous foraminifera of Um Sohryngkew River section, Meghalaya. *Journal of the Palaeontological Society of India*, 25, 53-74. [published in 1981]

Pessagno, E.A. Jr., 1967. Upper Cretaceous planktonic foraminifera from the Western Gulf coastal plain. *Palaeontographica Americana*, 5(37), 243-445.

Petrizzo, M.R., 2000. Upper Turonian-lower Campanian foraminifera from southern mid-high latitudes (Exmouth Plateau, NW Australia): biostratigraphy and taxonomic notes. *Cretaceous Research*, 21, 479-505.

Petters, S.W., 1983. Gulf of Guinea planktonic foraminiferal biochronology and geological history of the South Atlantic. *Journal of Foraminiferal Research*, 13, 32-59.

Rzehak, A., 1891. Die Foraminiferenfauna der alttertiären Ablagerungen von Bruderndorf in Nieder-Osterreich, mit Berüchsichtigung des angeblichen Kreidevorkommens von Leitzersdorf. *Annalen des K.K. Naturhistorischen Hofmuseums*, 6, 1-12.

Sacal, V., Debourle, A., 1957. Foraminifères d'Aquitaine 2e partie. Peneroplidae a Victoriellidae. *Mémoires de la Société Géologique de France*, 78, 1-87.

Said, R., Kenawy, A., 1956. Upper Cretaceous and Lower Tertiary foraminifera from northern Sinai, Egypt. *Micropaleontology*, 2, 105-173.

Sandidge, J.R., 1932. Significant foraminifera from the Ripley Formation of Alabama. *American Midland Naturalist*, 13, 190-202.

Schreiber, O.S., 1979. Heterohelicidae (Foraminifera) aus der Pemberger-Folge (Oberkreide) von Klein-Sankt Paul am Krappfeld (Kärnten). Beiträge *zur* Paläontologie Österreich, 6, 27-59.

Seiglie, G.A., 1958. Notas sobre algunas especies de Heterohelicidae del Cretacico superior de Cuba. *Boletín de la Asociatión Mexicana de Geólogos Petroleros*, 11, 51-62.

Sellier de Civrieux, J.M., 1952. Estudio de la microfauna de la seccion-tipo del Miembro Socuy de la Formacion Colon Distrito Mara, Estado Zulia. *Ministerio de Minas e Hidrocarburos Direccion de Geologia*, 2, 231-310.

Smith, C. C., Pessagno, E. A. Jr., 1973. Planktonic foraminifera and stratigraphy of the Corsicana Formation (Maestrichtian), north-central Texas. *Cushman Foundation for Foraminiferal Research, Special Publications*, 13, 5-68.

Thompson, L.B., 1991. Late Santonian to early Maastrichtian planktonic foraminiferal biostratigraphy and zonation of northeast Texas. *Micropaleontology Special Publication*, 5, 9-66.

Voorwijk, G.H., 1937. Foraminifera from the Upper Cretaceous of Habana, Cuba. *Proceedings of the Koninklijke Akademie van Wetenschappen te Amsterdam,* 40, 190-198.

Weiss, W., 1983. Heterohelicidae (seriale planktonische Foraminiferen) der tethyalen Oberkreide (Santon bis Maastricht). *Geologisches Jahrbuch,* A72, 3-93.

Wright, C.A., Apthorpe, M., 1976. Planktonic foraminiferids from the Maastrichtian of the northwest shelf, Western Australia. *Journal of Foraminiferal Research,* 6, 22-241.

Zapeda, M., 1998. Planktonic foraminiferal diversity, equitability and biostratigraphy of the uppermost Campanian-Maastrichtian, ODP Leg 122, Hole 762C, Exmouth Plateau, NW Australia, eastern Indian Ocean. *Cretaceous Research,* 19, 117-152.

In: Evolutionary Classification ...
Editors: M. Dan Georgescu and C. M. Henderson

ISBN: 978-1-63321-959-5
© 2015 Nova Science Publishers, Inc.

Chapter 3

NEW LATE CRETACEOUS (SANTONIAN-MAASTRICHTIAN) HETEROHELICID PLANKTIC FORAMINIFERA FROM THE PACIFIC AND INDIAN OCEANS AND THEIR BIOSTRATIGRAPHIC AND EVOLUTIONARY SIGNIFICANCE

M. Dan Georgescu[*]

Department of Geosciences, University of Calgary,
Calgary, Alberta, Canada

ABSTRACT

The study of the Late Cretaceous (Santonian-Maastrichtian) planktic foraminifera with gublerinid and pseudoguembelinid appearance and their taxonomic re-evaluation in evolutionary classification reveals the occurrence of five directional lineages, three of them new: *Magellanina*, *Leptobimodalia* and *Nederbragtina*; *Lipsonia* Georgescu and Abramovich 2008 and *Pseudoguembelina* Bronnimann and Brown 1953 are emended. Five new species are described, all of them being the initiating species in the directional lineage to which they belong: *Magellanina magellani*, *Lipsonia shatskyensis*, *Leptobimodalia leptobimodalis*, *Pseudoguembelina praecostulata*, and *Nederbragtina prima*. The new evolutionary framework indicates that the gublerinid architecture evolved in four lineages, whereas the tests with supplementary apertures in the posterior part of the last-formed chambers, which is a feature characteristic of the pseudoguembelinid test architecture, evolved independently in three lineages. Moreover, it is demonstrated that the supplementary apertures, which evolved in the directional lineages *Lipsonia* and *Nederbragtina* were abandoned during the evolution process, and the successor species in both lineages evolved adult stage with multichamber growth.

---

[*] Corresponding author: dgeorge@ucalgary.ca.

Keywords: Foraminifera, planktic, Late Cretaceous, new lineages, evolutionary classification

INTRODUCTION

Foraminifera are single-celled organisms that protect the cytoplasm with an organic or mineralized structure referred to as a test. Foraminiferal tests are quite common and fossilize relatively easy and therefore are frequent occurrences in the fossil record. The high rates of evolution of certain foraminiferal groups make them excellent tools in sedimentary rock dating and the preferences for certain ecological niches provide accurate indicators in reconstructing sedimentary paleoenvironments, paleobathymetry and its trends, paleoclimate and its fluctuations, etc. Foraminifera evolved in the Cambrian (circa 520 m.y. ago) and for the most time of their evolutionary history remained benthics; the group develop a planktic way of life starting in the early Jurassic (circa 183 m.y. ago), possibly in the latest Triassic (circa 201-202 m.y. ago). Irrespective of whether they are benthic or planktic, the foraminiferal classification is based in general on the test chemical and mineralogical composition, and gross test architecture, in which features such as chamber shape and arrangement, position and size of the aperture or apertures, *etc* play a primordial role; in this classification method, which is Linnaean or typological, the units of a lower rank are grouped into a unit of higher rank by the morphological resemblances in the test architecture.

The classical foraminiferal classification was challenged by the discovery of a distinct pattern in evolution, namely iterative evolution (Steineck and Fleisher 1978). According to the iterative evolution pattern similar morphological structures can occur in distant lineages, which are not in a direct evolutionary relationship. The extensive occurrence of this pattern among the Cretaceous planktic foraminifera indicates that the classification in use is not a natural one because species with the same morphological features that evolved independently are frequently grouped together in units of higher rank (i.e., genera). Solving this problem requires a new classification method, in which not only the morphological resemblances are to be taken into consideration, but also the differences resulting from the morphological divergence, which is the probably the most apparent effect of the evolution process. Therefore, grouping species into genera cannot be considered accurate in developing a natural classification framework; therefore, species should be grouped into lineages, accepting that the common ancestry is a *sine qua non* condition in the species grouping.

The development of a new classification framework for the Cretaceous planktic foraminifera is largely based on a significant influx of data from the test wall ultrastructure, ornamentation elements and their distribution over the chamber surface, pore size, shape and distribution patterns, and discrete test morphologic features such as periapertural structures, chamber extension, periphery shape, etc. Due to their small size, most of these data inaccessible through study with a classical optical microscope; therefore new data should be acquired with a more powerful tool, which is the Scanning Electron Microscope (SEM). Such data collected throughout the stratigraphic ranges of the Cretaceous planktic foraminiferal species, has led to a more accurate understanding of the morphological variability through time, taxonomic importance of the test features, and evolutionary relationships between the various species in the fossil record. The new evolutionary classification framework resulted in three breakthroughs in understanding the living world hierarchy. First, species grouping into lineages marked the transition from the static (e.g., genera) to dynamic (e.g., lineages) supraspecific units (Georgescu 2009). Second, the lineages can be separated into directional lineages (DL) and branched lineages (BL) based on their architecture (Georgescu 2010); the existence of two kinds of units at the same hierarchical level defines an opened system, which is opposite to the axiomatic character of the typological classification in which all the units at the same hierarchical level are considered equal (Georgescu, 2011). Third, the species of one lineage irrespective of their architecture (directional or branched) should be differently labelled as initiating species (IS), first descendant species (FDS), second descendant species (SDS), etc (Georgescu 2012a); such a system of species labelling proved useful and practical in lineage comparison.

The new methodologies in data acquisition and interpretation are herein applied to a relatively small group of Late Cretaceous (Santonian-Maastrichtian) planktic foraminifera from the Pacific and Indian Oceans; such taxa were overlooked in more than 40 years of studies of the Late Cretaceous planktic foraminiferal assemblages in the region. The outcomes of this study are multiple, and include a new taxonomic framework for the late Santonian-Maastrichtian gublerinid and pseudoguembelinid planktic foraminifera together with a significantly higher level of species predictability in the fossil record.

MATERIAL AND METHODS

The material used in this study was collected from Upper Cretaceous (Santonian-Maastrichtian) sediments from four sites and holes drilled under the auspices of the Deep Sea Drilling Project (DSDP), and Ocean Drilling Program (ODP): DSDP Site 305 (Shatsky Rise, Pacific Ocean), DSDP Site 463 (Mid-Pacific Mountains, Pacific Ocean), ODP Hole 762C (Exmouth Plateau, Indian Ocean), and ODP Hole 763B (Exmouth Plateau, Indian Ocean) (Figure 1). The stratigraphy, biostratigraphic framework, and foraminiferal assemblage characteristics are briefly presented for each site/hole.

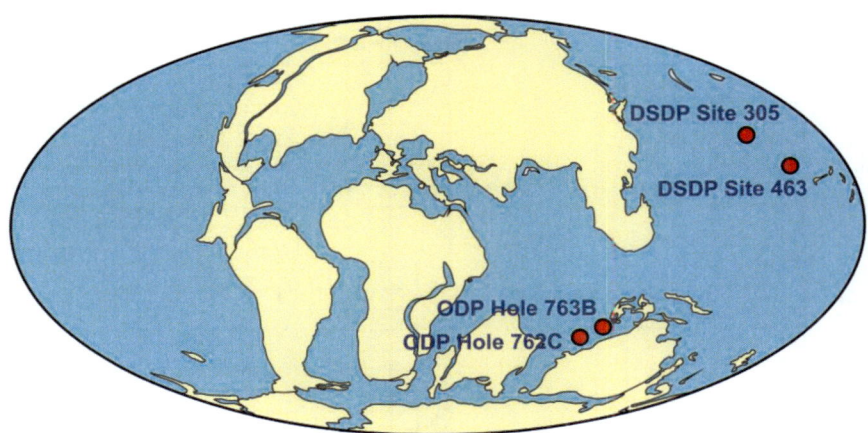

Figure 1. Geographical position of the DSDP sites and ODP Holes from which the fossil material considered in this study was collected. Base map is a reconstruction after Hay and others (1999).

Table 1. Selected planktic foraminiferal species distributions in the Campanian-Maastrichtian sediments at DSDP Site 305 (Shatsky Rise)

DSDP Leg 32, Site 305 (core number-core section, sample depth in centimeters)	Sample depth (meters below sea floor)	Stages	Biozonation	Magellanina magellani	M. acuta	Lipsonia shatskyensis	L. lipsonae	Leptobimodalia leptobimodalis	L. costellifera	L. kempensis	Pseudoguembelina praecostulata	P. costulata	P. excolata	Nederbragtina prima	N. palpebra	N. hariaensis
16-1 100-102 cm	140.00	Maastrichtian	P. hariaensis							x		x	x		x	x
16-2 100-102 cm	141.50									x		x	x		x	x
16-3 100-102 cm	143.00					x		x					x		x	x
16-4 100-102 cm	144.50		A. mayaroensis							x		x	x	x	x	
16-5 100-102 cm	146.00		R. fructicosa				x		x	x		x	x	x	x	
17-1 100-102 cm	149.50									x		x	x	x		
17-2 100-102 cm	151.00					x	x			x		x	x	x	x	
17-4 100-102 cm	154.00				x	x	x		x	x		x	x	x	x	
17-5 100-102 cm	155.50					x			x	x		x	x			
17-6 100-102 cm	157.00		G. stuarti			x			x			x	x			
18-1 100-102 cm	159.00					x						x	x	x		
18-2 100-102 cm	160.50								x			x		x		
18-3 100-102 cm	162.00											x		x		
18-4 100-102 cm	163.50				x							x		x		
18-5 100-102 cm	165.00				x							x		x		
18-6 100-102 cm	166.50				x				x			x				
19-1 100-102 cm	168.00				x							x				
19-2 100-102 cm	169.50				x				x			x				
19-3 100-102 cm	171.00		G. subpetaloidea		x				x			x				
19-4 100-102 cm	172.50				x				x		x	x				
19-5 100-102 cm	174.00								x			x				
19-6 100-102 cm	175.50				x						x	x				
20-1 100-102 cm	177.50				x				x	x	x	x				
20-2 100-102 cm	179.00				x				x	x	x	x				
21-1 50-52 cm	186.00	Campanian		x	x				x		x	x				
21-2 50-52 cm	188.00				x				x	x	x	x				
21-2 100-102 cm	188.50								x		x					
21-3 50-52 cm	189.50		R. calcarata						x		x					
21-4 50-52 cm	191.00								x		x					
21-5 50-52 cm	192.50								x							
23-1 100-102 cm	206.00								x							
23-2 100-102 cm	207.50			x					x		x					
23-3 100-102 cm	209.00								x		x					
23-4 100-102 cm	210.50			x					x		x					
23-5 100-102 cm	212.00								x		x					
23-6 100-102 cm	213.50			x					x							
24-1 100-102 cm	215.00								x							
24-2 100-102 cm	216.50			x					x		x					
24-3 100-102 cm	218.00		G. insignis	x					x		x					
24-4 100-102 cm	219.50								x		x					
24-5 100-102 cm	221.00								x							
25-1 100-102 cm	224.50								x							
25-2 100-102 cm	226.00								x							
25-3 100-102 cm	227.50			x					x		x					
25-4 100-102 cm	229.00			x					x		x					
25-5 100-102 cm	230.50								x		x					
25-6 100-102 cm	232.00								x		x					
26-1 100-102 cm	234.00								x		x					
26-2 100-102 cm	235.50								x		x					
26-3 100-102 cm	237.00								x		x					
26-4 100-102 cm	238.50		H. pacificus	x					x		x					
26-5 100-102 cm	240.00								x		x					
27-1 100-102 cm	243.00			x					x							
27-2 100-102 cm	244.50			x					x		x					
28-1 100-102 cm	252.50			x					x		x					

DSDP SITE 305 (SHATSKY RISE, PACIFIC OCEAN). The lithological succession consists of white soft nannofossil and foraminiferal chalks spanning the Campanian-Maastrichtian stratigraphic interval. The planktic foraminiferal zonation at this site was given by Georgescu (2012b, p. 9-11), and it is followed in this study. Foraminiferal tests are well-preserved although they appear affected by a low magnitude recrystallization. All five lineages and their component species studied occur in the Campanian-Maastrichtian sediments of DSDP Site 305 (Table 1). The occurrences at the same stratigraphic level (Sample 28-1, 100-102 cm) of the species *Magellanina magellani*, *Leptobimodalia leptobimodalis* and *Pseudoguembelina praecostulata* indicate that only the upper part of the early Campanian *Hendersonites pacificus* Biozone is represented at this site, and none of three last downhole occurrences can be considered evolutionary occurrences.

DSDP SITE 463 (MID-PACIFIC MOUNTAINS, PACIFIC OCEAN). A complete Coniacian-lower Maastrichtian succession consisting mostly of nannofossil and foraminiferal chalks was recognized at this sit by Georgescu (2012b, p. 14-17), and the succession of seven planktic foraminiferal biozones is followed herein. Only four of the studied lineages occur: *Magellanina*, *Leptobimodalia*, *Pseudoguembelina* and *Nederbragtina*. The lower portions of the stratigraphic ranges of the lineages *Magellanina*, *Leptobimodalia* and *Pseudoguembelina* are well-defined at this site in the Santonian-lower Camparian (Table 2). It appears evident from the occurrences at this site that *Leptobimodalia* evolved in the late Santonian (Sample 26-5, 53-58 cm) and *Pseudoguembelina* in the early Campanian (Sample 26-1, 50-52 cm). Only the initiating species of the *Nederbragtina* lineage is recorded at this site and this is due to the absence from the rock record of the middle-upper Maastrichtian deposits, and possibly the upper part of the lower Maastrichtian.

ODP HOLE 762C (EXMOUTH PLATEAU, INDIAN OCEAN). A complete upper Santonian-Maastrichtian lithological succession dominated by nannofossil chalks is recorded in the ODP Hole 762. The planktic foraminiferal zonation given by Georgescu (2012b, p. 18, 20-21) is followed and an additional five biozones that encompass the upper Campanian-Maastrichtian stratigraphic interval are identified (Table 3). Three of these lineages are recorded in these sediments: *Magellanina*, *Leptobimodalia* and *Nederbragtina*; only one of them, namely *Magellanina*, is represented by the initiating and descendant species. The small number of species relates to the shallower sedimentation, which is estimated to have occurred within the upper bathyal. This contrasts with the high diversity records from DSDP Sites 305 and 463, in which the sedimentation was bathyal. Such occurrences indicate that the five lineages included in this study were primarily inhabitants of open oceanic environments.

ODP HOLE 763B (EXMOUTH PLATEAU, INDIAN OCEAN). The upper Santonian-middle Campanian succession of nannofossil chalks with levels of chalky claystones of ODP Hole 763B yielded rare specimens of the initiating species of *Magellanina* and *Leptobimodalia* (Table 4). These occurrences appear controlled by local environmental factors, such as relative sea-level. Notably the sedimentation at this site was situated in the outer shelf throughout the upper Santonian-middle Campanian stratigraphic interval.

Table 2. Selected planktic foraminiferal species distributions in the upper Turonian-lower Maastrichtian sediments at DSDP Site 463 (Mid-Pacific Mountains)

Species columns (left table): *Magellanina magellani*, *M. acute*, *Leptobrodalia leptobrinodalis*, *L. costellifera*, *L. kempensis*, *Pseudoguembelina praecostulata*, *P. costulata*, *P. excolata*, *Neoinfragntna prima*

DSDP Leg 62, Site 463 (core number-core section, sample depth)	Sample depth (meters below sea floor)	Stages	Biozonation
7-4 20-22 cm	048.20		
8-1 50-52 cm	053.50		
8-2 50-52 cm	055.00		
8-3 15-17 cm	056.15		
9-1 50-52 cm	063.00		
9-2 50-52 cm	064.50		
9-3 47-49 cm	065.97		
10-1 50-52 cm	072.50		
10-2 50-52 cm	074.00		
10-3 50-52 cm	075.50		
10-4 50-52 cm	077.00		
10-5 50-52 cm	078.50		
10-6 50-52 cm	080.00		
11-1 50-52 cm	082.50		
11-2 53-55 cm	083.53		
11-3 34-36 cm	084.84		
11-4 50-55 cm	086.60		
11-5 50-52 cm	088.00		
12-2 52-54 cm	093.02		
12-3 50-52 cm	094.50	upper Campanian-lower Maastrichtian	
12-4 50-52 cm	096.00		
12-5 50-52 cm	097.50		
12-5 53-55 cm	099.53		
13-1 52-54 cm	101.02		
13-2 47-49 cm	102.47		
13-4 52-54 cm	105.52		
13-5 55-60 cm	107.00		
13-6 58-60 cm	108.58		*G. gansseri*
14-1 50-52 cm	110.50		
14-3 50-52 cm	113.50		
14-4 50-52 cm	115.00		
15-1 50-52 cm	120.00		
15-2 45-47 cm	121.45		
15-3 52-54 cm	123.02		
15-4 49-53 cm	124.49		
15-5 50-52 cm	126.00		
16-1 50-52 cm	129.50		
16-2 50-52 cm	131.00		
16-3 50-52 cm	132.50		
16-4 56-57 cm	134.06		
16-5 50-52 cm	135.50		
16-6 50-52 cm	137.00		
17-1 54-57 cm	139.04		
17-2 50-53 cm	140.50		
17-3 52-54 cm	142.02		
17-4 52-54 cm	143.52		
17-6 50-52 cm	146.50		
18-1 50-52 cm	148.50		
19-1 50-55 cm	158.00		
19-2 52-54 cm	159.52		
19-3 47-49 cm	160.97		
19-4 49-51 cm	162.49		
19-5 52-54 cm	164.02		
19-6 51-53 cm	165.51		
20-1 51-53 cm	167.51		
21-1 52-54 cm	177.02		*G. subpetaloidea*
21-2 52-54 cm	178.50		
21-3 52-54 cm	180.02		
21-4 50-54 cm	181.50		
21-5 38-40 cm	182.88		

Species columns (right table): *R. calcarata*, *G. insignis*, *H. pacificus*, *D. asymetrica*, *D. concava*

Sample	Sample depth	Stages	Biozonation
21-5 78-80 cm	183.28		*R. calcarata*
21-6 57-59 cm	184.57		
22-1 52-54 cm	186.52		
22-2 51-53 cm	188.01		
22-3 50-52 cm	189.50	Campanian	
22-4 52-54 cm	191.02		
22-4 10-12 cm	192.10		
23-1 50-52 cm	198.00		*G. insignis*
24-1 50-52 cm	200.00		
24-2 50-52 cm	201.50		
24-3 50-52 cm	203.00		
25-1 51-53 cm	205.51		
25-2 50-52 cm	207.00		
26-1 50-52 cm	215.00		*H. pacificus*
26-2 52-54 cm	216.52		
26-3 52-54 cm	218.02		
26-4 52-54 cm	219.52	u. Sant.	*D. asymetrica*
26-5 53-58 cm	221.03	Co-l.	
26-6 53-55 cm	222.53	Sant.	*D. concavata*
27-1 50-52 cm	224.50		
27-2 20-22 cm	225.70	u. Tur.	

Table 3. Selected planktic foraminiferal species distributions in the upper Santonian-Maastrichtian sediments at DSDP Hole 762C (Exmouth Plateau)

ODP Leg 122, Hole 762C (core number-core section, sample depth)	Sample depth (meters below sea floor)	Stages	Biozonation	Magellanina magellani	M. acuta	Leptobimodalia leptobimodalis	Nederbragtina hariaensis
43-1, 67-68 cm	555.17						x
43-2, 70-71 cm	556.70		N. hariaensis		x		x
43-3, 72-73 cm	558.22						
43-4, 68-69 cm	559.68						x
43-5, 66-67 cm	561.16						
44-1, 70-71 cm	564.70	upper Maastrichtian					
44-2, 72-73 cm	566.22						
44-3, 70-71 cm	567.70						
44-4, 70-71 cm	569.20						
44-5, 70-71 cm	570.70						
44-6, 63.5-65 cm	572.14						
45-1, 70.5-71.5 cm	574.21		A. mayaroensis				
45-2, 69.5-70.5 cm	575.70						
45-3, 70-71 cm	577.20				x		
45-4, 67-68 cm	578.67						
45-5, 68.5-69.5 cm	580.19						
46-1, 67.5-68.5 cm	583.61						
46-2, 65-66 cm	585.15				x		
46-3, 64.5-65.5 cm	586.65						
46-4, 73-74 cm	588.23						
47-1, 70-71 cm	593.20						
47-2, 71-72 cm	594.71	lower Maastrichtian					
47-3, 69.5-70.5 cm	596.20		G cuvillieri				
47-4, 71-72 cm	597.71						
47-5, 70-71 cm	599.20						
47-6, 70-71 cm	600.70						
48-1, 70-71 cm	602.70				x		
48-2, 71-72 cm	604.21						
48-3, 71-72 cm	605.71		G. subpetaloidea				
48-4, 70-71 cm	607.20						
48-5, 70.5-71.5 cm	608.71						
48-6, 72-73 cm	610.22						
49-1, 70-71 cm	612.20						
49-2, 71-72 cm	613.71				x		
49-3, 71-72 cm	615.21				x		
49-4, 70.5-71.5 cm	616.71						
49-5, 64.5-65.5 cm	618.15						
50-1, 70-71 cm	621.70						
50-2, 72-73 cm	623.22						
50-3, 72-73 cm	624.72			x			
50-4, 69-70 cm	626.19						
50-5, 63-64 cm	627.63						
50-6, 74.5-75.5 cm	629.25	upper Campanian					
51-1, 77.5-78.5 cm	631.28						
51-2, 68-69 cm	632.68				x		
51-3, 71.5-72.5 cm	634.22		G. rajagopalani		x		
51-4, 69.5-70.5 cm	635.70						
51-5, 75-76 cm	637.25						
51-6, 65.5-66.5 cm	638.66						
52-1, 73.5-74.5 cm	640.74						
52-2, 70-71 cm	642.20			x			
52-3, 73.5-74.5 cm	643.74						
52-4, 72-73 cm	645.22						
52-5, 70-71 cm	646.70						
52-6, 73-74 cm	648.23					x	
53-1, 69.5-70.5 cm	650.20						
53-2, 71-72 cm	651.71			x			
53-3, 73-74 cm	653.23			x			
53-4, 60-61 cm	654.60			x	x		
53-5, 70-71 cm	656.20						
53-6, 62.5-63.5 cm	657.63						
54-1, 70-71 cm	659.70			x			
54-2, 68-69 cm	661.29			x			
54-3, 69-70 cm	662.80						
54-4, 71-72 cm	664.32						
54-5, 65-66 cm	665.76			x			
55-1, 69-70 cm	669.19						
55-2, 70-71 cm	670.70			x			
55-3, 68-69 cm	672.18						
56-1, 70-71 cm	678.69						
56-2, 70-71 cm	680.20			x			
56-3, 67-68 cm	681.67	lower-middle Campanian	G. ventricosa	x	x		
56-4, 71-75 cm	683.24						
56-5, 71.5-72.5 cm	684.72						
56-6, 71-72.5 cm	686.21			x			
57-1, 73.5-74.5 cm	688.24			x			
57-2, 67-68 cm	689.67			x			
58-1, 69-70.5 cm	697.69			x	x		
58-2, 71.5-72.5 cm	699.22			x			
58-3, 72-73 cm	700.72						
58-4, 75-76 cm	702.25			x			
58-5, 69-70 cm	703.69			x	x		
59-1, 69-70 cm	707.19			x			
59-2, 69-70 cm	708.69			x	x		
59-3, 68-69 cm	710.18						
59-4, 68-69 cm	711.68			x	x		
59-5, 62-63 cm	713.12						
59-6, 29-30 cm	714.29				x		
60-1 20-21 cm	716.20				x		
60-1 120-121 cm	717.20				x		
60-2 20-21 cm	717.70			x	x		
60-2 120-121 cm	718.70				x		
60-3 20-21 cm	719.20				x		
60-3 120-121 cm	720.20				x		
61-1 20-21 cm	725.70			x	x		
61-1 120-121 cm	726.70				x		
61-2 20-21 cm	727.20			x	x		
61-2 120-121 cm	728.20			x	x		
62-1 20-21 cm	735.20			x	x		
62-1 120-121 cm	736.20			x	x		
62-2 20-21 cm	736.70				x		
62-2 120-121 cm	737.70				x		
62-3 20-21 cm	738.20			x	x		
62-3 120-121 cm	739.20				x		
62-4 20-21 cm	739.70				x		
62-4 120-121 cm	740.70			x	x		
63-1 20-21 cm	744.70		G. elevata equivalent		x		
63-1 120-121 cm	745.70				x		
63-2 20-21 cm	746.20			x	x		
63-2 120-121 cm	747.20			x	x		
64-1 18.5-19.5 cm	754.19						
64-1 125-126 cm	755.25						
64-2 20-21 cm	755.70				?		
64-2 120-121 cm	756.70						
64-3 20-21 cm	757.20						
64-3 120-121 cm	758.20						
65-1 20-21 cm	760.20						
66-1 21.5-22.5 cm	765.22						
66-1 118-119 cm	766.18						
66-2 20-21 cm	766.70	upper Santonian			?		
66-2 120-121 cm	767.70		D. asymetrica		x		
66-3 20-21 cm	768.20				x		
66-3 120-121 cm	769.20				x		
66-4 20-21 cm	769.70				x		
66-4 110-111 cm	770.60						

Table 4. Selected planktic foraminiferal species distributions in the upper Santonian-middle Campanian sediments at DSDP Hole 763B (Exmouth Plateau)

ODP Leg 122, Hole 763B (core number-core section, sample depth)	Sample depth (meters below sea floor)	Stages	Biozonation	Magellanina magellani	Leptobimodalia leptobimodalis
8-1, 71-72 cm	247.71				
8-2, 67.5-68.5 cm	249.18				
8-3, 70-71 cm	250.70				
8-4, 43-44 cm	251.93				
9-1, 71-72 cm	257.21				
9-2, 70-71 cm	258.70				
9-3, 70-71 cm	260.20			x	
9-4, 67-68 cm	261.67				
9-5, 70-71 cm	263.20				
9-6, 68-69 cm	264.37				
10-1, 68.5-69.5 cm	266.66				
10-2, 69-70 cm	268.19				
10-3, 68-69 cm	269.68				
10-4, 70-71 cm	271.20				
10-5, 67.5-69 cm	272.68		G. ventricosa		
10-6, 71-72.5 cm	274.21				
11-1, 69-70 cm	276.19				
11-2, 71-72 cm	277.71	lower-middle Campanian			
11-3, 70-71 cm	279.20				
11-4, 70-71 cm	280.70				
11-5, 71-72 cm	282.21				
12-1, 67-68 cm	285.67				
12-2, 70-71 cm	287.20				x
12-3, 60-61 cm	288.60				
12-4, 70-71 cm	290.20				
13-1, 71-72 cm	295.21				
13-2, 69-70 cm	296.69				
13-3, 70-71 cm	298.20				
13-4, 72-73 cm	299.72				
13-5, 70-71 cm	301.20				
13-6, 73-74 cm	302.73				
14-1, 70-71 cm	304.70				
14-2, 68-69 cm	306.18				
14-3, 63-64 cm	307.63				
14-4, 70-71 cm	309.20				
14-5, 70-71 cm	310.70				
15-1, 71-72 cm	314.21				
15-2, 77-78 cm	315.77		G. elevata equivalent		
15-3, 67-68 cm	317.17				x
15-4, 61-62 cm	318.61				x
15-5, 70-71 cm	320.20				
15-6, 70-71 cm	321.70				
16-1, 74-75 cm	323.74				x
16-2, 76-77 cm	325.26				
16-3, 70-71 cm	326.70				x
16-4, 71-72 cm	328.21				
16-5, 72-73 cm	329.72				
16-6, 36-37 cm	330.86				
17-1, 70-71 cm	333.20	u. Sant.	D. asymetrica		
17-2, 70-71 cm	334.70				

NEW MORPHOLOGIC STRUCTURES AND TERMS

Two new morphological features are described in the Cretaceous planktic foraminifers with chambers alternately added with respect to the test growth axis. Accordingly new morphological terms are herein proposed for the two features related to the chamber ornamentation and periapertural structures.

The earliest representatives of the heterohelicid group were smooth, and only tests lacking ornamentation occur in the upper Albian-lower upper Cenomanian stratigraphic interval. Costate ornamentation evolved in the late Cenomanian, and remained the dominant ornamentation throughout the uppermost Cenomanian-Maastrichtian stratigraphic interval. Little attention was given to the costae orientation over the chamber surface despite the high frequency in the fossil record of the heterohelicid tests with costate ornamentation; it was generally accepted that the costae are longitudinally arranged, more or less parallel to the test axis of growth and periphery. New data on the late Santonian-Maastrichtian indicate that there are two distinct kinds of distribution of the costae forming the ornamentation, which are herein named unimodal and bimodal. *Unimodal ornamentation* is that where the costae are parallel to the periphery in the test marginal regions, and parallel to the test growth axis in the proximity of the central suture (Figure 2:1). *Bimodal ornamentation* is that in which the costae are parallel to the periphery in the test marginal zone, and oblique to the test growth axis in the proximity of the central suture (Figure 2:2).

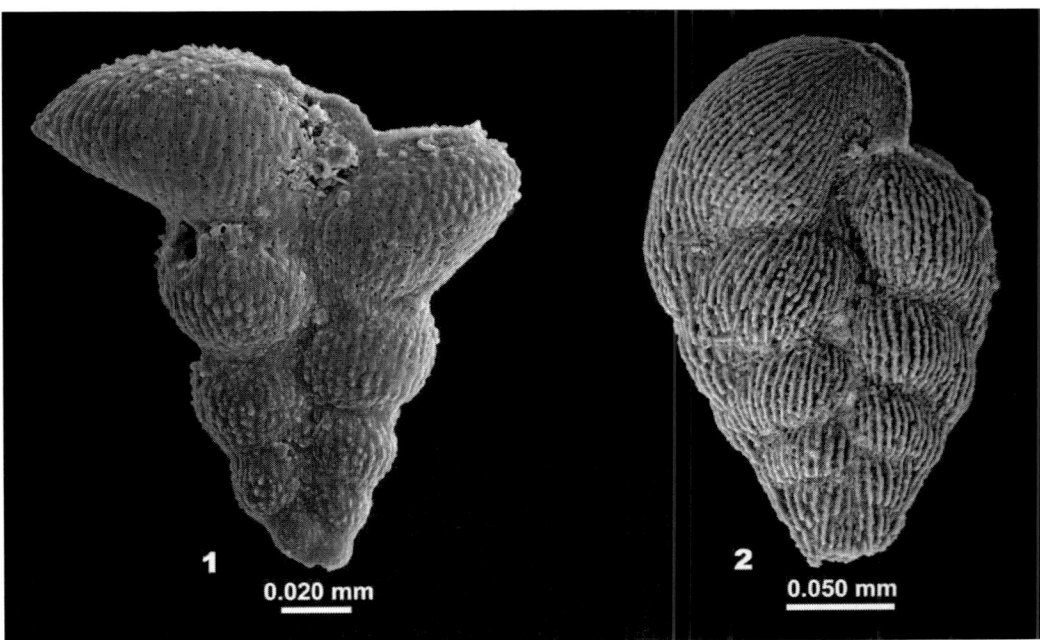

Figure 2. Two specimens of Spiroplecta americana (Ehrenberg 1843) (1) and Leptobimodalia leptobimodalis - new species and directional lineage (2) illustrating the unimodal and bimodal ornamentation respectively. Specimen provenance: Sample 1595b, Ehrenberg Collection (Naturkundemuseum, Berlin) (1), and Sample 62-463-26-3, 52-54 cm (2).

A four-fold classification of the periapertural structures in the heterohelicid group was given by Georgescu (2010, p. 62-63): archaeoflanges, orthoflanges, metaflanges, and leptoflanges. The new data complement this terminology, and a new type is herein recognized and named: *retroflanges*. The characteristic of the new type of heterohelicid periapertural structures is that they are wider towards the test growth axis and narrow towards the periphery. Retroflanges can be simple (Figure 3:1) or rimmed (Figure 3:2).

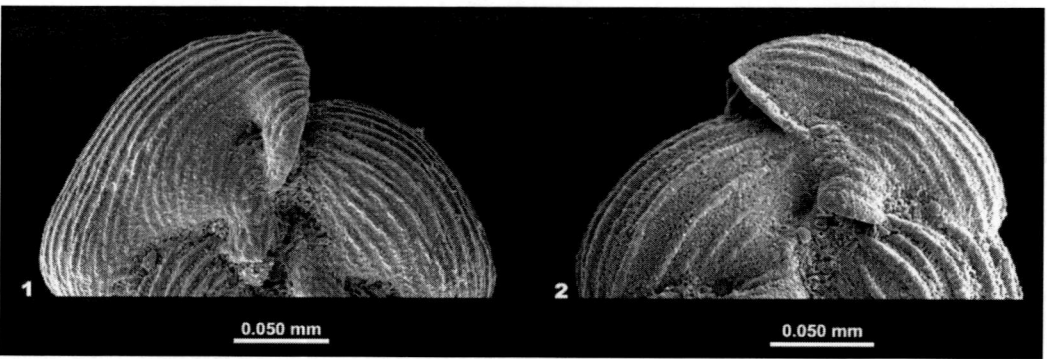

Figure 3. Two specimens of *Pseudoguembelina* sp. exhibiting retroflanges, which are simple (1) and rimmed (2); specimens from the Shatsky Rise (Pacific Ocean). Specimen provenance: Sample 32-305-17-4, 80-95 cm (1), and Sample 32-305-16-5, 60-76 cm (2).

SYSTEMATIC DESCRIPTIONS

Evolutionary classification units above species level are after Georgescu (2010). Species kinds within the lineages are after Georgescu (2012a): IS - initiating species, FDS - first descendant species, and SDS - second descendant species.

Directional Lineage: *Magellanina* – new
(Figure 4)

Species included. IS: *Magellanina magellani* - new species and FDS: *M. acuta* (de Klasz 1953).

Diagnosis. Gublerinid directional lineage leptocostate in the IS, and with five quasi-longitudinal bands of leptocostate and reticulate ornamentation in the FDS.

Description. Test with the chambers alternately added with respect to the test growth axis in the IS, and multichamber growth in the adult stage in the FDS; the adult stage consists of the progressive chamber, which is followed by one pair of chambers, one on each side of the progressive chamber. Sutures are distinct and depressed, distinctly curved towards the test anterior par in the adult stage. Test is compressed in edge view, with rounded or rarely subangular periphery. Aperture is a small arch-shaped opening at the base of the last-formed chamber in the IS and early stage of the FDS; the biaperturate chamber of the FDS is the only one with more than two apertures. Two symmetrically developed metaflanges border the aperture, one on each test side. Chamber surface is throughout ornamented with closely spaced discontinuous and longitudinal leptocostae in the IS, and five quasi-longitudinal bands of leptocostae and reticulate ornamentation in the FDS; the leptocostate ornamentation occurs in the peripheral regions and central test portion. whereas the reticulate one occurs in the central portion of the chambers. Leptocostae thickness is of 0.0020-0.0027 mm in the IS, and 0.0030-0.0034 mm in the FDS. Test wall is calcitic, hyaline, simple and perforate; pores are circular or elliptical, and with a diameter or maximum dimension of 0.0006-0.0014 mm in the IS, and 0.0006-0.0046 mm in the FDS; the largest pores in this lineage are recorded in the reticulately ornamented regions of the FDS.

KEY FEATURES	DL: *Planoheterohelix*	DL: *Magellanina* - new	
	SDS: *P. planata* (Cushman 1938)	IS: *M. magellani* - new species	FDS: *M. acuta* (de Klasz 1953)
	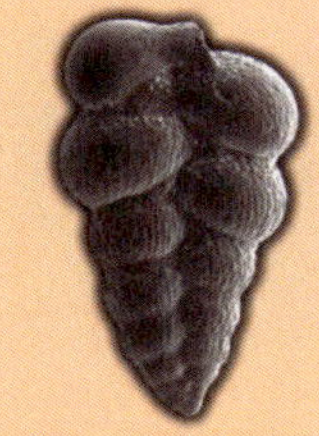 No scale is implied.	 No scale is implied.	 No scale is implied.
Chamber arrangement	Alternately added with respect to the test growth axis throughout	Alternately added with respect to the test growth axis throughout	Alternately added with respect to the test growth axis in the early stage, and with multichamber growth in adult
Aperture	Single, a medium high to high arch at the base of the last-formed chamber	Single, a medium high arch at the base of the last-formed chamber	Single in the early stage, and multiple in the adult
Periapertural structures	Dominant orthoflanges, and occasionally metaflanges	Metaflanges	Metaflanges
Ornamentation	Unimodal and leptocostate, 0.0018-0.0025 mm in thickness	Unimodal and leptocostate, 0.0020-0.0027 mm in thickness	Five quasi-longitudinal bands, leptocostate at the periphery and central part of the test (leptocostae thickness is of 0.0030-0.0034 mm), and reticulate over the chambers
Porosity	Circular, 0.0006-0.0008 mm in diameter	Circular or elliptical in shape, 0.0006-0.0014 mm in diameter or maximum dimension	Circular or elliptical in shape, 0.0006-0.0008 mm in diameter or maximum dimension in the leptocostate regions and 0.0020-0.0049 mm in the reticulate regions

Figure 4. The key features of species of *Magellanina* - new directional lineage (green background), and its ancestor, SDS of *Planoheterohelix*.

Remarks. Georgescu and others (2008) reviewed the gublerinid planktic foraminifera, and described the genus *Praegublerina* as branched lineage, which included *P. pseudotessera* (Cushman 1938), *P. robusta* (de Klasz 1953) and *P. acuta* (de Klasz 1953); the former species is the IS of *Praegublerina*. New data from the Pacific and Indian Oceans indicate that the two descendant species recognized by Georgescu and others (2008) evolved independently, and therefore *Praegublerina* should be redefined as DL consisting of *P. pseudotessera* - IS and *P. robusta* - FDS. *Magellanina* differs from *Praegublerina* by having (i) IS with periapertural structures consisting of metaflanges rather than orthoflanges, (ii) FDS ornamented with five quasi-longitudinal bands of leptocostae and reticulate ornamentation rather than only leptocostae, and (iii) FDS with fewer chambers in the adult stage with multichamber growth; the two directional lineages apparently initiated through divergent evolution from *Planoheterohelix planata* (Cushman 1938) in the early Campanian; such evolutionary relationships between *Planoheterohelix* as ancestor, and *Magellanina* and *Praegublerina* as descendants are indicated by the leptocostate ornamentation, pore size and distribution, and general test architecture between the ancestral species *P. planata* and the IS of the two descendant DL, namely *M. magellani* and *P. pseudotessera* respectively.

Derivation. Directional lineage named after Fernão de Magalhães (Ferdinand Magellan), the first European who crossed the Pacific Ocean.

Age. Late Santonian-Maastrichtian.

Geographical distribution. Cosmopolitan.

IS: *Magellanina magellani* - new species
(Figure 5:1-12)

Holotype. Specimen WKB 010110 (Figure 5:1-2).

Dimensions of the holotype. Length: L=0.198 mm; width: W=0.158 mm; thickness: T=0.070 mm; W/L=0.798; T/L=0.354.

Paratypes. Nine specimens, WKB 010111-010115.

Dimensions. L=0.143-0.198 mm (mean 0.167 mm); W=0.100-0.158 mm (mean 0.122 mm); T=0.048-0.070 mm (mean 0.049 mm); W/L=0.629-0.909 (mean 0.731); T/L=0.246-0.354 (mean 0.295). Ranges are based on the average measurements of the holotype and paratypes.

Material. Around 100 specimens.

Type locality. DSDP Site 463 (Mid-Pacific Mountains, Pacific Ocean); geographical coordinates: 21° 21' N, 174° 40' E.

Type level. White nannofossil chalk of the upper part of the *G. elevata* Biozone equivalent of late early Campanian age; Sample 62-463-25-1, 51-53 cm.

Derivation. As for the directional lineage.

Diagnosis. Flattened leptocostate species, with the main aperture bordered by metaflanges, and larger pores in the middle portion of the test.

Description. Test consists of a small proloculus (0.0100-0.0120 mm) followed by 10-13 chambers with gradual size increase, and alternately added with respect to the test growth axis. Chambers are subrectangular in shape, and the last-formed one or two chambers may be reniform. Sutures are distinct and depressed, straight to slightly curved, and oblique to the test growth axis. Test is compressed in edge view; periphery is rounded or more rarely subangular. Aperture is a small arch-shaped opening at the base of the last-formed chamber; two symmetrically developed metaflanges border the aperture, one on each test side. Chamber surface is ornamented with closely spaced discontinuous leptocostae (0.0020-0.0027 mm in thickness); a periapertural pustulose area consisting of scattered dome-like pustules occurs in the chamber anterior portion. Test wall is calcitic, hyaline, simple, and perforate; pores are simple, and circular or elliptical in shape with a diameter or maximum dimension of 0.0006-0.0014 mm.

Remarks. *Magellanina magellani* differs from *P. planata* by having (i) smaller sizes, (ii) periapertural structures consisting of metaflanges rather than orthoflanges, (iii) occasionally subangular periphery, and (iv) larger pores (0.0006-0.0014 mm rather than 0.0006-0.0008 mm); the similarities between the two species indicate that *M. magellani* is descendant from *P. planata*. This species is the ancestor of *M. magellani* of which it differs mainly by the absence of the adult stage with multichamber growth.

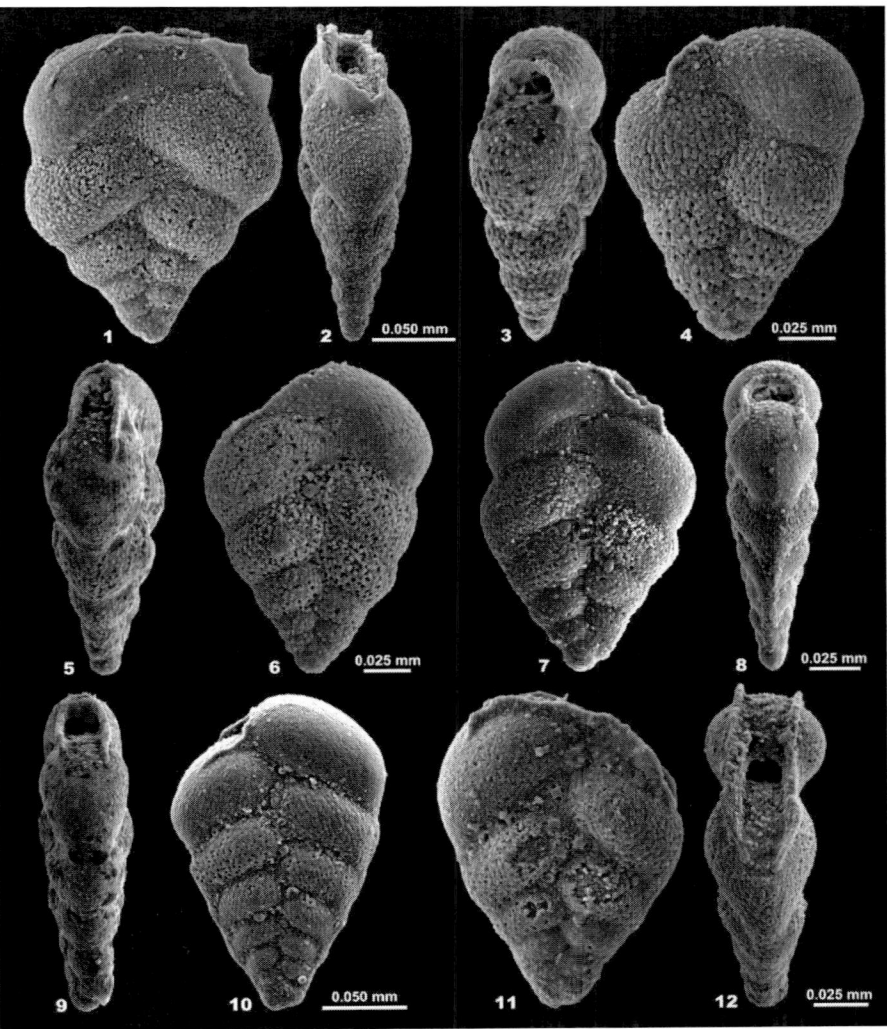

Figure 5. Specimens of *Magellanina magellani* - new directional lineage and new species from DSDP Site 463 (Mid-Pacific Mountains, Pacific Ocean). 1-2 Holotype, Sample 62-463-25-1, 51-53 cm (late early Campanian, upper part of the *G. elevata* Biozone equivalent) (WKB 010110). 3-4 Paratype, Sample 62-463-25-1, 51-53 cm (late early Campanian, upper part of the *G. elevata* Biozone equivalent). 5-6 Paratype, Sample 62-463-25-1, 51-53 cm (late early Campanian, upper part of the *G. elevata* Biozone equivalent). 7-8 Paratype, Sample 62-463-24-1, 50-52 cm (middle Campanian, *G. ventricosa* Biozone); note the subangular periphery. 9-10 Paratype, Sample 62-463-24-1, 50-52 cm (middle Campanian, *G. ventricosa* Biozone). 11-12 Paratype, Sample 62-463-24-1, 50-52 cm (middle Campanian, *G. ventricosa* Biozone).

Age. Late Santonian-Campanian.

Geographical distribution. Pacific Ocean (Mid-Pacific Mountains and Shatsky Rise).

Directional lineage: *Lipsonia* Georgescu and Abramovich 2008 – emended
(Figure 6)

Lipsonia Georgescu and Abramovich 2008, p. 119.

Species included. IS: *L. shatskyensis* - new species and FDS: *L. lipsonae* Georgescu and Abramovich 2008.

Diagnosis. Directional lineage consisting of tests with sutures lined by sutural ridges, multichamber growth stage in the FDS, and one supplementary aperture in the last-formed chamber in the IS.

Description. Test with chambers alternately added with respect to the growth axis in the IS, and with multichamber growth in the FDS. Earlier chambers of the stage with alternate chamber addition are subglobular, then subrectangular in shape; there are up to 10 chamber sets, and 45 to 61 petaloid chambers in total in the adult stage with multichamber growth of the FDS. Sutures are curved, and lined by well-developed phaneroridges that are absent in the earlier portion of the test due to the ornamentation change resulted from the successive addition of layers of calcite during the ontogeny. Test is compressed in edge view; periphery is rounded, and with the phaneroridges connected across the periphery by transverse keels in the IS. Aperture is a low to medium high arch at the base of the last-formed chamber in the stage with alternating addition of chambers, and multiple in the adult stage with multichamber growth; two symmetrically developed metaflanges border each aperture, one on each side of the test. One small supplementary aperture occurs in the posterior part of the last-formed chamber in the IS; the supplementary aperture is covered by a backward lid-like wall extension. Chamber surface is smooth, but vestiges of the leptocostate ornamentation occur frequently in the sutural ridges especially in the IS; two prominent longitudinal ridges occur in the posterior portion of the test in the IS. Test wall is calcitic, hyaline, simple and perforate; pores are circular with a diameter of 0.0008-0.0014 mm in the IS, and 0.0008-0.0017 mm in the FDS. Vuggy pores with a maximum dimension of 0.0036-0.0081 mm occur in the posterior portion of the test in the FDS.

Remarks. The existence of two intervals in which the planktic foraminifera evolved phaneroridges in the Late Cretaceous (Santonian-Maastrichtian) was demonstrated by Georgescu and Abramovich (2008) with the description of *L. lipsonae* from the Maastrichtian sediments of the Shatsky Rise. The new data from Pacific Ocean help in defining the DL *Lipsonia*, which evolved in the late Campanian. *Lipsonia* is almost a homoeomorph of the Santonian lineages *Sigalia* Reiss 1957 and *Proliferania* Georgescu 2010. Notably neither *Sigalia* nor *Proliferania* crosses the Santonian/Campanian boundary and the earliest occurrence of *Lipsonia* is in the late Campanian; therefore there is a gap of circa 8.0 m.y. between these lineages that evolved through convergent and iterative evolution. The IS in the DL *Lipsonia* presents one small supplementary aperture in the posterior part of the last-formed chamber; such a feature does not occur in any species of the Santonian lineages.

Age. Maastrichtian.

Geographical distribution. Pacific Ocean (Shatsky Rise), USA (California) and Mexico (Baja California).

KEY FEATURES	DL: *Gublerina*	DL: *Lipsonia* Georgescu and Abramovich 2008	
	IS: *G. rajagopalani* (Govindan 1972)	IS: *L. shatskyensis* - new species	FDS: *L. lipsonae* Georgescu and Abramovich 2008
	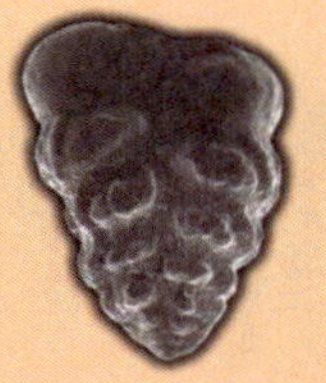 No scale is implied.	 No scale is implied.	 No scale is implied.
Chamber arrangement	Alternately added with respect to the test growth axis throughout	Alternately added with respect to the test growth axis throughout	Alternately added with respect to the test growth axis in the early stage, and with multichamber growth in adult
Sutures	Flush or slightly depressed	Lined with phaneroridges	Lined with phaneroridges
Apertures	Single, a low to medium high arch at the base of the last-formed chamber	Main aperture is a low to medium high arch at the base of the last-formed chamber: one supplementary aperture occurs in the posterior part of the last-formed chamber	Single in the early stage, and multiple in the adult
Periapertural structures	Metaflanges	Metaflanges	Metaflanges
Ornamentation	Coarsely pycnocostate and with irregular agglomerations of ornamentation elements over the earlier chambers; vestiges of the ancestral leptocostate ornamentation occur occasionally	Smooth; vestiges of the ancestral leptocostate ornamentation occur in the phaneroridges	Smooth
Porosity	Circular, 0.0006-0.0010 mm in diameter	Circular, 0.0008-0.0014 mm in diameter	Circular or elliptical in shape, 0.0008-0.0017 mm in diameter or maximum dimension; vuggy, irregular in shape occur over the earlier portion of the test

Figure 6. The key features of species of *Lipsonia* Georgescu and Abramovich 2008 - emended (green background), and its ancestor, IS of *Gublerina*.

IS: *Lipsonia shatskyensis* - new species
(Figure 7:1-12)

Holotype. Specimen WKB 010116 (Figure 7:1-3).

Dimensions of the holotype. L=0.468 mm; W=0.310 mm; T=0.117 mm; W/L=0.662; T/L=0.250.

Paratypes. Five specimens, WKB 010117-010121.

Dimensions. L=0.322-0.468 mm (mean 0.405 mm); W=0.241-0.314 mm (mean 0.296 mm); T=0.096-0.164 mm (mean 0.118 mm); W/L=0.662-0.781 (mean 0.734); T/L=0.250-0.385 (mean 0.292). Ranges are based on the average measurements of the holotype and paratypes.

Material. Around 50 specimens.

Type locality. DSDP Site 305 (Shatsky Rise, Pacific Ocean); geographical coordinates: 32° 00`N, 157° 51`E.

Type level. White-yellowish nannofossil chalk of the upper part of *G. stuarti* Biozone (early Maastrichtian age); Sample 32-305-17-6, 100-102 cm.

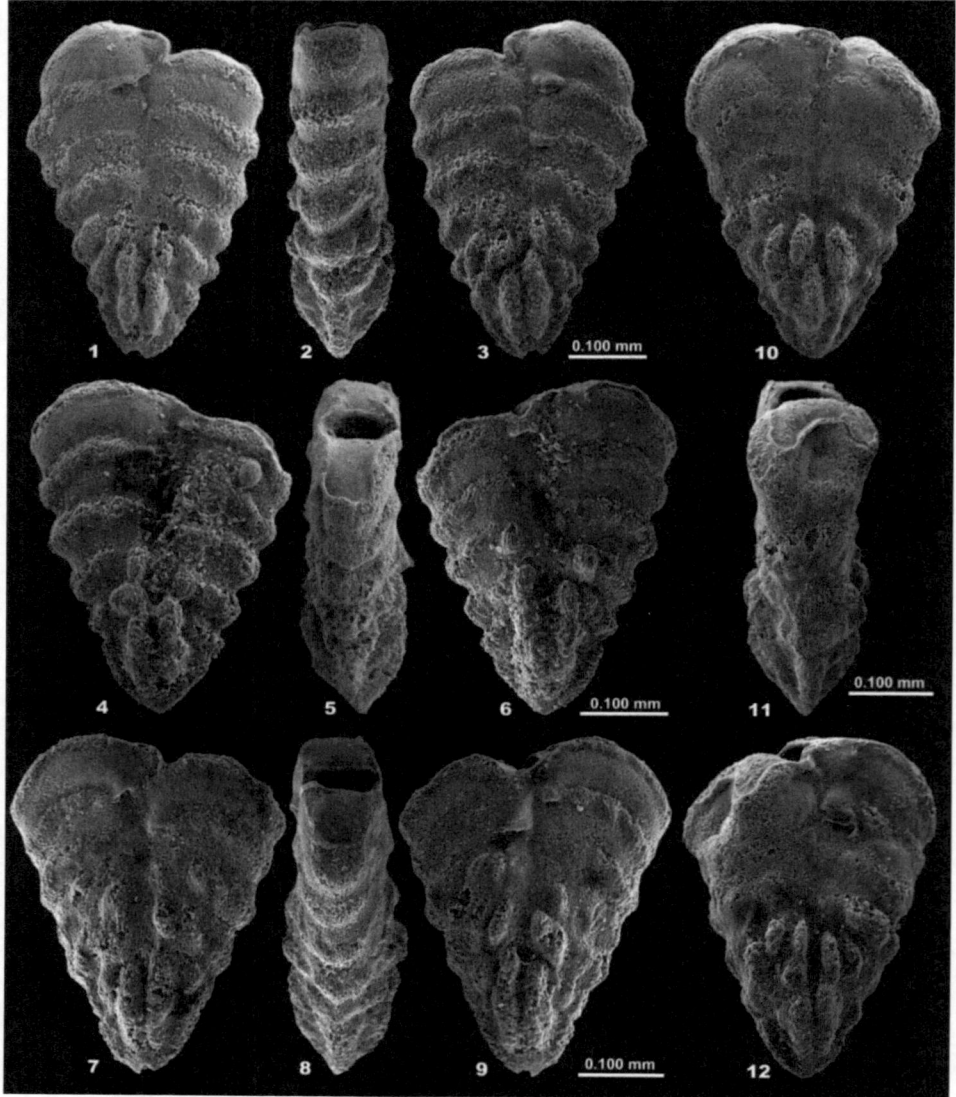

Figure 7. Specimens of *Lipsonia shatskyensis* - new species from the DSDP Site 305 (Shatsky Rise, Pacific Ocean). 1-3 Holotype, Sample 32-305-17-6, 100-102 cm (early Maastrichtian, upper part of the *G. stuarti* Biozone) (WKB 010116). 4-6 Paratype, 32-305-17-4, 100-102 cm (early Maastrichtian, upper part of the *G. stuarti* Biozone). 7-9 Paratype, Sample 32-305-17-6, 100-102 cm (early Maastrichtian, upper part of the *G. stuarti* Biozone). 10-12 Paratype, Sample 32-305-17-6, 100-102 cm (early Maastrichtian, upper part of the *G. stuarti* Biozone).

Derivation. Species named after the *Shatsky* Rise where the type locality is situated.

Diagnosis. *Lipsonia* without multichamber growth in the adult stage, and with the last-formed chamber with one supplementary aperture covered by a backward lid-like extension.

Description. Test consists of a small proloculus followed by 14-17 chambers with gradual size increase, which are added alternately with respect to the test growth axis; chamber shape is subrectangular. Sutures are lined by phaneroridges, which are more prominent in the adult stage of the test, curved towards the test anterior part, and oblique to the test growth axis; the central suture is depressed. Test is compressed in edge view. Periphery is rounded; sutural ridges of the same chamber on the test sides connect across the periphery resulting in a transverse-keeled appearance. Aperture is a low to medium high arch at the base of the last-formed chamber; two symmetrically developed metaflanges border the aperture. One small supplementary aperture occurs in the posterior part of the last-formed chamber in the proximity of the central suture; it is covered by a short backward lid-like wall extension. Chamber surface is smooth, but vestiges of the ancestral leptocostate ornamentation occur sporadically in the phaneroridges. Two strong longitudinal ridges symmetrically developed with respect to the test growth axis occur in the posterior side of the test each lateral side. Test wall is calcitic, hyaline, simple, and perforate; pores are circular with a diameter of 0.0008-0.0014 mm.

Remarks. *Lipsonia shatskyensis* differs from its descendant, *L. lipsonae*, mainly by lacking the multichamber growth in the adult stage. This species evolved from *Gublerina rajagopalani* Govindan 1972 from which it differs by having (i) sutures lined by phaneroridges rather than depressed, (ii) two strong longitudinal ridges in the posterior portion of the test on both lateral sides, and (iii) one supplementary aperture in the posterior part of the last-formed chamber. *Lipsonia shatskyensis* has significant morphological resemblances with the late Santonian species *Sigalia carpathica* (Salaj and Samuel 1963); the two species differ by the occurrence of the strong longitudinal ridges in the posterior test portion, and one supplementary aperture in the posterior part of the last-formed chamber in *L. shatskyensis*. The stratigraphic ranges of the two species do not overlap, and their strong resemblance is the result of iterative and convergent evolution.

Age. Maastrichtian.

Geographical distribution. Pacific Ocean (Shatsky Rise).

Directional Lineage: *Leptobimodalia* – new
(Figure 8)

Species included. IS: *L. leptobimodalis* - new species, FDS: *L. costellifera* (Masters 1976) and SDS: *L. kempensis* (Esker 1968).

Diagnosis. Directional lineage with the tests consisting of chambers ornamented with discontinuous leptocostae with bimodal orientation.

Description. Test consists of proloculus, which is followed by chambers alternately added with respect to the test growth axis; chambers gradually increase in size. Earlier chambers are subrectangular in shape in the test early portion; the last-formed one or two are reniform, and with one bulbous to elongate backward projection. Sutures are distinct and depressed, straight to slightly curved, and oblique to nearly perpendicular to the test growth axis. Test is compressed in edge view, with rounded to subangular periphery in the IS and FDS, and a wall flexure resulting in a rimmed aspect in the SDS. Aperture is a low to medium high arch at the base of the last-formed chamber; two symmetrically retroflanges with leptocostate ornamentation border the aperture, one on each test side. Supplementary apertures occur at the posterior end of the last one to four chamber backward extensions in the FDS and SDS; they are protected by an elongate lid-like wall extension. Chamber surface is ornamented with closely spaced discontinuous leptocostae that gradually thicken from 0.0017-0.0036 mm in the IS to 0.0027-0.0043 mm in the SDS; the leptocostae are parallel to the periphery in the marginal region, and oblique to the test growth axis in the proximity of the central suture; thickened ornamentation frequently occurs over the earlier chamber in the SDS. Test wall is calcitic, hyaline, simple and perforate; pores are situated in the space between the leptocostae, and are circular or elliptical, and have a diameter or maximum dimension of 0.0005-0.0010 mm throughout the directional lineage.

KEY FEATURES	DL: *Planoheterohelix*	DL: *Leptobimodalia* - new		
	SDS: *P. planata* (Cushman 1938)	IS: *L. leptobimodalis* - new species	FDS: *L. costellifera* (Masters 1976)	SDS: *L. kempensis* (Esker 1968)
	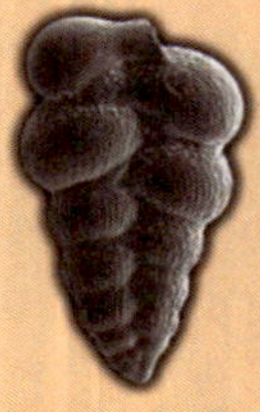 No scale is implied.	 No scale is implied.	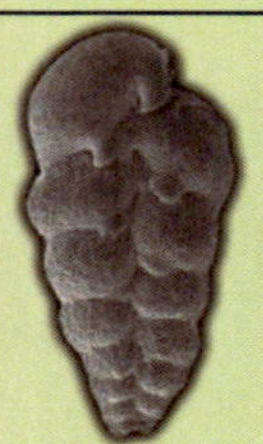 No scale is implied.	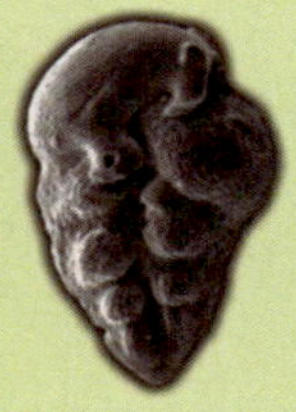 No scale is implied.
Chamber arrangement	Alternately added with respect to the test growth axis throughout	Alternately added with respect to the test growth axis throughout	Alternately added with respect to the test growth axis throughout	Alternately added with respect to the test growth axis throughout
Main aperture	Single, a medium high to high arch at the base of the last-formed chamber	Single, a low to medium high arch at the base of the last-formed chamber	Single, a low arch at the base of the last-formed chamber	Single, a low arch at the base of the last-formed chamber
Supplementary apertures	Absent	Absent	On the last-formed one to four chambers	On the last-formed one to four chambers
Periapertural structures	Dominant orthoflanges, and occasionally metaflanges	Simple retroflanges	Simple and rimmed retroflanges	Rimmed retroflanges
Ornamentation	Unimodal and leptocostate, 0.0018-0.0025 mm in thickness	Bimodal and leptocostate, 0.0017-0.0036 mm in thickness	Bimodal and leptocostate, 0.0023-0.0046 mm in thickness	Bimodal and leptocostate, 0.0027-0.0047 mm in thickness
Porosity	Circular, 0.0006-0.0008 mm in diameter	Circular or elliptical, 0.0005-0.0010 mm in diameter or maximum dimension	Circular or elliptical, 0.0008-0.0012 mm in diameter or maximum dimension	Circular or elliptical, 0.0007-0.0014 mm in diameter or maximum dimension

Figure 8. The key features of species of *Leptobimodalia* - new directional lineage (green background), and its ancestor, SDS of *Planoheterohelix*.

Remarks. The new DL *Leptobimodalia* of the Santonian-Maastrichtian is almost a homoeomorph of the Turonian-early Coniacian BL *Huberella* (Georgescu 2007; Georgescu and others 2011). It differs from the latter mainly by having (i) tests ornamented with leptocostae exhibiting bimodal rather than unimodal arrangement, and (ii) development of supplementary apertures in the posterior portion of the last-formed chambers of the FDS and SDS. A gap of circa 1.0 m.y. (late Coniacian) separates the extinction event of *Huberella* from the evolutionary occurrence of *Leptobimodalia*. The differences in the last-formed chamber backward extension, which is subtriangular to subtrapezoidal in *H. huberi* (Georgescu 2007) and bulbous in the new species *L. leptobimodalis* indicate that there is no evolutionary relationship between the two.

Derivation. The name results from the combination of the Greek prefix *lepto-* (=thin), and the English word *bimodal*; to them the Latin suffix *-ia* is added.

Age. Santonian-Maastrichtian.

Geographical distribution. Cosmopolitan.

IS: *Leptobimodalia leptobimodalis* - new species
(Figure 9:1-11)

Pseudoguembelina costulata Masters. Nederbragt, 1991, p. 358, pl. 8, fig. 2.

Holotype. Specimen WKB 010122 (Figure 9:1-2).

Dimensions of the holotype. L=0.240 mm; W=0.156 mm; T=0.080 mm; W/L=0.650; T/L=0.333.

Paratypes. Twelve specimens, WKB 010123-010134.

Dimensions. L=0.240-0.285 mm (mean 0.259 mm); W=0.140-0.196 mm (mean 0.162 mm); T=0.071-0.137 mm (mean 0.097 mm); W/L=0.540-0.688 (mean 0.622); T/L=0.280-0.484 (mean 0.367). Ranges are based on the average measurements of the holotype and paratypes.

Material. Around 120 specimens.

Type locality. DSDP Site 305 (Shatsky Rise, Pacific Ocean); geographical coordinates: 32° 00`N, 157° 51`E.

Type level. White-yellowish nannofossil chalk of the *Radotruncana calcarata* Biozone (late Campanian age); Sample 32-305-21-2, 100-102 cm.

Derivation. As for the directional lineage.

Diagnosis. *Leptobimodalia* with backward chamber extension, but without supplementary aperture at its end.

Description. The test consists of the proloculus followed by 12-15 chambers with a gradual size increase, which are alternately added with respect to the test growth axis. Chamber shape is subrectangular in the earlier portion of the test; the last-formed one or two chambers are reniform, and with one distinct bulbous backward extension. Sutures are distinct and depressed, straight to slightly curved, and almost perpendicular to the test growth axis. Test is compressed in edge view, with rounded periphery and without peripheral structures. Aperture is a low to medium high arch at the base of the last-formed chamber; two symmetrically developed retroflanges with leptocostate ornamentation border the aperture, one on each side of the test. Chamber surface is ornamented with leptocostae with a width of 0.0017-0.0036 mm; the leptocostae are parallel to the periphery in the test marginal region, and oblique to the test axis of growth in the proximity of the central suture. Test wall is calcitic, hyaline, simple, and perforate; pores are circular or elliptical, have a diameter or maximum dimension of 0.0005-0.0010 mm, and are situated in the space between the leptocostae.

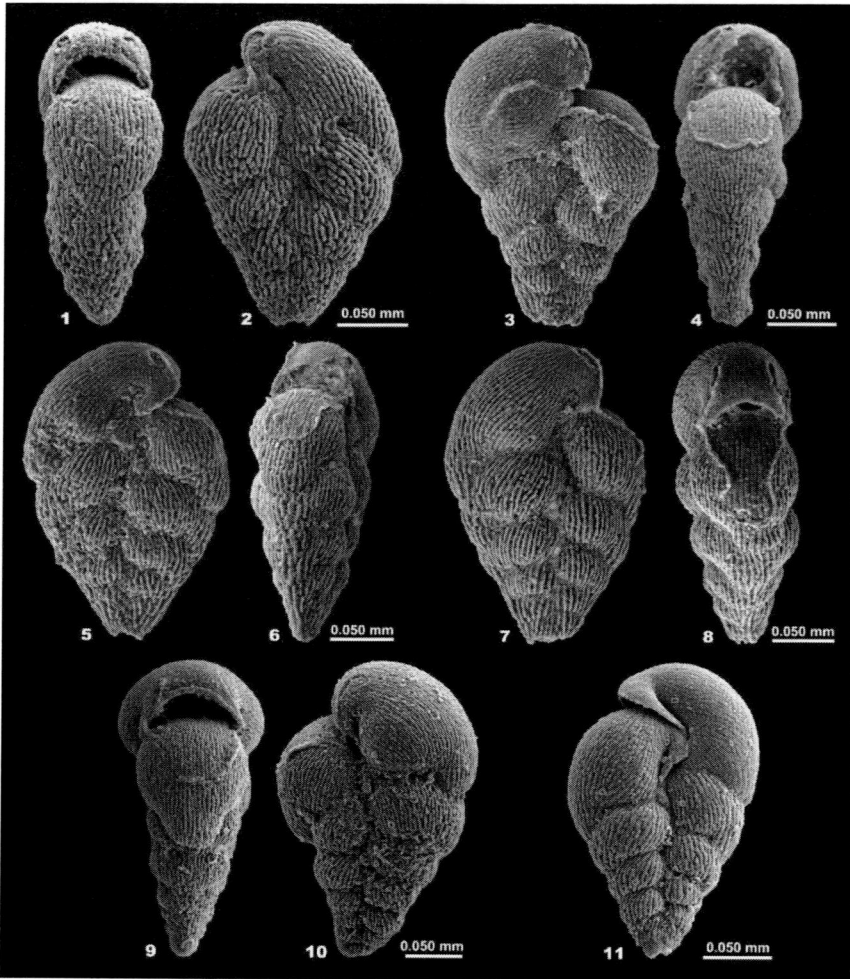

Figure 9. Specimens of *Leptobimodalia leptobimodalis* - new species from the DSDP Site 305 (Shatsky Rise) and Site 463 (Mid-Pacific Mountains). 1-2 Holotype, Sample 32-305-21-2, 100-102 cm (late Campanian, *R. calcarata* Biozone) (WKB 010122). 3-4 Paratype, Sample 62-463-26-3, 52-54 cm (lower Campanian, lower part of the *H. pacificus* Biozone). 5-6 Paratype, Sample 62-463-26-3, 52-54 cm (lower Campanian, lower part of the *H. pacificus* Biozone). 7-8 Paratype, Sample 62-463-26-3, 52-54 cm (lower Campanian, lower part of the *H. pacificus* Biozone). 9-10 Paratype, Sample 32-305-25-1, 100-102 cm (early Campanian, lower part of *G. insignis* Biozone). 11 Paratype, Sample 32-305-25-4, 100-102 cm (early Campanian, lower part of *G. insignis* Biozone).

Remarks. *Leptobimodalia leptobimodalis* differs from *Huberella huberi* Georgescu 2007 by having ornamentation consisting of leptocostae with bimodal rather unimodal orientation. The two species have stratigraphic ranges that do not overlap: the LO of *H. huberi* is at the Turonian/Coniacian boundary whereas the FO of *L. leptobimodalis* is in the early Santonian; a time interval of circa 2.2-2.3 m.y. separates the two bioevents. The species differs from its descendant *L. costellifera* (Masters) mainly by lacking the supplementary aperture in the posterior part of the last-formed one to four chambers. There are significant morphological resemblances between *L. leptobimodalis* and *H. yucatanensis* Georgescu and others 2011 of the latest Turonian-early Coniacian; the two differ by (i) the bimodal ornamentation and (ii) thinner leptocostae (0.0017-0.0036 mm rather than 0.0026-0.0041 mm) in the former species.

Age. Santonian-Campanian.

Geographical distribution. Pacific Ocean (Mid-Pacific Mountains and Shatsky Rise), northern Africa (Tunisia).

Directional Lineage: *Pseudoguembelina* Brönnimann and Brown 1953 - emended
(Figure 10)

Pseudoguembelina BRÖNNIMANN and BROWN 1953, p. 150.
Pseudoguembelina Brönnimann and Brown 1953. Montanaro Gallitelli 1957, p. 139.
Pseudoguembelina Brönnimann and Brown 1953. Loeblich and Tappan 1964, p. C656.
Pseudoguembelina Brönnimann and Brown 1953. Saavedra 1965, p. 344.
Pseudoguembelina Brönnimann and Brown 1953. Brown 1969, p. 36.
Pseudoguembelina Brönnimann and Brown 1953. El-Naggar 1971, p. 446.
Pseudoguembelina Brönnimann and Brown 1953. Masters 1977, p. 369.
Pseudoguembelina Brönnimann and Brown 1953. Weiss 1983, p. 54.
Pseudoguembelina Brönnimann and Brown 1953. Caron 1985, p. 24.
Pseudoguembelina Brönnimann and Brown 1953. Loeblich and Tappan 1987, p. 457.
Pseudoguembelina Brönnimann and Brown 1953. Nederbragt 1989, p. 113.
Pseudoguembelina Brönnimann and Brown 1953. Nederbragt 1991, p. 358.

Species included. IS: *P. praecostulata* - new species, FDS: *P. costulata* (Cushman 1938) and SDS: *P. excolata* (Cushman 1926).

Diagnosis. Directional lineage with the tests with pycnocostate ornamentation, which developed supplementary apertures in the last-formed one to four chambers of the FDS and SDS.

Description. Tests consists of the proloculus followed by chambers added alternately with respect to the test growth axis; chambers gradually increase in size. Earlier chambers are subglobular, then subrectangular, and the last-formed one to four reniform with one distinct backward extension. Sutures are depressed and often discrete, straight to slightly curved, and oblique to the test growth axis. Test is compressed in edge view; periphery is rounded and simple, without peripheral structures. Main aperture is a low to medium high arch at the base of the last-formed chamber. Two symmetrically developed retroflanges, one on each test side, border the aperture; retroflanges are simple in the IS, and rimmed in the SDS, whereas both morphologies occur in the FDS. Supplementary apertures occur under the last-formed one to four chambers backward extension in the FDS and SDS; each of them is protected by an elongate lid-like wall extension. Chamber surface is ornamented with pycnocostae, which has a thickness of 0.0028-0.0052 mm in the IS, 0.0054-0.0118 mm in the FDS, and 0.0055-0.0187 mm in the SDS, and are thicker over the earlier portion of the test in the SDS; pycnocostae orientation is bimodal in the IS and FDS, and unimodal in the SDS due to ornamentation thickening. Test wall is calcitic, hyaline and perforate; pores are circular, elliptical, or irregular, and have a diameter or maximum dimension of 0.0009-0.0013 mm throughout the lineage; pores are situated in the space between the pycnocostae, and are not longitudinally aligned.

Remarks. *Pseudoguembelina* is emended, and accommodates a directional lineage that consists of tests with the pycnocostate ornamentation; pycnocostae exhibit a bimodal arrangement. The DL *Pseudoguembelina* differs from the DL *Leptobimodalia* mainly by the ornamentation consisting of pycnocostae rather than leptocostae; the two lineages have parallel evolution and develop supplementary apertures in the posterior part of the last-formed chambers of the FDS and SDS.

Age. Campanian-Maastrichtian.

Geographical distribution. Cosmopolitan.

KEY FEATURES	DL: *Leptobimodalia* - new IS: *L. leptobimodalis* - new species	DL: *Pseudoguembelina* Brönnimann and Brown 1953		
		IS: *P. praecostulata* - new species	FDS: *P. costulata* (Cushman 1938)	SDS: *P. excolata* (Cushman 1926)
	 No scale is implied.	 No scale is implied.	 No scale is implied.	 No scale is implied.
Chamber arrangement	Alternately added with respect to the test growth axis throughout	Alternately added with respect to the test growth axis throughout	Alternately added with respect to the test growth axis throughout	Alternately added with respect to the test growth axis throughout
Main aperture	Single, a low to medium high arch at the base of the last-formed chamber	Single, a low to medium high arch at the base of the last-formed chamber	Single, a low to medium high arch at the base of the last-formed chamber	Single, a low arch at the base of the last-formed chamber
Supplementary apertures	Absent	Absent	On the last-formed one to four chambers	On the last-formed one to four chambers
Periapertural structures	Simple retroflanges	Simple retroflanges	Simple and rimmed retroflanges	Rimmed retroflanges
Ornamentation	Bimodal and leptocostate, 0.0017-0.0036 mm in thickness	Bimodal and pycnocostate, pycnocostae are 0.0028-0.0052 mm in thickness	Bimodal and pycnocostate, pycnocostae are 0.0054-0.0118 mm in thickness	Bimodal and pycnocostate, pycnocostae are 0.0055-0.0187 mm in thickness
Porosity	Circular or elliptical, 0.0005-0.0010 mm in diameter or maximum dimension	Circular, elliptical, or with irregular shape 0.0009-0.0013 mm in diameter or maximum dimension		

Figure 10. The key features of species of *Pseudoguembelina* Brönnimann and Brown 1953 - emended (green background), and its ancestor, IS of *Leptobimodalia*.

IS: *Pseudoguembelina praecostulata* - new species
(Figure 11:1-13)

Holotype. Specimen WKB 010135 (Figure 11:1-3).

Dimensions of the holotype. L=0.300 mm; W=0.152 mm; T=0.097 mm; W/L=0.507; T/L=0.323.

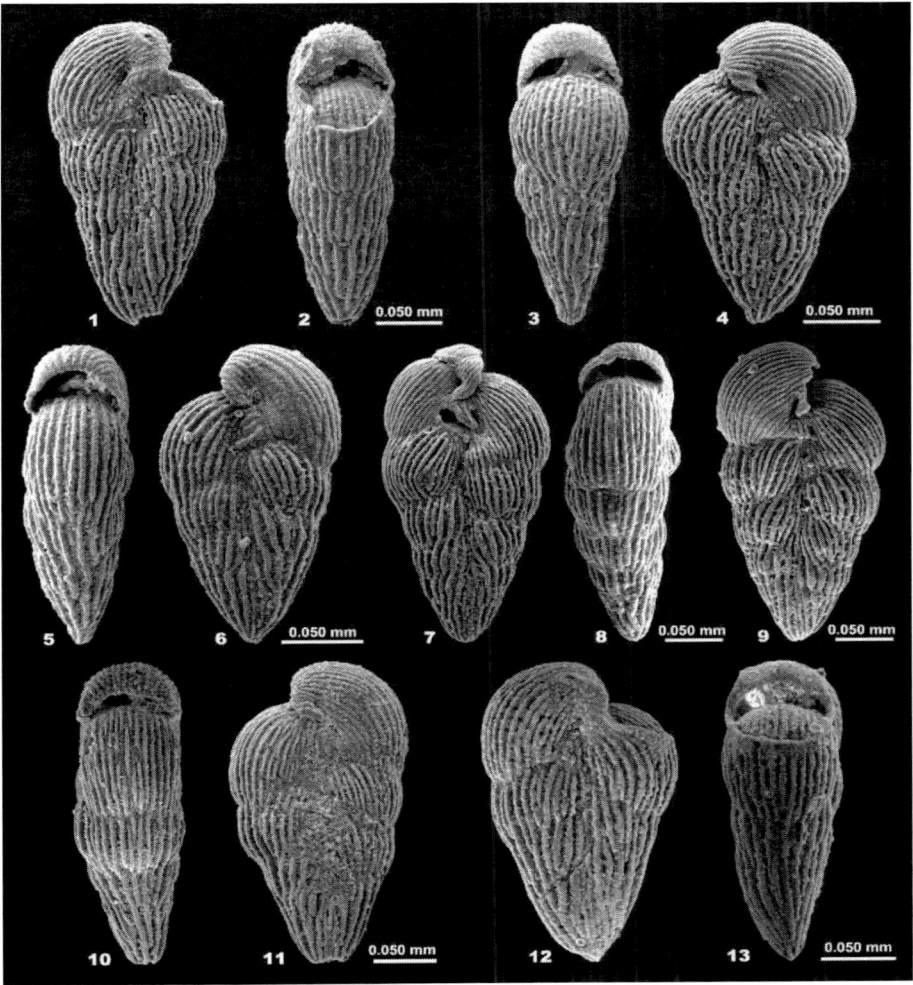

Figure 11. Specimens of *Pseudoguembelina praecostulata* - new species from the DSDP Site 305 (Shatsky Rise) and Site 463 (Mid-Pacific Mountains). 1-2 Paratype, Sample 32-305-28-1, 100-102 cm (early Campanian, lower part of the *H. pacificus* Biozone). 3-4 Paratype, Sample 32-305-28-1, 100-102 cm (early Campanian, lower part of the *H. pacificus* Biozone). 5-6 Paratype, Sample 32-305-27-2, 106-107 cm (early Campanian, lower part of the *H. pacificus* Biozone). 7-8 Holotype, Sample 32-305-27-2, 106-107 cm (early Campanian, lower part of the *H. pacificus* Biozone) (WKB 010135). 9 Paratype, Sample 32-305-27-2, 106-107 cm (early Campanian, lower part of the *H. pacificus* Biozone). 10-11 Hypotype, Sample 62-463-25-1, 51-53 cm middle Campanian, lower part of the *G. insignis* Biozone). 12-13 Hypotype, Sample 62-463-25-1, 51-53 cm middle Campanian, lower part of the *G. insignis* Biozone).

Paratypes. Seven specimens, WKB 010136-010141.

Dimensions. L=0.215-0.300 mm (mean 0.262 mm); W=0.115-0.152 mm (mean 0.136 mm); T=0.067-0.097 mm (mean 0.081 mm); W/L=0.468-0.556 (mean 0.520); T/L=0.275-0.346 (mean 0.313). Ranges are based on the average measurements of the holotype, paratypes, and two hypotypes.

Material. Around 150 specimens.

Type locality. DSDP Site 305 (Shatsky Rise, Pacific Ocean); geographical coordinates: 32° 00`N, 157° 51`E.

Type level. White-yellowish nannofossil chalk of the *H. pacificus* Biozone (early Campanian age); Sample 32-305-27-2, 106-107 cm.

Derivation. The Latin prefix *prae-* (=before) is added to the pre-existing species name *costulata*, thereby indicating species order in the stratigraphic succession.

Diagnosis. *Pseudoguembelina* without supplementary apertures in the posterior part of the last-formed chambers.

Description. Test is biserial throughout consisting of a small proloculus followed by 12-14 chambers that gradually increase in size; the chambers are added alternately with respect to the test growth axis. Chambers in the earlier portion of the test are subglobular and often hardly visible due to the well-developed ornamentation, then with subrectangular shape, and the last-formed one to two are reniform, and with one distinct rounded backward extension. Sutures are shallow, depressed, straight to slightly curved, and oblique to the test growth axis. Test is compressed in edge view, with rounded and simple periphery. Aperture is a low or more rarely medium high arch at the base of the last-formed chamber; two symmetrically developed retroflanges with pycnocostate ornamentation border the aperture, one on each side of the test; retroflanges often define false supplementary apertures along the anterior portion of the central suture. Chamber surface is ornamented with continuous pycnocostae with a width of 0.0028-0.0052 mm; the pycnocostae are longitudinal in the periphery region, and oblique to the test growth axis in the proximity of the central suture. Test wall is calcitic, hyaline, simple, and perforate; pores are circular or elliptical, with the diameter or maximum dimension 0.0009-0.0013 mm, and are situated in the space between the pycnocostae.

Remarks. *Pseudoguembelina praecostulata* differs from its descendant *P. costulata* mainly by lacking supplementary apertures in the posterior part of the last-formed chambers, and having the main aperture in the shape of a low rather than medium high arch. It differs from *L. leptobimodalis* mainly having the chamber surface and periapertural structures ornamented with pycnocostae rather than leptocostae, and larger pores (0.0009-0.0013 mm rather than 0.0006-0.0010 mm). *Heterohelix labellosa* Nederbragt 1991 has general resemblances with *P. praecostulata* but it was reported from a higher stratigraphic level, namely upper Campanian-Maastrichtian; the medium high main aperture (Nederbragt 1991, pl. 2, fig. 4) indicates that such specimens should be considered juveniles of *P. costulata* rather than a distinct species.

Age. Campanian.

Geographical distribution. Pacific Ocean (Shatsky Rise, Mid-Pacific Mountains).

Directional Lineage: *Nederbragtina* – new
(Figure 12)

Species included. IS: *N. prima* - new species, FDS: *N. palpebra* (Brönnimann and Brown 1953) and SDS: *N. hariaensis* (Nederbragt 1991).

Diagnosis. Directional lineage consisting of leptocostate tests that evolved supplementary apertures in the FDS and irregular adult stage with chamber proliferation in the SDS.

Description. Test with the chambers alternately added with respect to the test growth axis throughout in the IS and FDS, and with multichamber growth in the adult stage in the SDS.

KEY FEATURES	DL: *Mihaia*	DL: *Nederbragtina* - new		
	SDS: *M. reussi* (Cushman 1938)	IS: *N. prima* - new species	FDS: *N. palpebra* (Brönnimann and Brown 1953)	SDS: *N. hariaensis* (Nederbragt 1991)
	 No scale is implied.	 No scale is implied.	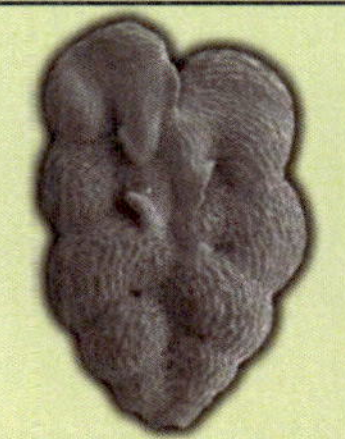 No scale is implied.	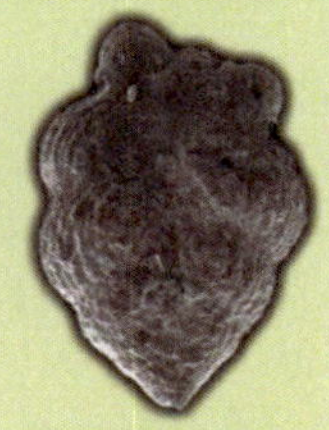 No scale is implied.
Chamber arrangement	Alternately added with respect to the test growth axis throughout	Alternately added with respect to the test growth axis throughout	Alternately added with respect to the test growth axis throughout	Adult stage with multichamber growth
Main aperture	Single, a low to medium high to high arch at the base of the last-formed chamber	Single, a low to medium high arch at the base of the last-formed chamber	Single, a low to medium high arch at the base of the last-formed chamber	Multiple in the adult stage
Supplementary apertures	Absent	On the last-formed one to two chambers	On the last-formed two to five chambers	Absent
Periapertural structures	Metaflanges	Rimmed retroflanges	Rimmed retroflanges	Rimmed retroflanges
Ornamentation	Unimodal and leptocostate, 0.0032-0.0053 mm in thickness	Bimodal and leptocostate, 0.0033-0.0068 mm in thickness	Bimodal and leptocostate, 0.0061-0.0130 mm in thickness	Bimodal and leptocostate, 0.0071-0.0147 mm in thickness
Porosity	Circular, 0.0006-0.0012 mm in diameter	Circular or elliptical, 0.0010-0.0021 mm in diameter or maximum dimension	Circular or elliptical, 0.0010-0.0025 mm in diameter or maximum dimension	Circular or elliptical, 0.0012-0.0028 mm in diameter or maximum dimension

Figure 12. The key features of species of *Nederbragtina* - new directional lineage (green background) and its ancestor, SDS of *Planoheterohelix*.

Earlier chambers are subglobular, then subrectangular, and the last-formed ones are often petaloid. Sutures are distinct, depressed, straight to curved, and perpendicular to oblique to the test growth axis. Test is compressed in edge view; periphery is rounded and simple, without peripheral structures. Main aperture is a low to medium high arch situated at the base of the last-formed chamber in the IS and FDS; the adult stage with multichamber growth in SDS presents biaperturate (e.g., progressive chamber), and monoaperturate (e.g., relapsed) chambers. The main apertures are bordered by two symmetrically developed rimmed or not retroflanges, one on each side of the test. Small-sized supplementary apertures occur in the posterior portion of the last-formed one to two chambers in the IS, are well-developed in the last-formed two to five chambers, rarely one in the FDS, and absent in the SDS; the supplementary apertures are situated at the end of the retroflange. Chamber surface is ornamented with leptocostae with a thickness of 0.0033-0.0068 mm in the IS, 0.0061-0.0130 mm in the FDS, and 0.0071-0.0147 mm in the SDS; leptocostae are thicker over the earlier chambers due to the addition of successive layers of calcite during the ontogenetic development. Test wall is calcitic, hyaline, simple and perforate; pores are circular or elliptical, with a diameter or maximum dimension of 0.0010-0.0021 mm in the IS, 0.0010-0.0025 mm in the FDS, and 0.0012-0.0028 mm in the SDS; vuggy pores with irregular shape and maximum dimension of 0.0031-0.0045 mm occur occasionally in the proximity of the central suture in the FDS.

Remarks. *Nederbragtina* accommodates a directional lineage, which includes three species that have similar earlier stages, and ornamentation consisting of leptocostae. It differs from the DL *Lipsonia* mainly by having (i) leptocostate ornamentation rather than smooth chamber surface, and (ii) depressed sutures rather than lined with phaneroridges. Supplementary apertures in the backward portion of the last-formed chambers are developed in the FDS of *Nederbragtina*, whereas in *Lipsonia* they occur only in the IS; in addition the adult stage with multichamber growth is developed in the SDS in *Nederbragtina*, whereas it evolved in the FDS of *Lipsonia*. *Nederbragtina* can be easily distinguished from *Pseudoguembelina* mainly by the ornamentation that consists of leptocostae rather than pycnocostae.

Derivation. Directional lineage named after Dr Alexandra J. Nederbragt as appreciation for her outstanding contributions in the study of the Cretaceous planktic foraminifera.

Age. Late Campanian-Maastrichtian.

Geographical distribution. Cosmopolitan.

IS: *Nederbragtina prima* - new species
(Figure 13: 1-12)

Holotype. Specimen WKB 010142 (Figure 1-3).

Dimensions of the holotype. L=0.391 mm; W=0.313 mm; T=0.138 mm; W/L=0.801; T/L=0.353.

Paratypes. Five specimens, WKB 010143-010147.

Dimensions. L=0.353-0.428 mm (mean 0.382 mm); W=0.257-0.313 mm (mean 0.280 mm); T=0.138-0.183 mm (mean 0.159 mm); W/L=0.671-0.801 (mean 0.732); T/L=0.353-0.442 (mean 0.416). Ranges are based on the average measurements of the holotype, paratypes, and three hypotypes.

Material. Around 50 specimens.

Type locality. DSDP Site 305 (Shatsky Rise, Pacific Ocean); geographical coordinates: 32° 00`N, 157° 51`E.

Type level. White-yellowish nannofossil chalk of the *G. stuarti* Biozone (late Campanian age); Sample 32-305-18-2, 100-102 cm.

Derivation. Species name comes from the Romanian word *prima* (=first), which indicates it is the oldest in the history of the directional lineage *Nederbragtina*.

Diagnosis. *Nederbragtina* without multichamber growth in the adult stage, and with incipiently developed supplementary apertures in the posterior portion of the last-formed chambers.

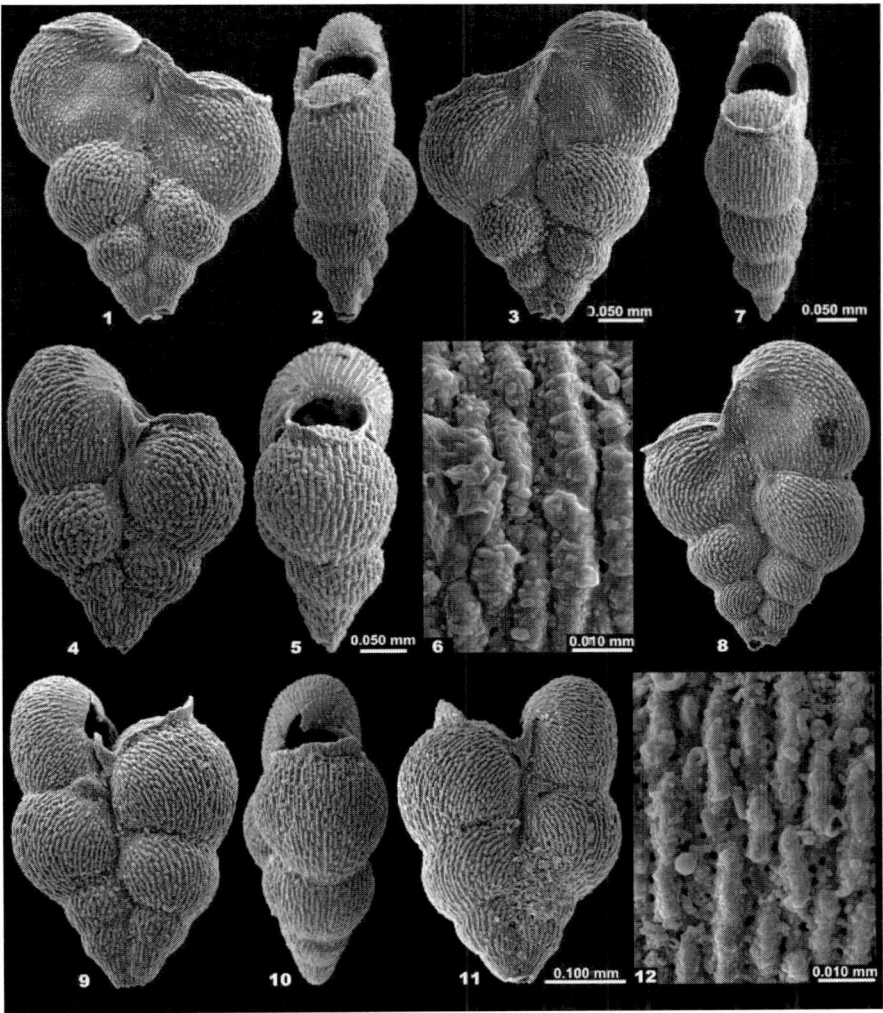

Figure 13. Specimens of *Nederbragtina prima* - new species from the DSDP Site 305 (Shatsky Rise). 1-3 Holotype, Sample 32-305-18-2, 100-102 cm (late Campanian, lower part of the *G. stuarti* Biozone) (WKB 010142). 4-6 Paratype, Sample 32-305-18-2, 100-102 cm (late Campanian, lower part of the *G. stuarti* Biozone); note the inflated chambers in edge view. 7-8 Paratype, Sample 32-305-18-3, 100-102 cm (late Campanian, lower part of the *G. stuarti* Biozone). 9-12 Paratype, Sample 32-305-18-4, 100-102 cm (late Campanian, lower part of the *G. stuarti* Biozone).

Description. The test consists of the proloculus followed by 11-13 chambers that gradually increase in size; chambers are alternately added with respect to the test growth axis throughout. Earlier chambers are subglobular, then with subrectangular shape, and the last-formed one to two petaloid. Sutures are distinct, depressed throughout, straight or slightly curved, and oblique to the test growth axis. Last-formed chambers are slightly compressed in edge view; periphery is rounded and simple, without peripheral structures. Main aperture is a low to medium high arch at the base of the last-formed chamber; two symmetrically developed rimmed retroflanges border the aperture, one on each side of the test. Small-sized supplementary apertures occur at the junction between the retroflanges and previous chambers in the last-formed one or two chambers; the supplementary apertures are situated at the end of the retroflanges. Chamber surface is ornamented with leptocostae, which are longitudinal and continuous in the peripheral region, and oblique to the test growth axis and discontinuous in the proximity of the central suture; leptocostae are more prominent over the earlier chambers, and have a width is of 0.0033-0.0068 mm. Test wall is calcitic, hyaline, simple, and perforate; pores are circular or elliptical, with a diameter or maximum dimension of 0.0010-0.0021 mm.

Remarks. *Nederbragtina prima* resembles *Planulitella sphenoides* (Masters 1976) in general test appearance; the main difference is in the occurrence of supplementary apertures on the last-formed chambers in the former species; there is gap of circa 5.5 m.y. between the extinction of the latter and the evolutionary occurrence of the former. *Nederbragtina prima* exhibits significant variability in the chamber inflation; the specimens with higher T/L ratio (Figure 13:4-6, 9-12) also present better developed supplementary apertures resulting from the retroflanges rimming; such specimens apparently indicate the initiation of the FDS of the *Nederbragtina* directional lineage, namely *N. palpebra* (Brönnimann and Brown 1953).

Age. Late Campanian-Maastrichtian.

Geographical distribution. Pacific Ocean (Shatsky Rise. Mid-Pacific Mountains).

CONCLUSION

The study of gublerinid and pseudoguembelinid planktic foraminifera of the Late Cretaceous (Santonian-Maastrichtian) in the Pacific and Indian Oceans led to the recognition of five lineages: *Magellanina* - new directional lineage, *Lipsonia* Georgescu and Abramovich 2008 - emended, *Leptobimodalia* - new directional lineage, *Pseudoguembelina* Bronnimann and Brown 1953 - emended, and *Nederbragtina* - new directional lineage. The five lineages are of paramount importance in understanding the evolutionary relationships between representatives of the gublerinid and pseudoguembelinid planktic foraminifera.

Magellanina is a newly described directional lineage of the Campanian and Maastrichtian, and includes the new species *M. magellani* as IS and *M. acuta* (de Klasz 1953) as FDS. The tests are small-sized, and ornamented with five quasi-longitudinal bands of leptocostate and reticulate ornamentation in the FDS. The IS has chambers alternately added with respect to the test growth axis, and multichamber growth occurs in the FDS. *Magellanina* evolved from *Planoheterohelix planata* (Cushman 1938), a species of the heterohelicid stem group (Figure 9).

Lipsonia Georgescu and Abramovich 2008 is emended as a directional lineage in the evolutionary classification, and includes two species; its stratigraphic range is re-evaluated as late Campanian-Maastrichtian. The new species *L. shatskyensis* is the IS, and *L. lipsonae* the FDS; the IS presents chambers alternately added with respect to the test growth axis throughout, and a well-developed multichamber growth stage evolved in the FDS. The DL *Lipsonia* evolved phaneroridges, which are often indistinct over the test's earlier portion due to the successive additions of calcite layers. Notably one supplementary aperture situated at the posterior end of the last-formed chamber and in the proximity of the central suture occurs in the IS. The ancestor of the DL *Lipsonia* is the IS of *Gublerina* (Figure 14).

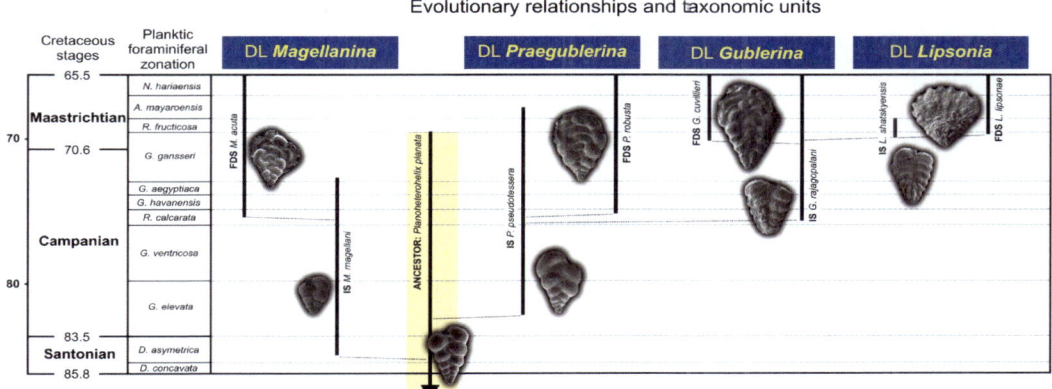

Figure 14. Evolutionary relationships between the planktic foraminiferal species that evolved gublerinid-like tests. Ages are after Gradstein and others (2004). Planktic foraminiferal zonation is after Robaszynski and Caron (1995).

An overall view on the Campanian-Maastrichtian gublerinid planktic foraminiferal evolution shows that although two lineages, namely *Magellanina* and *Praegublerina*, initiated their evolution in the early Campanian it was not until the late Campanian when the gublerinid chamber proliferation in the adult stage evolved independently in the two lineages (Figure 14). The ornamentation of the FDS in the two lineages is different: mixed leptocostate and reticulate in *M. acuta* and leptocostate throughout in *P. robusta*; in addition *P. robusta* has more chambers in the adult stage with multichamber growth when compared to *M. acuta* (Georgescu and others 2008). The independent evolution in the two lineages is reflected not only in the ornamentation and adult stage with multichamber growth of the FDS, but also in the IS of the two lineages: chamber divergence in the adult stage of *P. pseudotessera* (Cushman 1938) is more apparent when compared to *M. magellani*, which appears more conservative with respect to the development of this feature. The two other lineages in which the gublerinid adult stage of the FDS evolved are *Gublerina*, which was initiated in the late Campanian, and *Lipsonia* in the Maastrichtian. It is interesting that together with *Praegublerina* the three lineages form an evolutionary continuum, in which the initiating species are in direct evolutionary relationships (Figure 9). The main difference between *Gublerina* and *Lipsonia* is in the ornamentation, which is concentrated over the chamber surface in the former, and in the phaneroridges in the latter.

Leptobimodalia is a newly described directional lineage, which initiated its evolution in the early Santonian, and consists of three species: the new species *L. leptobimodalis* is the IS, *L. costellifera* (Masters 1976) the FDS and *L. kempensis* (Esker 1968) the SDS. The main evolutionary change in this lineage is the development of a chamber with a backward extension, which bears supplementary apertures in the last-formed one to four chambers only in the FDS and SDS; the IS lacks such supplementary apertures. The ornamentation is bimodal and leptocostate with an evident trend of thickening in the SDS. *Leptobimodalia* evolved from the heterohelicid stem, namely from *P. planata* (Figure 15).

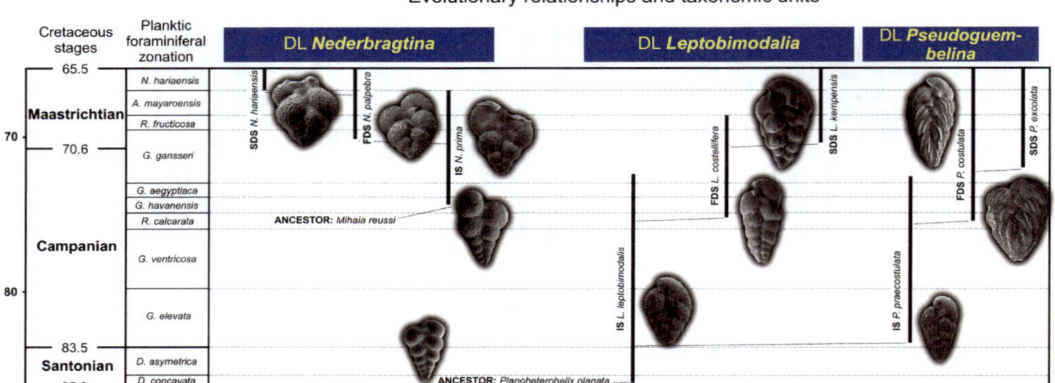

Figure 15. Evolutionary relationships between the planktic foraminiferal species that evolved pseudoguembelinid-like tests. Ages are after Gradstein and others (2004). Planktic foraminiferal zonation is after Robaszynski and Caron (1995).

Pseudoguembelina Brönnimann and Brown 1953 is emended and accommodates a directional lineage consisting of three species: the new species *P. praecostulata* is the IS, *P. costulata* (Cushman 1938) the FDS, and *P. excolata* (Cushman 1926) the SDS. The DL *Pseudoguembelina* is the only known lineage that evolved supplementary apertures and is ornamented with pycnocostae. The evolution of *Pseudoguembelina* took place in the early Campanian from *L. leptobimodalis* (Figure 15) and supplementary apertures in the last-formed one to four chambers occur only in the FDS and SAD.

Nederbragtina evolved in the late Campanian from *Mihaia reussi* (Cushman 1938) of the heterohelicid stem and continued its evolution into the Maastrichtian; this new directional lineage consists of three species: the new species *N. prima* is the IS, *N. palpebra* (Brönnimann and Brown 1953) the FDS, and *N. hariaensis* (Nederbragt 1991) the SDS. The IS has a test consisting entirely of chambers that are added with respect to the test growth axis throughout ontogeny; there are supplementary apertures in the IS and FDS (smaller in the former and well-developed in the latter), and SDS evolved to the adult stage with multichamber growth. *Nederbragtina* consists of completely leptocostate tests, in which the ornamentation is slightly thicker over the earlier portion of the test due to the addition of successive layers of calcite. There are no known descendants from *Nederbragtina*.

All the new species described are the IS of the five lineages; this shows the high degree of species lumping in the typological classification of the Cretaceous planktic foraminifera, in which the species with features that are intermediate between two generally accepted morphologies are often overlooked. Such practice prevents us from having a more accurate view on the fossil record. For example, it is quite evident from the high resolution test observations that *P. planata* and *L. costellifera* are related, but it was impossible to document the evolution from the former species as ancestor to the latter; the possible occurrence of a leptocostate species with bulbous chamber backward extensions in the adult stage, but without supplementary apertures could be inferred, then documented and finally formalized with the description of *L. leptobimodalis*. Therefore, it is concluded that the evolutionary classification herein developed for the Cretaceous planktic foraminifera, has a major characteristic of science, which is predictability. By contrast, predictability is absent in the classical typological classification, in which a genus is a conglomerate of species grouped together according to the opinion of a certain author or group of authors. Notably this was observed for the first time by Gandolfi (1955, p. 78): "Some transitional forms could be postulated before they were actually found."

ACKNOWLEDGMENTS

The DSDP/ODP headquarters are thanked for providing the samples used in this study. Dr L.V. Hills (University of Calgary) is thanked for the presubmittal review of the manuscript. Dr M. Schoel (Microscope and Imaging Facility, University of Calgary) is thanked for the professional and enthusiastic help during the SEM operations.

REFERENCES

Brown, N.K., Jr., 1969. Heterohelicidae Cushman, 1927, amended, a Cretaceous planktonic foraminiferal family. In: *Proceedings of the First International Conference on Planktonic Microfossils, Geneva 1967* (P. Brönnimann, P. and H.H. Renz, Eds). Leiden: E.J. Brill, 2, 21-67.

Brönnimann, P., Brown, N.K. Jr., 1953. Observations on some planktonic Heterohelicidae from the Upper Cretaceous of Cuba. *Contributions from the Cushman Foundation for Foraminiferal Research*, 4, 150-156.

Caron, M., 1985. Cretaceous planktic foraminifera. In: *Plankton Stratigraphy* (H.M. Bolli, J.B. Saunders, K. Perch-Nielsen, Eds). Cambridge: Cambridge University Press, 17-86.

Cushman, J.A., 1926. Some foraminifera from the Mendez Shale of eastern Mexico. *Contributions from the Cushman Laboratory for Foraminiferal Research*, 2, 16-26.

Cushman, J.A., 1938. Cretaceous species of *Gümbelina* and related genera. *Contributions from the Cushman Laboratory for Foraminiferal Research*, 14, 2-27.

El-Naggar, Z.R., 1971. On the classification, evolution and stratigraphical distribution of the Globigerinacea. In: *Proceedings of the II Planktonic Conference, Roma 1970* (A. Farinacci, Ed.). Roma: Edizioni Tecnoscienza, 421-476.

Esker, G.C., 1968. A new species of *Pseudoguembelina* from the Upper Cretaceous of Texas. *Contributions from the Cushman Foundation for Foraminiferal Research*, 19, 168-169.

Gandolfi, R., 1955. The genus *Globotruncana* in northeastern Colombia. *Bulletins of American Paleontology*, 36 (155), 1-118.

Georgescu, M.D., 2007. A new planktonic heterohelicid foraminiferal genus from the Upper Cretaceous (Turonian). *Micropaleontology*, 53, 212-220.

Georgescu, M.D., 2009. Upper Albian-lower Turonian non-schackoinid planktic foraminifera with elongate chambers: morphology reevaluation, taxonomy and evolutionary classification. *Revista Española de Micropaleontología*, 41, 255-293.

Georgescu, M.D., 2010. Origin, taxonomic revision and evolutionary classification of the late Coniacian-early Campanian (Late Cretaceous) planktic foraminifera with multichamber growth in the adult stage. *Revista Española de Micropaleontología*, 42, 59-118.

Georgescu, M.D., 2011. New data on the evolutionary classification of the Late Cretaceous (late Coniacian-Santonian) planktic foraminifera with elongate chambers. *Revista Española de Micropaleontología*, 43, 39-54.

Georgescu, M.D., 2012a. Iterative evolution, taxonomic revision and evolutionary classification of the praeglobotruncanid planktic foraminifera, Cretaceous (late Albian-Santonian). *Revista Española de Micropaleontología*, 43, 173-207.

Georgescu, M.D., 2012b. Morphology, taxonomy, stratigraphic distribution and evolutionary classification of the schackoinid planktic foraminifera (late Albian-Maastrichtian, Cretaceous). In: *Deep-Sea Marine Biology, Geology, and Human Impact* (D.R. Bailey and S.E. Howard, Eds), 1-62. New York: Nova Publishers.

Georgescu, M.D., Abramovich, S., 2008. A new serial planktic foraminifer (Family Heterohelicidae Cushman, 1927) from the Upper Maastrichtian of the equatorial Central Pacific. *Journal of Micropaleontology*, 27, 117-123.

Georgescu, M.D., Quinney, A.E., Anderson, K.D., 2011. New data on the taxonomy, evolution and biostratigraphic significance of the Turonian-Coniacian (Late Cretaceous) planktic foraminifer *Huberella* Georgescu 2007. *Micropaleontology*, 57, 247-254.

Georgescu, M.D., Saupe, E.E., Huber, B.T., 2008. Morphometric and stratophenetic basis for phylogeny and taxonomy in Late Cretaceous gublerinid planktonic foraminifera. *Micropaleontology*, 54, 397-424. [published in 2009]

Govindan, A., 1972. Upper Cretaceous planktonic foraminifera from the Ponicherry area, south India. *Micropaleontology*, 19, 160-193.

Gradstein, F.M., Ogg, J.G., Smith, A.G. (Eds.), 2004. *A Geologic Time Scale 2004*. Cambridge: Cambridge University Press, 589 pp.

Hay, W.H., DeConto, R.M., Wold, C.N., Wilson, K.M., Voigt, S., Schulz, M., Wold, A.R., Dullo, W, Ronov, A.B., Balukhovsky, A.N., Söding, E., 1999. Alternative global Cretaceous paleogeography. In: *Evolution of the Cretaceous Ocean-Climate System* (E. Barrera and C.C. Johnson, Eds). *The Geological Society of America Special Publication*, 332, 1-47.

Klasz, I. de, 1953. On the foraminiferal genus *Gublerina* Kikoïne. *Geologica Bavarica*, 17, 245-251.

Loeblich, A.R. Jr., Tappan, H., 1964. Sarcodina Chiefly "Thecamoebians" and Foraminifera. In: *Treatise on Invertebrate Paleontology. Part C* (R.C. Moore, Ed.). The Geological Society of America and The University of Kansas Press, 900 p.

Loeblich, A.R. Jr., Tappan, H., 1987. *Foraminiferal Genera and Their Classification*. New York: Van Nostrand Reinhold Company, 970 p.

Masters, B.A., 1976. Planktic foraminifera from the Upper Cretaceous Selma Group, Alabama. *Journal of Paleontology*, 50, 318-330.

Masters, B.A., 1977. Mesozoic planktonic foraminifera. A world–wide review and analysis. In: *Oceanic Micropaleontology* (A.T.S. Ramsay, Ed.). London-New York-San Francisco: Academic Press, 1, 301-731.

Montanaro Gallitelli, E., 1957. A revision of the foraminiferal family Heterohelicidae. In: *Studies in foraminifera* (A.R. Jr. Loeblich, Ed.). Washington, D.C.: *United States National Museum History Bulletin*, 215, 133-154.

Nederbragt, A.J., 1989. Chamber proliferation in the Cretaceous planktonic foraminifera Heterohelicidae. *Journal of Foraminiferal Research*, 19, 105-114.

Nederbragt, A.J., 1991. Late Cretaceous biostratigraphy and development of Heterohelicidae (planktic foraminifera). *Micropaleontology*, 37, 329-372.

Reiss, Z., 1957. Notes on foraminifera from Israel. *Sigalia* - a new genus of foraminifera. *Bulletin of the Research Council of Israel*, 6b, 239-244.

Robaszynski, F., Caron, M., 1995. Foraminifères planctoniques du Crétacé: commentaire de la zonation Europe-Mèditerranée. *Bulletin de la Société Géologique de France*, 6, 681-692.

Saavedra, J.L. 1965. La evolución de los Globigerináceos. *Boletín de la Real Sociedad Española de Historia Natural*, 63, 317-349.

Salaj, J. and Samuel, O. 1963. Mikrobiostratigrafia srednej a vrchnej Kreidy z východnej Časti Bradloveho Pasma. *Geologicke Prace*, 30, 93-112.

Steineck, P.L., Fleisher, R.L., 1978. Towards the classical evolutionary classification of the Cenozoic Globigerinacea (Foraminiferida). *Journal of Faleontology*, 53, 618-635.

Weiss, W., 1983. Heterohelicidae (seriale planktonische Foraminiferen) der tethyalen Oberkreide (Santon bis Maastricht). *Geologisches Jahrbuch*, A72, 3-93.

Chapter 4

Taxonomic Revision of *Planoglobulina* Cushman 1927 as Directional Lineage in Evolutionary Classification

M. Dan Georgescu[*]

Department of Geosciences, University of Calgary,
Calgary, Alberta, Canada

Abstract

The taxonomic revision of *Planoglobulina* Cushman 1937 in evolutionary classification results in the definition of the homonym directional lineage. The initiating species is *P. sphaeralis*, a new species described from the upper part of the middle Campanian-Maastrichtian sediments of Alabama and Texas; this species presents the chambers alternately added with respect to the test growth axis throughout the ontogenetic development. First descendant species is *P. acervulinoides* (Egger 1899), which presents the adult stage with chamber proliferation. *Planoglobulina* is the only known lineage of Cretaceous planktic foraminifera with chambers alternately added with respect to the test growth axis with adult stage with multichamber growth and ornamentation consisting of longitudinal pycnocostae.

Keywords: Foraminifera, planktic, Campanian, Maastrichtian, new species, evolutionary classification

Introduction

Evolution of an adult stage with multichamber growth is an iterative process that occurred in fourteen lineages of foraminifera with chambers alternately added with respect to the test growth axis. The occurrence of such chamber proliferation in the adult stage was for the first time considered of importance in taxonomy by Egger (1899) in the description of the

---

[*] Corresponding author: dgeorge@ucalgary.ca.

species *Gümbelina acervulinoides* (Figure 1); Cushman (1927) used the proliferating adult stage as morphological criterion at generic level in the genus *Planoglobulina* and further demonstrated that this genus independently evolved from *Ventilabrella* Cushman 1928 from lower stratigraphic levels (Santonian-lower Campanian) for which *Planoglobulina* was often confused or synonymized. One distinct attempt to synonymize *Planoglobulina* and *Ventilabrella* was made by Montanaro Gallitelli (1957) based on a discrepancy in the synonymies of *P. acervulinoides* by Cushman (1938, 1946), but was not accepted by Brown (1969, p. 39) and Georgescu (2010, p. 102).

   Planoglobulina is herein revised in evolutionary classification based on new high resolution data collected from generally well-preserved tests collected from locations worldwide. The new data indicate that *Planoglobulina* is a directional lineage, which consists of two species: *P. sphaeralis* – new species and *P. acervulinoides*.

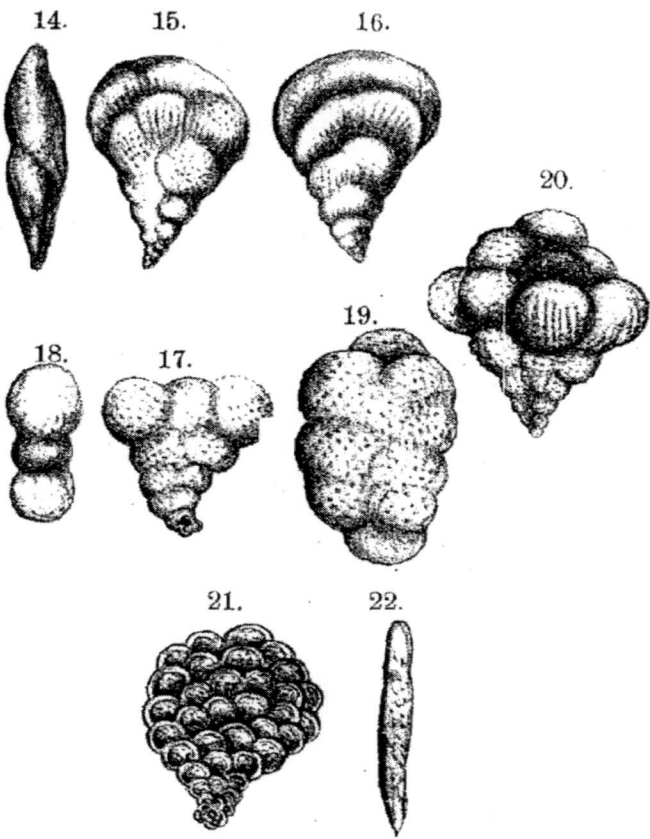

Figure 1. Original illustrations by Egger (1899) of the species *Gümbelina acervulinoides* showing the original numbering; note the polytaxic nature of the group of illustrated specimens.

MATERIAL AND PROVENANCE

Most of the studied fossil material used in the taxonomic revision of the directional lineage *Planoglobulina* was collected from Campanian-Maastrichtian sediments of ten Deep Sea Drilling Project (DSDP)/Ocean Drilling Program (ODP) sites and holes: DSDP Hole 111A (Orphan Knoll, North Atlantic Ocean), DSDP Site 152 (Nicaragua Rise, Caribbean Region), DSDP Site 305 (Shatsky Rise, Central Pacific Ocean), DSDP Site 356 (São Paulo Plateau, South Atlantic Ocean), DSDP Site 357 (Rio Grande Rise, South Atlantic Ocean), DSDP Site 384 (*J*-Anomaly Ridge, North Atlantic Ocean), ODP Hole 761B (Wombat Plateau, East Indian Ocean), ODP Hole 763B (Exmouth Plateau, East Indian Ocean) and ODP Holes 1050C and 1052E (Blake Nose, North Atlantic Ocean). The sediments at these locations consist mostly of white and yellow nannofossil chalks, which were accumulated in deep oceanic conditions (bathyal). Only tests of the first descendant species of *Planoglobulina* were recorded at the ten sites; wherever they were encountered the tests of *P. acervulinoides* are well-preserved and allowed high resolution observations on delicate test structures, ultrastructure, ornamentation and porosity despite the low magnitude recrystallization.

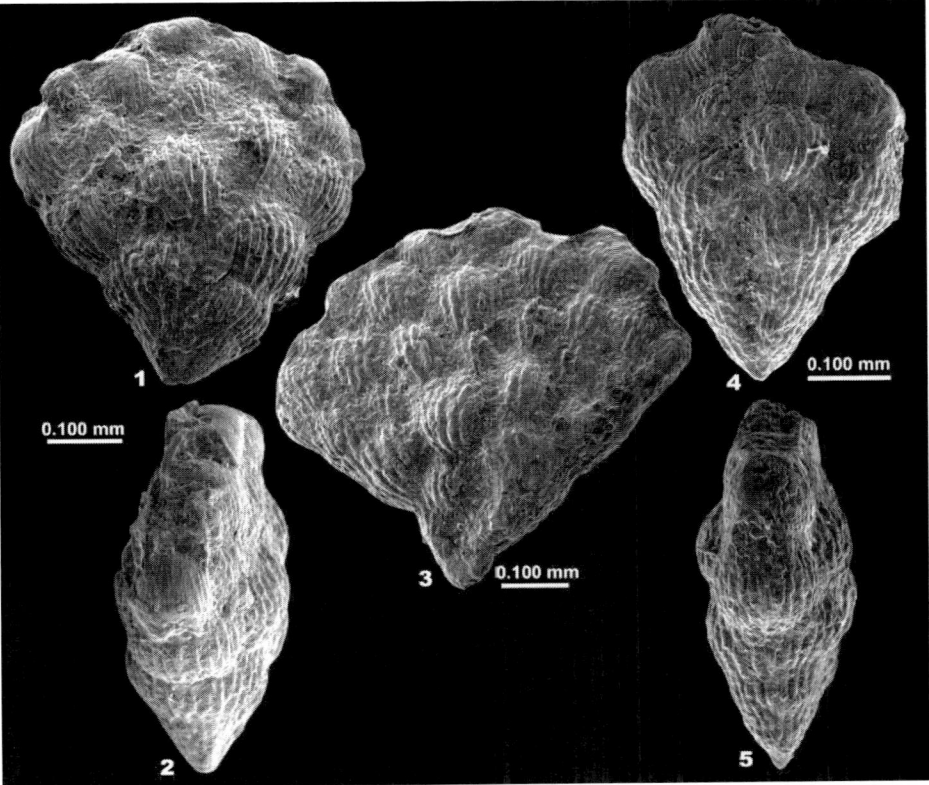

Figure 2. Illustrations of the three specimens of *Planoglobulina acervulinoides* (Egger 1899) from the Loeblich and Tappan Topotype Collection, National Museum of Natural History, Smithsonian Institution, Washington, D.C. **1-2** Specimen from Bruderndorf, Niederöstereich (USNM 480989). **3-5** Specimens from Eisenärtz, Bavaria (USNM 477905).

Planoglobulina sphaeralis, which is the initiating species of the directional lineage, is herein recognized and documented based on the material from the McGugan Collection of the University of Calgary. The material was collected by Dr. Alan McGugan during a field trip in Alabama (USA) from the sections described in detail by Scott and others (1968). The samples were collected from the Demopolis Chalk, Ripley Formation and Prairie Bluff Chalk, covering the middle Campanian-lower Maastrichtian stratigraphic interval; the age dating of these samples confirm the stratigraphic framework of Liu (2007). The foraminiferal tests from these lithostratigraphic units are very well preserved and in pristine condition allowing high quality observations on the test morphology.

One collection specimen is of paramount importance in the taxonomic revision of the directional lineage *Planoglobulina*. It was discovered in the Loeblich and Tappan Topotype Collection, National Museum of Natural History, Smithsonian Institution, Washington, D.C. (Figure 2:1-2). This specimen of *P. acervulinoides* (USNM 480989) was collected from the Maastrichtian glauconitic sands of Bruderndorf, Niederöstereich and sent to A.R. Jr. Loeblich and H.N. Tappan by V. Pokorný when the material of the type locality of the species *Pseudotextularia elegans* of A. Rzehak was curated. Two additional specimens of the same species from the Loeblich and Tappan Topotype Collection (USNM 477905) were collected from Buchecker Beds near Eisenärtz, Bavaria (Figure 2:3-5).

SYSTEMATIC DESCRIPTIONS

Evolutionary classification units at lineage and species level are from Georgescu (2010, 2011, 2013). Species abbreviations: IS-initiating species, FDS-first descendant species.

Directional Lineage: *Planoglobulina* Cushman 1927a - emended

Planoglobulina CUSHMAN 1927a, p. 77.
Planoglobulina Cushman 1927. Cushman 1927b, 157.
Planoglobulina Cushman 1927. Cushman 1927c, p. 61.
Planoglobulina Cushman 1927. Cushman 1933, p. 212.
Planoglobulina Cushman 1927. Galloway 1933, p. 346.
Planoglobulina Cushman 1927. Cushman 1950, p. 257.
Planoglobulina Cushman 1927. Montanaro Gallitelli 1957, p. 141.
Planoglobulina Cushman 1927. Loeblich and Tappan 1964, p. C655.
Planoglobulina Cushman 1927. Saavedra 1965, p. 344.
Planoglobulina Cushman 1927. Govindan 1972, p. 171.
Planoglobulina Cushman 1927. Martin 1972, p. 141.
Planoglobulina Cushman 1927. Smith and Pessagno 1973, p. 20.
Planoglobulina Cushman 1927. Masters 1976, p. 321.
Platystaphyla MASTERS 1976, p. 325.
Planoglobulina Cushman 1927. Aliyulla 1977, p. 203.
Platystaphyla Masters 1976. Masters 1977, p. 363.
Planoglobulina Cushman 1927. Smith 1978, p. 316.

Planoglobulina Cushman 1927. Pandey 1980, p. 63.
Planoglobulina Cushman 1927. Weiss 1983, p. 49.
Planoglobulina Cushman 1927. Aliyulla in Ali-zade and others 1988, p. 128.
Planoglobulina Cushman 1927. Loeblich and Tappan 1987, p. 455.
Planoglobulina Cushman 1927. Nederbragt 1989a, p. 113.
Planoglobulina Cushman 1927. Nederbragt 1991, p. 354.

Species included. IS: *P. sphaeralis* - new species and FDS: *P. acervulinoides* (Egger 1899).

Emended diagnosis. This is a Campanian-Maastrichtian directional lineage with pycnocostate ornamentation with chambers alternately added with respect to the test growth axis in the IS and evolved multichamber growth in the FDS.

Emended description. Test with the proloculus followed by chambers alternately added with respect to the test growth axis throughout in the IS and an adult stage with multichamber growth in the FDS. Chambers are globular throughout the ontogeny and with a variable overlap rate. Sutures are distinct and depressed, straight to slightly curved between all the chambers of the test. The adult stage with multichamber growth begins with the biaperturate progressive chamber, which is followed by up to six chamber sets in which the number increases by one; irregularities in the number in the successively added chamber sets occur frequently. Test shape is subtriangular in edge view, with the chambers gradually increasing in size in the IS and lanceolate in the FDS; periphery is rounded to broadly rounded and simple, without peripheral structures. Chamber surface is ornamented with longitudinal pycnocostae, which are 0.0023-0.0084 mm in the IS and 0.0070-0.0149 mm in the FDS. Test wall is calcitic, hyaline. Simple and perforate; pores have a circular outline and with a diameter of 0.0004-0.0017 mm in the IS and circular to elliptical in shape and with a diameter or maximum dimension of 0.0017-0.0034 mm in the FDS.

Remarks. *Planoglobulina* is herein reviewed in the evolutionary classification and includes a directional lineage with the chamber surface ornamented with pycnocostae that evolved one adult stage with multichamber growth in the FDS. It differs from the directional lineage *Ehrenbergites* Georgescu 2013 mainly by having the test surface ornamented with pycnocostae rather than leptocostae and adult stage in the FDS with fewer chamber sets; in addition the two directional lineages evolved at different times and from different ancestors [*Ehrenbergites* in the early Santonian *Planoheterohelix planata* (Cushman 1938) and *Planoglobulina* in the late middle Campanian from *Muhaia reussi* (Cushman 1938)]. *Planoglobulina* is the only known lineage of the Campanian-Maastrichtian stratigraphic interval that evolved pycnocostate ornamentation and chamber proliferation in the adult stage in the same plane in which the champers were added in the juvenile one.

Age. Late middle Campanian-Maastrichtian.

Geographic distribution. Cosmopolitan.

IS: *Planoglobulina sphaeralis* – new species
(Figures 3:1-9, 4:1-10)

Heterohelix aff. *striata* (Ehrenberg). Yassini 1979, p. 22, pl. 1, figs 11-13.

Planoglobulina carseyae (Plummer). Yassini 1979, p. 23, pl. 2, figs 6-8.

Heterohelix striata (Ehrenberg). Petters 1983, p. 43, pl. 1, figs 14-15. Zapeda 1998, p. 137, fig. 11: 5.

Heterohelix globulosa (Ehrenberg). Keller and others 2002, pl. 1, fig. 12 (only).

Holotype. Specimen WKB 010154 (Figure 3:1-3).

Holotype dimensions. Length: L=0. 289 mm; width: W=0.213 mm; thickness: T=0.130 mm; W/L=0.737; T/L=0.450.

Paratypes. Six specimens, WKB 010155-010160.

Material. Around 100 specimens.

Dimensions. L=0.176-0.326 mm; W=0.127-0.247 mm; T=0.083-0.153 mm; W/L=0.595-0.984; T/L=0.405-0.667. Ranges based on the average measurements of 16 specimens: holotype, six paratypes and nine hypotypes.

Type locality. Stop 7 of Scott and others (1968, p. 54-56), Montgomery County, Alabama, USA.

Type level. Light-grey and yellowish-grey Demopolis Chalk (see Scott and others 1968, p. 56 for the complete description).

Derivation. From the Latin word *sphaeralis* (=spherical).

Diagnosis. *Planoglobulina* with chambers alternately added with respect to the test growth axis throughout the ontogenetic development.

Description. The test consists of 8-14 globular chambers, which are alternately added with respect to the test growth axis throughout the ontogeny; chambers are spherical in shape, variable degree of overlapping, and gradual size increase. Sutures are distinct, depressed and straight to slightly curved. Periphery is broadly rounded and simple, without peripheral structures. Aperture has the shape of a medium high arch and is situated at the base of the last-formed chamber; two small-sized symmetrically developed metaflanges and one imperforate band between the metaflanges border the aperture. Chamber surface is ornamented with continuous longitudinal pycnocostae with a thickness of 0.0023-0.0084 mm. Wall is calcareous, hyaline, simple and perforate; pores are circular in shape and with a diameter of 0.0004-0.0017 mm.

Remarks. The morphological similarities in the test general appearance, globular to spheroidal chambers and periapertural structures consisting of symmetrically developed metaflanges, one on each side of the test, and one imperforate band bordering the aperture between the flanges indicate the *Planoglobulina sphaeralis* evolved from *Mihaia reussi* (Cushman 1938).

The evolution from *M. reussi* to *P. sphaeralis* is apparent in an increase in pore diameter from 0.0006-0.0012 mm in the ancestor to 0.0004-0.0017 mm in the descendant. The evolution of pycnocostae in *P. sphaeralis* from the leptocostate *M. reussi* resulted in a significant increase in thickness of these ornamentation structures from 0.0032-0.0046 mm in the ancestor to 0.0023-0.0084 mm in the descendant.

Age. Late middle Campanian-Maastrichtian.

Geographic distribution. USA (Alabama, Texas), eastern Atlantic Ocean (Gulf of Guinea), northern Africa (Tunisia) and Middle East (Jordan).

Figure 3. Specimens of *Planoglobulina sphaeralis* – new species from the central part of the Montgomery County, Alabama (McGugan Collection). All specimens are from the middle Campanian Demopolis chalk from Stop 7 of Scott and others (1968). 1-3 Holotype (WKB 010154). 4-5 Paratype. 6-7 Paratype. 8-9 Paratype.

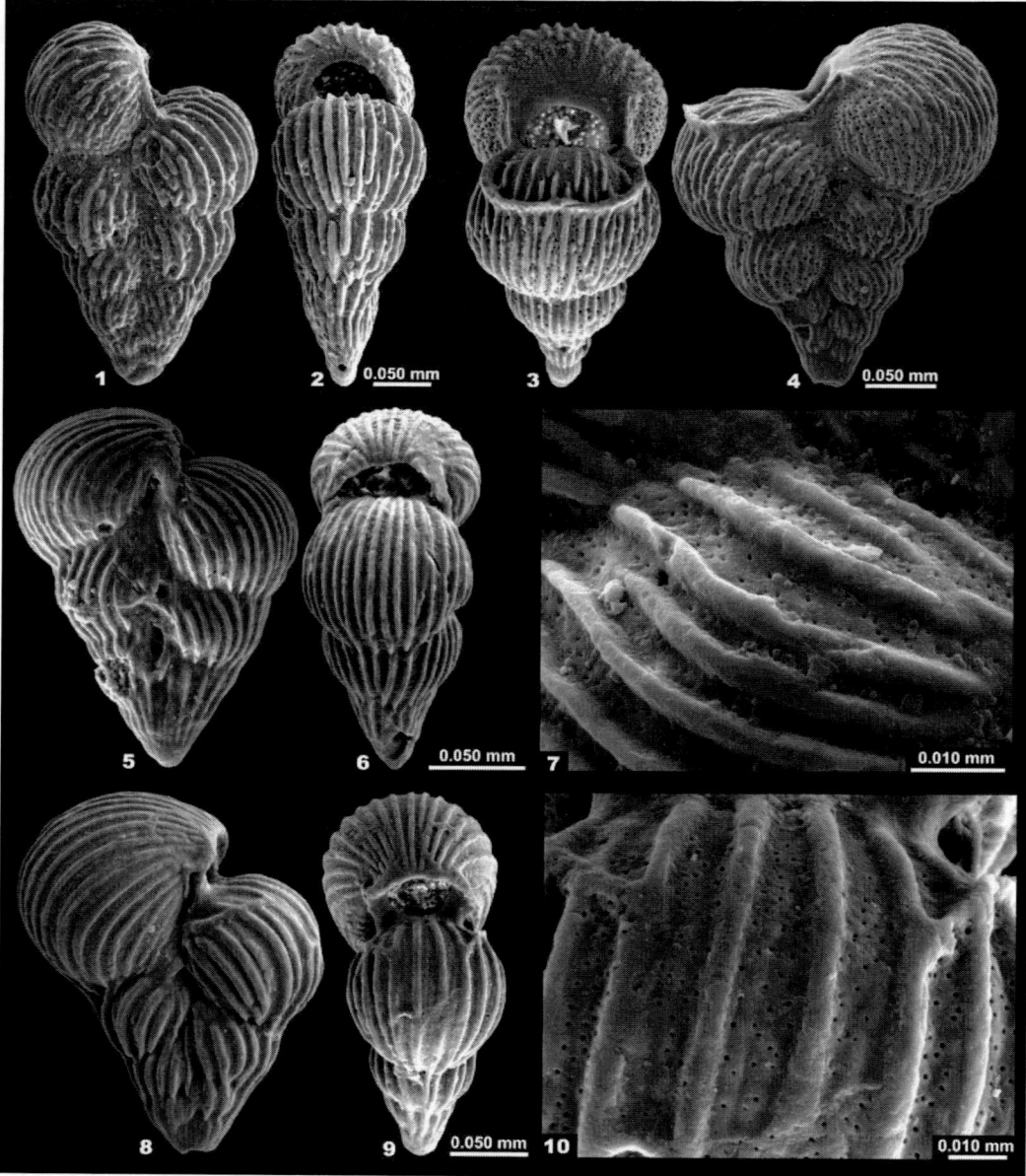

Figure 4. Specimens of *Planoglobulina sphaeralis* – new species. 1-2 Paratype from the middle Campanian Demopolis Chalk of the Montgomery County, Alabama collected from the Stop 7 of Scott and others (1968); specimen from the McGugan Collection. 3-4 Paratype from the early Maastrichtian Prairie Bluff Chalk from Stop 4 of Scott and others (1968); specimen from the McGugan Collection. 5-7 Paratype from the early Maastrichtian Prairie Bluff Chalk from Stop 4 of Scott and others (1968); specimen from the McGugan Collection. 8-10 Paratype from the Maastrichtian Kemp Clay from the Mullinax-1 well of Texas (Sample 27-1).

FDS: *Planoglobulina acervulinoides* (Egger 1899)
(Figures 5:1-10, 6:1-9, 7:1-9)

Gümbelina acervulinoides EGGER 1899, p. 36, pl. 14, figs 19-20 (only).

Pseudotextularia acervulinoides (Egger). Cushman 1926, p. 17, pl. 2, fig. 5. Wille-Janoschek 1966, p. 119, pl. 8, fig. 8. Sturm in Faupl and others 1970, p. 113, pl. 8, fig. 3.

Planoglobulina acervulinoides (Egger). Cushman 1927b, p. 158, pl. 27, fig. 3. Voorwijk 1937, p. 195, pl. 1, fig. 19. Cushman 1938, p. 23, pl. 4, figs 5-8. Cushman 1946, p. 111, pl. 47, figs 12-15. Kikoïne 1948, pl. 2, fig. 6. Cushman 1949, p. 8, pl. 3, fig. 27. Djafarov and others 1951, p. 111, pl. 16, figs 3-4. Noth 1951, p. 62, pl. 7, fig. 1. Itzhaki 1952, p. 188, figs 9-11. Said and Kenawy 1956, p. 140, pl. 3, fig. 45. Sacal and Debourle 1957, p. 13, pl. 3, fig. 11. Olsson 1960, p. 28, pl. 4, fig. 12. Said and Kerdany 1961, p. 334, pl. 2, fig. 15. Kavary and Frizzell 1963, p. 64, pl. 13, figs 11-12. Salaj and Samuel 1963, p. 234, pl. 26, figs 1-2, pl. 37, fig. 14. Said and Sabry 1964, p. 394, pl. 3, fig. 30. Pringgoprawiro 1965, p. 37, pl. 6, fig. 2. Ansary and Tewfik 1968, p. 43, pl. 3, fig. 15. Lehmann 1966, p. 315, pl. 2, fig. 5. Pessagno 1967, p. 271, pl. 87, fig. 14. Dupeuble 1969, p. 158, pl. 4, fig. 16. Funnell and others 1969, p. 22, pl. 1, figs 7-8, text-fig. 4. Hanzliková 1969, p. 41, pl. 8, figs 1-2. Bertels 1970, p. 32, pl. 2, fig. 6. Rahhali 1970, p. 68, text-fig. 7: l-o. Cita and Gartner 1971, pl. 5, fig. 3. Martin 1972, p. 81, pl. 3, figs 3-6. Pessagno and Longoria in Shipboard Scientific Party 1972, pl. 10, fig. 4, pl. 11, fig. 2. Wright and Apthorpe 1976, pl. 1, fig. 2. Linares-Rodríguez 1977, pl. 48, figs 8-9. Pandey 1980, p. 64, pl. 3, figs 11-12. Peryt 1980, p. 46, pl. 5, fig. 5. Butt 1981, pl. 19, fig. E. Weiss 1983, p. 50, pl. 4, figs 5-7. Abdel-Kireem 1986, p. 223, pl. 2, fig. 1. Jansen and Kroon 1987, p. 565, pl. 8, fig. 10. Nederbragt 1989a, pl. 3, fig. 5. Nederbragt 1989b, p. 200, pl. 4, figs 4-6, pl. 5, figs 1-2. Malmgren 1991, pl. 1, fig. 12. Nederbragt 1991, p. 356, pl. 6, figs 5-6, pl. 7, fig. 1. Gawor-Biedowa 1992, p. 73, pl. 11, fig. 13. Georgescu 1995, p. 405, pl. 2, figs 4-6, 9. Mancini and others 1996, fig. 6: 9. Georgescu 1997, fig. 6: 8. Abramovich and others 2003, pl. 4, fig. 11. Howe and others 2003, pl. 7, figs 29-30. Ohmert 2011, pl. 3, fig. 5. Pérez-Rodriguez and others 2012, fig. 7: C.

Pseudotextularia (Gümbelina) acervulinoides Egger. Liebus 1927, p. 375, pl. 14, fig. 2 (only).

Ventilabrella carseyae PLUMMER 1931, p. 178, pl. figs 7-8 (only).

Ventilabrella carseyae Plummer. Sandidge 1932, p. 362, pl. 31, fig. 29. Jennings 1936, p. 28, pl. 3, fig. 13. Cushman 1938, p. 26, pl. 4, figs 20-24. Cushman and Hedberg 1941, p. 93, pl. 22, fig. 18. Cushman and Todd 1943, p. 65, pl. 11, fig. 18. Cushman 1946, p. 112, pl. 48, figs 1-5. Cushman 1947, p. 14, pl. 4, figs 9-10. Cushman 1949, p. 8, pl. 3, fig. 28. Sellier de Civrieux 1952, p. 271, pl. 6, figs 17-18. Hamilton 1953, p. 235, pl. 30, fig. 8. Bettenstaedt and Wicher 1955, pl. 1, fig. 5 (right). Salaj and Samuel 1966, p. 230, pl. 37, fig. 8. Salaj 1980, pl. 16, fig. 2.

Pseudotextularia elegans acervulinoides Egger. Glaessner 1936, p. 102, pl. 1, figs 7-8, 10 (only).

Pseudotextularia elegans (Rzehak). Bettenstaedt and Wicher 1955, pl. 1, fig. 6 (left). Hiltermann in Koch in Bartenstein and others 1962, p. 337, pl 46, fig. 12.

Planoglobulina carseyae (Plummer). Montanaro Gallitelli 1957, pl. 32, fig. 13. Olsson 1960, p. 28, pl. 4, fig. 13. Skinner 1962, p. 42, pl. 5, fig. 18. Kavary and Frizzell 1963, p. 64, pl. 13, figs 11-12. Perlmutter and Todd 1965, p. 114, pl. 2, fig. 15. Pessagno 1967, p. 271, pl.

87, figs 10, 15-16. Hanzliková 1969, p. 41, pl. 8, figs 6-8, 29-30. Neagu 1970, p. 61, pl. 14, figs 13-14. Todd 1970, p. 152, pl. 5, fig. 6. Govindan 1972, p. 171, pl. 1, fig. 17. Hanzliková 1972, p. 94, pl. 23, figs 20-22. Martin 1972, p. 83, pl. 4, figs 4-7. Smith and Pessagno 1973, p. 21, pl. 5, figs 3-12. Webb 1973, pl. 1, fig. 9. Peryt 1980, p. 46, pl. 5, fig. 4. Petters 1983, p. 45, pl. 1, fig. 12. Weiss 1983, p. 51, pl. 5, figs 1-4. Nederbragt 1989a, pl. 3, fig. 4. Nederbragt 1989b, p. 200, pl. 5, figs 3-5. Malmgren 1991, pl. 1, figs 14-15. Nederbragt 1991, p. 356, pl. 7, figs 2-3. Ayyad and others 1996, fig. 6: g. Abramovich and others 2002, pl. 1, fig. 5. Campbell and others 2004, fig. 13: S-T.

Pseudotextularia (*Racemiguembelina*) *fructicosa* (Egger). Berggren 1962, p. 22, pl. 6, fig. 6.

Pseudotextularia carseyae (Plummer). Brown 1969, pl. 4, figs 8-9. Masters 1977, p. 381, pl. 6, figs 1-2. Hofker 1978, pl. 1, fig. 7.

Planoglobulina brazoensis MARTIN 1972, p. 82, pl. 3, fig. 7, pl. 4, figs 1-2.

Planoglobulina brazoensis Martin. Pessagno and Longoria in Shipboard Scientific Party 1972, pl. 11, fig. 1. Smith and Pessagno 1973, p. 20, pl. 4, figs 5-10, pl. 5, figs 1-2. Weiss 1983, p. 51, pl. 4, figs 1-4. Abdel-Kireem 1986, p. 223, pl. 2, fig. 4. Jansen and Kroon 1987, p. 565, pl. 9, figs 1-3. Almogi-Labin and others 1990, p. pl. 1, fig. 7. Gawor-Biedowa 1992, p. 73, pl. 11, fig. 12. Ayyad and others 1996, fig. 6: f. Abramovich and others 2002, pl. 1, fig. 4. Keller and others 2002, pl. 2, fig. 11. Abramovich and others 2003, pl. 4, fig. 10. Korchagin 2011, pl. 1, fig. 7. Ohmert 2011, pl. 3, fig. 4.

Platystaphyla brazoensis (Martin). Masters 1977, p. 365, pl. 4, figs 3-4.

Heterohelix pseudocarseyae Pandey 1980, p. 62, pl. 3, figs 6-8.

Ventilabrella brazoensis (Martin). Salaj 1983, pl. 1, fig. 47.

Pseudoplanoglobulina carseyae (Plummer). Georgescu 1995, p. 405, pl. 2, figs 7-8. Georgescu 1997, fig. 6: 7.

Diagnosis. *Planoglobulina* with chamber proliferation in the adult stage.

Description. Test with two distinct growth stages: the juvenile stage consists of the proloculus followed by chambers alternately added with respect to the test growth axis and the adult stage with chamber proliferation. The adult stages begins with the biaperturate progressive chambers, which is followed by up to six chamber sets that increase in number by one but irregular additions on chambers in the adult stage occur frequently. Sutures are distinct and depressed, straight to slightly curved between all the chambers of the test. In edge view the test is the thickest in the region of the progressive chamber, which confers it a lanceolate appearance; periphery is rounded to broadly rounded and simple, without peripheral structures. Aperture is single at the base of the last-formed chamber in the juvenile stage and multiple, rounded, without transverse walls and on the apertural face in the adult stage with multichamber growth; apertures are bordered by narrow metaflanges. Chamber surface is ornamented with longitudinal pycnocostae with a thickness of 0.0070-0.0149 mm; the ornamentation is less developed over the last-formed chambers of the test. Test wall is calcitic, hyaline, simple and perforate; pores are simple, with a circular or elliptical outline and a diameter or maximum dimension of 0.0017-0.0034 mm.

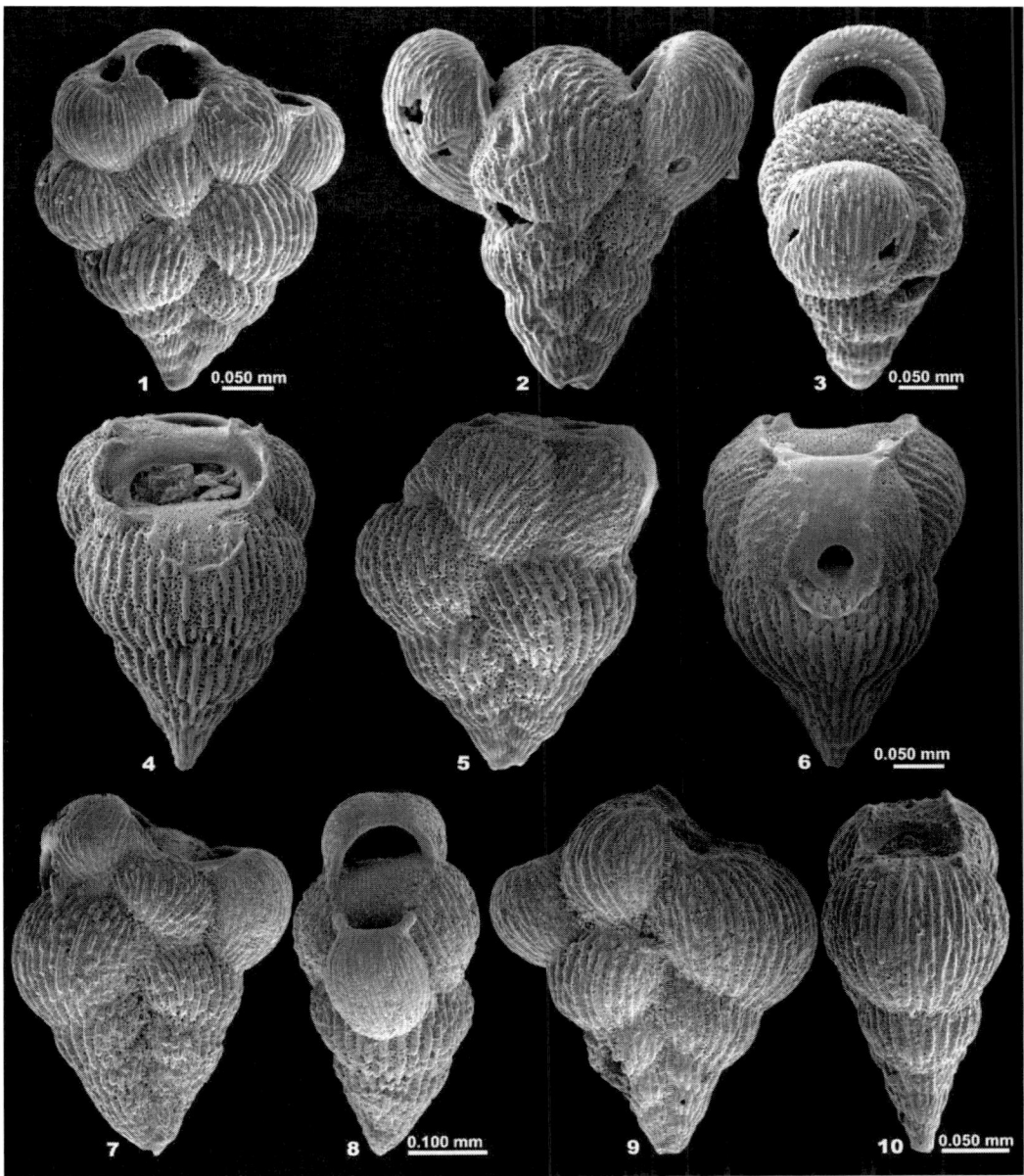

Figure 5. Specimens of *Planoglobulina acervulinoides* (Egger). 1 Hypotype from the middle Campanian Demopolis Chalk of the Montgomery County, Alabama collected from the Stop 7 of Scott and others (1968); specimen from the McGugan Collection. 2-3 Hypotype from the early Maastrichtian Prairie Bluff Chalk from Stop 4 of Scott and others (1968); specimen from the McGugan Collection. 4-6 Hypotype from the Maastrichtian Kemp Clay from the Mullinax-3 well of Texas (Sample 6-1). 7-8 Hypotype from the late Maastrichtian of Orphan Knoll, North Atlantic Ocean (Sample 12-111A-11-2, 123-137 cm). 9-10 Hypotype from the late Maastrichtian of Orphan Knoll, North Atlantic Ocean (Sample 12-111A-11-2, 5-19 cm).

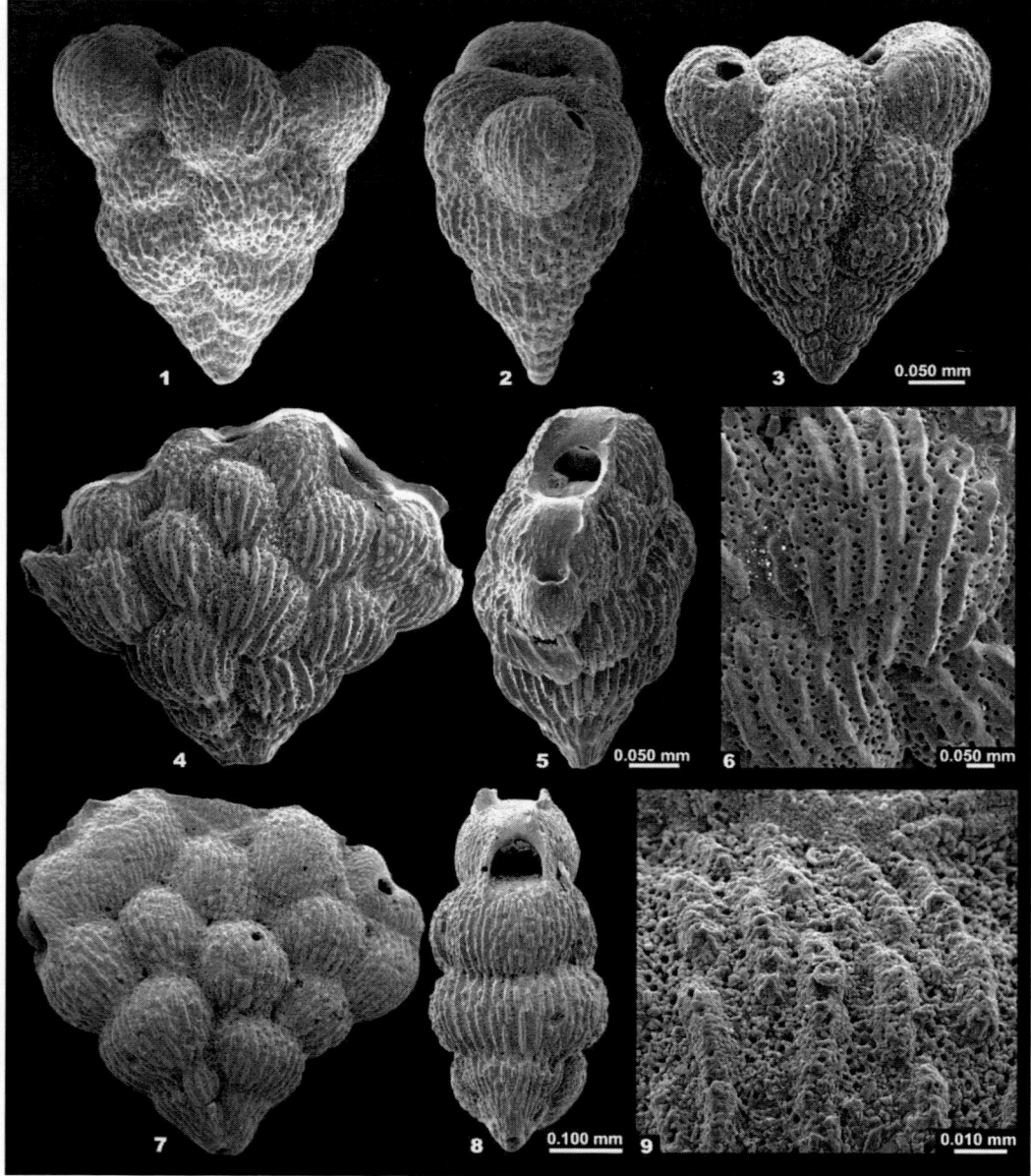

Figure 6. Specimens of *Planoglobulina acervulinoides* (Egger). 1-3 Hypotype from the Maastrichtian of Blake Nose, Western North Atlantic Ocean (Sample 171B-1050C-15-3, 71-73 cm). 4-6 Hypotype from the Maastrichtian Kemp Clay from the surroundings of the Limestone City, Texas; Sample Huber, National Museum of Natural History, Smithsonian Institution, Washington, D.C. 7-9 Hypotype from the late Maastrichtian of Orphan Knoll, North Atlantic Ocean (Sample 12-111A-11-2, 5-19 cm).

Remarks. *Planoglobulina acervulinoides* differs from its ancestor *P. sphaeralis* by having the adult stage with multichamber growth rather than having the test with chambers alternately added with respect to the test growth axis throughout, thicker leptocostae (0.0070-0.0149 mm rather than 0.0023-0.0084 mm) and larger pores with a diameter or maximum dimension of 0.0017-0.0034 mm rather than 0.0004-0.0017 mm. It differs from *Ehrenbergites* *riograndensis* (Martin 1972) mainly by having the test surface ornamented with pycnocostae rather than leptocostae.

Age. Late middle Campanian-Maastrichtian.

Geographic distribution. Cosmopolitan.

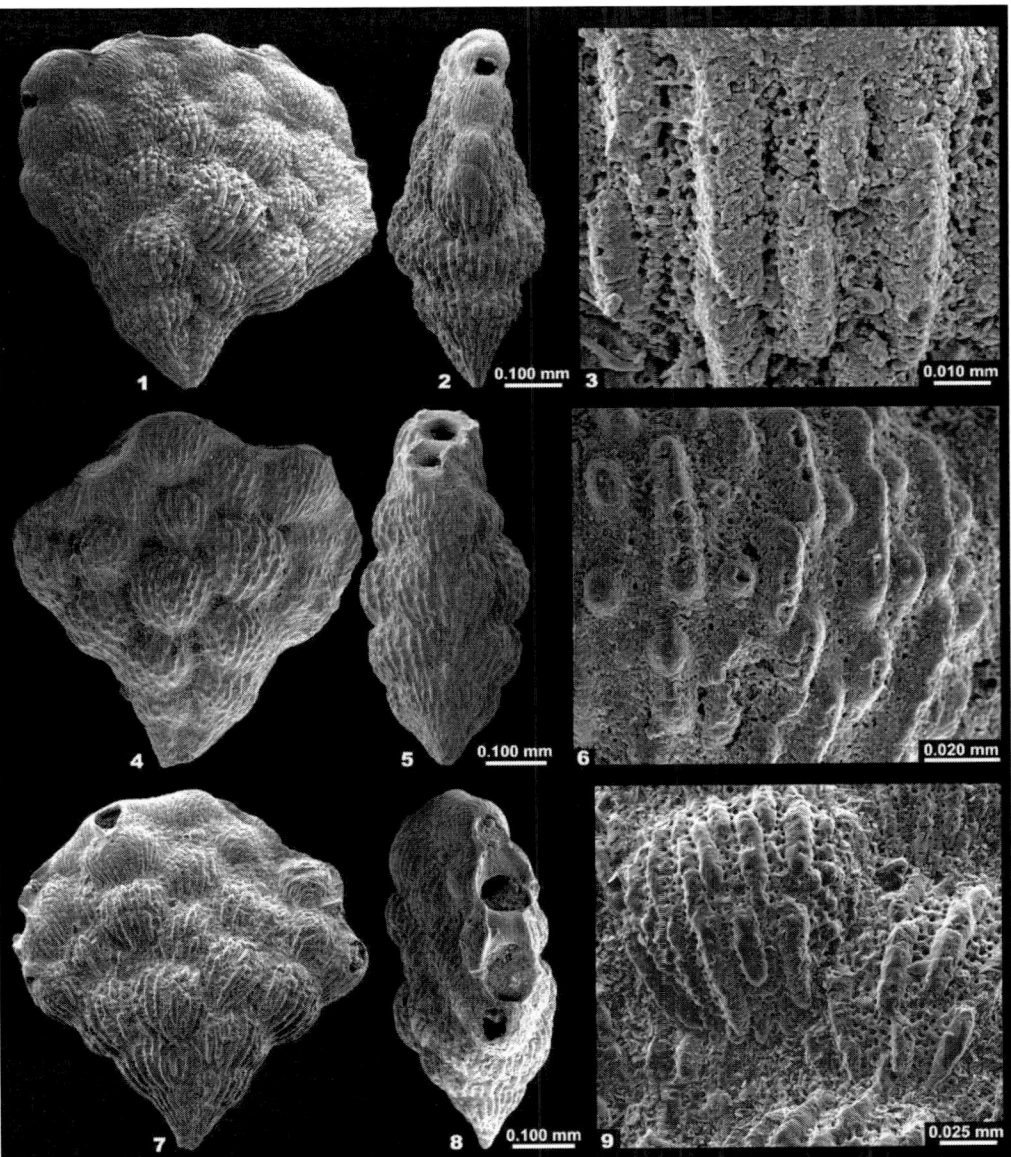

Figure 7. Specimens of *Planoglobulina acervulinoides* (Egger). 1-3 Hypotype from the late Maastrichtian of Orphan Knoll, North Atlantic Ocean (Sample 12-111A-11-2, 5-19 cm). 4-6 Hypotype from the late Maastrichtian of Orphan Knoll, North Atlantic Ocean (Sample 12-111A-11-2, 5-19 cm). 7-9 Hypotype from the Maastrichtian of Blake Nose, Western North Atlantic Ocean (Sample 171B-1050C-15-3, 71-73 cm).

CONCLUSION

The taxonomic revision of the Cretaceous planktic foraminifer *Planoglobulina* Cushman 1927 results in the definition of the homonym directional lineage consisting of two species. The initiating species is herein described from the upper middle Campanian-Maastrichtian sediments of the USA (Alabama, Texas): *P. sphaeralis*. *Planoglobulina sphaeralis* has the tests consisting of chambers alternately added with respect to the growth axis throughout the ontogenetic development. The first descendant species of this directional lineage is *P. acervulinoides* (Egger 1899), which evolved an adult stage with multichamber growth stage and no further descendants from it are known. In parallel with the development of the adult stage with multichamber growth there is a gradual increase in pycnocostae thickness and pore size. The similarities in the general test architecture, globular to spherical chamber shape and periapertural structures consisting of symmetrically developed metaflanges, one on each test side, indicate that the directional lineage *Planoglobulina* evolved from *Mihaia reussi* (Cushman 1938).

The study in stratigraphic context of the adult stage with multichamber growth shows that it presents a wide morphological variability. Specimens with well-developed adult proliferating stage are herein reported from early stages of the directional lineage evolution indicating that the initiating and first descendant species coexisted for nearly the entire history of *Planoglobulina*.

REFERENCES

Abdel-Kireem, M.R., 1986. Planktonic foraminifera and stratigraphy of the Tanjero Formation (Maastrichtian), northeastern Iraq. *Micropaleontology*, 32, 215-231.

Abramovich, S., Keller, G., Adatte, T., Stinnesbeck, W., Hottinger, L., Stueben, D., Berner, Z., Ramanivosoa, B., Randiriamanantenasoa, A., 2002. Age and paleoenvironment of the Maastrichtian to Paleocene of the Mahajanga Basin, Madagascar: a multidisciplinary approach. *Marine Micropaleontology*, 47, 17-70.

Abramovich, S., Keller, G., Stüben, D., Berner, Z., 2003. Characterization of late Campanian and Maastrichtian planktonic foraminiferal depth habitats and vital activities based on stable isotopes. *Palaeogeography, Palaeoclimatology, Palaeoecology*, 202, 1-29.

Ali-zade, A., Aliev, G.A., Aliev, M.M., Aliyulla, K., Khalilov, A.G., 1988. Cretaceous fauna of Azerbaijan. Baku: Akademia Nauk Azerbaydzhanskoy SSR, Institut Geologia im. Akad. I. M. Gubkina. Izdatelstvo "Elm", 447 p. [in Russian]

Aliyulla, K., 1977. Upper Cretaceous and foraminiferal development in the Lesser Caucasus (*Azerbaijan*). Baku: Akademia Nauk Azerbaydzhanskoy SSR, Institut Geologia im. Akad. I. M. Gubkina. Izdatelstvo "Elm", 229 p. [in Russian]

Almogi-Labin, A., Flexer, A., Honigstein, A., Rosenfeld, A., Rosenthal, E., 1990. Biostratigraphy and tectonically controlled sedimentation of the Maastrichtian in Israel and adjacent countries. *Revista Española de Micropaleontología*, 5, 41-52.

Ansary, S.E., Tewfik, N.M., 1968. Planktonic foraminifera and some benthonic species from the subsurface Upper Cretaceous of Ezz El Orban area, Gulf of Suez. *Journal of Geology of the United Arab Republic*, 10, 37-76.

Ayyad, S.N., Abed, M.M., Abu Zied, R.H., 1996. Biostratigraphy and correlation of Cretaceous rocks in Gebel Arif El–Naga, northeastern Sinai, Egypt, based on planktonic foraminifera. *Cretaceous Research*, 17, 263-291.

Berggren, W. A., 1962. Some planktonic foraminifera from the Maestrichtian and type Danian of southern Scandinavia. *Stockholm Contributions in Geology*, 9, 1-106.

Bertels, A., 1970. Los foraminiferos planctonicos de la Cuenca Cretacico-Terciaria en Patagonia septentrional (Argentina), con consideraciones sobre la estratigrafia de Fortin general roca (Provincia de Rio Negro). *Ameghiniana*, 7, 1-47.

Bettenstaedt, F., Wicher, C.A., 1955. Stratigraphic correlation of Upper Cretaceous and Lower Cretaceous in the Tethys and Boreal by the aid of microfossils. *Proceedings of the Fourth World Petroleum Congress, Geology and Geophysics*, I/D, 493-515.

Brown, N.K., Jr., 1969. Heterohelicidae Cushman, 1927, amended, a Cretaceous planktonic foraminiferal family. In: *Proceedings of the First International Conference on Planktonic Microfossils, Geneva 1967* (P. Brönnimann, P. and H.H. Renz, Eds). Leiden: E.J. Brill, 2, 21-67.

Butt, A., 1981. Depositional environments of the Upper Cretaceous rocks in the northern part of the Eastern Alps. *Cushman Foundation for Foraminiferal research Special Publication*, 20, 1-121.

Campbell, R. J., Howe, R. W., Rexilius, J. P., 2004 Middle Campanian–lowermost Maastrichtian nannofossil and foraminiferal biostratigraphy of the northwestern Australian margin. *Cretaceous Research*, 25, 827-864.

Cita, M.B., Gartner, S., Jr., 1971. Deep sea Upper Cretaceous from the western North Atlantic. In: *Proceedings of the II Planktonic Conference, Roma 1967* (A. Farinacci, Ed.). Roma: Edizioni Tecnoscienza, 1, 287-319.

Cushman, J.A., 1926. Some foraminifera from the Mendez Shale of eastern Mexico. *Contributions from the Cushman Laboratory for Foraminiferal Research*, 2, 16-28.

Cushman, J.A., 1927a. Some new genera of foraminifera. *Contributions from the Cushman Laboratory for Foraminiferal Research*, 3, 77-81.

Cushman, J.A., 1927b. Some characteristic Mexican fossil foraminifera. *Journal of Paleontology*, 1, 147-172.

Cushman, J.A., 1927c. An outline of a re-classification of the foraminifera. *Contributions from the Cushman Laboratory for Foraminiferal Research*, 3, 1-105.

Cushman, J.A., 1928. Additional genera of foraminifera. *Contributions from the Cushman Laboratory for Foraminiferal Research*, 4, 1-10.

Cushman, J.A., 1933. *Foraminifera, their classification and economic use*; 2nd edition. Cambridge, Massachusetts: Harvard University Press, 349 p.

Cushman, J.A., 1938. Cretaceous species of *Gümbelina* and related genera. *Contributions from the Cushman Laboratory for Foraminiferal Research*, 14, 2-28.

Cushman, J.A., 1946. Upper Cretaceous foraminifera of the Gulf coastal region of the United States and adjacent areas. *United States Geological Survey Professional Paper*, 206, 1-241.

Cushman, J.A., 1947. A foraminiferal fauna from the Santa Anita Formation of Venezuela. *Contributions from the Cushman Foundation for Foraminiferal Research*, 23, 1-18.

Cushman, J.A., 1949. The foraminiferal fauna of the Upper Cretaceous Arkadelphia Marl in Arkansas. *United States Geological Survey Professional Paper*, 221A, 1-17.

Cushman, J.A., 1950. *Foraminifera, their classification and economic use*; 4th edition. Cambridge, Massachusetts: Harvard University Press, 605 p.

Cushman, J.A., Hedberg, H.D., 1941. Upper Cretaceous foraminifera from Santander del Norte, Colombia, S.A. *Contributions from the Cushman Laboratory for Foraminiferal Research*, 17, 79-100.

Cushman, J.A., Todd, R., 1943. Foraminifera of the Corsicana Marl. *Contributions from the Cushman Laboratory for Foraminiferal Research*, 19, 49-72.

Djafarov, D.I., Agalarova, D.A., Khalilov, D.M., 1951. *Dictionary of microfauna of the Cretaceous deposits of Azerbaijan*. Baku: Gosudarstvenoe Nauchno-technicheskoe Izdatelstvo Neftianoi i Gorno-toplivnoi Lineraturyi Azerbaijanskoe Otdelenie, 128 p. [in Russian]

Dupeuble, P.A., 1969. Foraminifères planctoniques (Globotruncanidae et Heterohelicidae) du Maastrichtien supérieur en Aquitaine Occidentale. In: *Proceedings of the First International Conference on Planktonic Microfossils, Geneva 1967* (P. Brönnimann, P. and H.H. Renz, Eds). Leiden: E.J. Brill, 2: 153-161.

Egger, J.G., 1899. Foraminiferen und Ostrakoden aus den Kreidemergeln der Oberbayerischen Alpen. *Abhandlungen der Mathematisch-Physikalischen Klasse der Königlich Bayerischen Akademie der Wissenschaften*, 21, 3-230. [published in 1902]

Faupl, P., Grün, W., Lauer, G., Maurer, R., Papp, A., Schnabel, W., Sturm, M., 1970. Zur Typisierung des Sieveringer Schichten im Flysch des Wienerwaldes. *Jahrbuch der Geologisches Bundesanstalt*, 113, 73-158.

Funnell, B.M., Friend, J.K., Ramsay, T.S., 1969. Upper Maastrichtian planktonic foraminifera from Galicia Bank, west of Spain. *Palaeontology*, 12, 19-41.

Galloway, J.J., 1933. *A Manual of Foraminifera*. Bloomington, Indiana: The Principia Press, 483 p.

Gawor-Biedowa, E., 1992. Campanian and Maastrichtian foraminifera from the Lublin Upland, Eastern Poland. *Palaeontologica Polonica*, 52, 1-187.

Georgescu, M.D., 1995. Upper Cretaceous Heterohelicidae in the Romanian Western Black Sea offshore. *Revista Española de Micropaleontología*, 27, 91-106.

Georgescu, M.D., 1997. Upper Jurassic–Cretaceous planktonic biofacies succession and the evolution of the western Black Sea Basin. In: *Regional and petroleum geology of the Black Sea and surrounding region* (A.G. Robinson, Ed.). *The American Association of Petroleum Geologists Memoir*, 68, 169-182.

Georgescu, M.D., 2010. Origin, taxonomic revision and evolutionary classification of the late Coniacian-early Campanian (Late Cretaceous) planktic foraminifera with multichamber growth in the adult stage. *Revista Española de Micropaleontología*, 42, 59-118.

Georgescu, M.D., 2011. Iterative evolution, taxonomic revision and evolutionary classification of the praeglobotruncanid planktic foraminifera, Cretaceous (late Albian-Santonian). *Revista Española de Micropaleontología*, 43, 173-207. [published in 2012]

Georgescu, M.D., 2013. Revised evolutionary systematics of the Cretaceous planktic foraminifera described by C.G. Ehrenberg. *Micropaleontology*, 59, 1-49.

Glaessner, M.F., 1936. Die Foraminiferengattungen *Pseudotextularia* und *Amphimorphina*. *Problemyi Paleontologhyi*, 1, 97-130.

Govindan, A., 1972. Upper Cretaceous planktonic foraminifera from the Pondicherry are, south India. *Micropaleontology*, 18, 160–193.

Hamilton, E.L., 1953. Upper Cretaceous, Tertiary, and Recent Planktonic Foraminifera from Mid-Pacific flat-topped seamounts. *Journal of Paleontology*, 27, 204-237.

Hanzliková, E., 1969. The foraminifera of the Frýdek Formation (Senonian). *Sborník Geologických Věd, Paleontologie*, 11, 7-79.

Hanzliková, E., 1972. Carpathian Upper Cretaceous Foraminiferida of Moravia (Turonian-Maastrichtian). *Rozpravy Ústředního Ústavu Geologického*, 39, 1-160.

Hiltermann, H., Koch, W., 1962. Oberkreide des nördlich Mitteleuropa. In: *Leitfossilien der Mikropaläontologie. Ein Abriss herausgegeben von einem Arbeitskreis deutscher Mikropaläontologen* (H. Bartenstein and others, Eds). Berlin-Nikolassee: Gebrüder Borntraeger, 229-338.

Hofker, J., 1978. Analysis of a large succession of samples through the upper Maastrichtian and the lower Tertiary of drill hole 47.2, Shatsky Rise, Pacific, Deep Sea Drilling Project. *Journal of Foraminiferal Research*, 8, 46-75.

Howe, R.W., Campbell, R.J., Rexilius, J.P., 2003. Integrated uppermost Campanian-Maastrichtian calcareous nannofossil and foraminiferal biostratigraphic zonation of the northwestern margin of Australia. *Journal of Micropaleontology*, 22, 29-62.

Itzhaki, J., 1952. Séries de variabilité de *Pseudotextularia* (Rzehak) d'après la forme du test et ses tendances évolutives. *Comptes Rendus Sommaires de la Société Géologique de France*, 10, 187-189.

Jansen, H., Kroon, D., 1987. Maestrichtian foraminifers from Site 605, Deep Sea Drilling Project Leg 93, northwest Atlantic. In: *Initial Reports of the Deep Sea Drilling Project, Volume 93* (van Hinte, J.E. and others, Eds). Washington, D.C.: United States Government Printing Office, 93, 555-575.

Jennings, P.H., 1936. A microfauna from the Monmouth and Basal Rancocas Groups of New Jersey. *Bulletins of American Paleontology*, 23(78), 1-76.

Kavary, E., Frizzell, D.L., 1963. Upper Cretaceous and Lower Cenozoic oraminifera from west central Iran. *University of Missouri School of Mines and Metallurgy Bulletin*, 102, 1-89.

Keller, G., Adate, T., Stinnesbeck, W., Luciani, V., Karoui-Yaakoub, N. and ZaghbibTurki, D., 2002. Paleoecology of the Cretaceous–Tertiary mass extinction in planktonic foraminifera. *Palaeogeography, Palaeoclimatology, Palaeoecology*, 178, 257-297.

Kikoïne, J., 1948. Les Heterohelicidae du Crétacé supérieur pyrénéen. *Bulletin de la Société Géologique de France*, 18, 15-35.

Korchagin, O.A., Upper Campanian-lower Maastrichtian planktonic foraminifers and biostratigraphy of the Moni Formation, Southern Cyprus. *Stratigraphy and Geological Correlation*, 19, 526-544.

Lehmann, R., 1966. Description des Globotruncanidés et Hétérohelicidés d'une faune maestrichtienne du Prérif (Maroc). *Eclogae Geologicae Helvetiae*, 59, 309-317.

Liebus, A., 1927. Neue Beiträge zur Kenntnis des Eozänfauna des Krappfeldes in Kärnten. *Jahrbuch der Geologisches Bundesanstalt*, 3, 333-392.

Linares-Rodríguez, D., 1977. *Foraminiferos planctonicos del Cretacico superior de las Cordilleras Beticas (sector central)*. Universidad de Málaga, Departamento de Geología, 410 p.

Liu, K., 2007. Sequence stratigraphy and orbital cyclostratigraphy of the Mooreville Chalk (Santonian-Campanian), northeastern Gulf of Mexico area, USA. *Cretaceous Research*, 28, 405-418.

Loeblich, A.R. Jr., Tappan, H., 1964. Sarcodina Chiefly "Thecamoebians" and Foraminifera. In: *Treatise on Invertebrate Paleontology. Part C* (R.C. Moore, Ed.). The Geological Society of America and The University of Kansas Press, 900 p.

Loeblich, A.R. Jr., Tappan, H., 1987. *Foraminiferal Genera and Their Classification.* New York: Van Nostrand Reinhold Company, 970 p.

Malmgren, B.A., 1991. Biogeographic patterns in terminal Cretaceous planktonic foraminifera from Tethyan and warm transitional waters. *Marine Micropaleontology*, 18, 73-99.

Mancini, E.A., Puckett, T.M., Tew, B.H., 1996. Integrated biostratigraphic and sequence stratigraphic framework for Upper Cretaceous strata of the eastern Gulf Coastal Plain, USA. *Cretaceous Research*, 17, 645-669.

Martin, S.E., 1972. Reexamination of the Upper Cretaceous planktonic foraminiferal genera *Planoglobulina* Cushman and *Ventilabrella* Cushman. *Journal of Foraminiferal Research*, 2, 73-92.

Masters, B.A., 1976. Planktic foraminifera from the Upper Cretaceous Selma Group, Alabama. *Journal of Paleontology*, 50, 318-330.

Masters, B.A., 1977. Mesozoic planktonic foraminifera. A world–wide review and analysis. In: *Oceanic Micropaleontology* (A.T.S. Ramsay, Ed.). London-New York-San Francisco: Academic Press, 1, 301-731.

Montanaro Gallitelli, E., 1957. A revision of the foraminiferal family Heterohelicidae. In: *Studies in foraminifera* (A.R. Jr. Loeblich, Ed.). Washington, D.C.: *United States National Museum History Bulletin*, 215, 133-154.

Neagu, T., 1970. Micropaleontological and stratigraphical study of the Upper Cretaceous deposits between the upper valleys of the Buzău and Rîul Negru Rivers. *Memoriile Institului de Geologie şi Geofizică*, 12, 1-109.

Nederbragt, A.J., 1989a. Chamber proliferation in the Cretaceous planktonic foraminifera Heterohelicidae. *Journal of Foraminiferal Research*, 19, 105-114.

Nedebragt, A.J., 1989b. Maastrichtian Heterohelicidae (planktic foraminifera) from the West North Atlantic. *Journal of Micropaleontology*, 8, 183-206.

Nedebragt, A.J., 1991. Late Cretaceous biostratigraphy and development of Heterohelicidae (planktic foraminifera). *Micropaleontology*, 37, 329-372.

Noth, R., 1951. Foraminiferen aus Unter- und Oberkreide des Österreichischen anteils an Flysch, Helvetikum und Vorlandvorkommen. *Jahrbuch der Geologischen Bundesanstaldt, Sonderband*, 3, 1-91.

Ohmert, W., 2011. Radiolarien-Faunen und Stratigraphie der Plattenau-Formation (Campanium bis Maastrichtium) im Helvetikum von Bad Tölz. *Zittelliana*, 51, 37-95.

Olsson, R. K., 1960. Foraminifera of Latest Cretaceous and Earliest Tertiary age in the New Jersey coastal plain. *Journal of Paleontology*, 34, 1-58.

Pandey, J., 1980. Cretaceous foraminifera of Um Sohryngkew River section, Meghalaya. *Journal of the Palaeontological Society of India*, 25, 53-74. [published in 1981]

Pérez-Rodriguez, I., Lees, J.A., Larrasoaña, J.C., Arz, J.A., Aremillas, I., 2012. Planktonic foraminiferal and calcareous nanofossil biostratigraphy and magnetostratigraphy of the uppermost Campanian and Maastrichtian at Zumaia, northern Spain. *Cretaceous Research*, 37, 100-126.

Perlmutter, N. M., Todd, R., 1965. Correlation and foraminifera of the Monmouth Group (Upper Cretaceous) Long Island, New York. *United States Geological Survey Professional Paper*, 483-I, 1-21.

Peryt, D., 1980. Planktic foraminifera zonation of the Upper Cretaceous in the middle Vistula River Valley, Poland. *Palaeontologia Polonica*, 41, 3-101.

Pessagno, E.A. Jr., 1967. Upper Cretaceous planktonic foraminifera from the Western Gulf coastal plain. *Palaeontographica Americana*, 5(37), 243-445.

Petters, S.W., 1983. Gulf of Guinea planktonic foraminiferal biochronology and geological history of the South Atlantic. *Journal of Foraminiferal Research*, 13, 32-59.

Plummer, H.J., 1931. Some Cretaceous foraminifera in Texas. *University of Texas Bureau of Economic Geology and Technology Bulletin*, 3101, 109-203.

Pringgoprawiro, H., 1965. Some significant Upper Cretaceous from Groisbach, Morzger Hügel, and Michelstetten, Austria. *Institute of Technology Bandung*, 60, 21-65.

Rahhali, I., 1970. Foraminifères benthoniques et pelagiques du Crétacé supérieur du synclinal d'El-Koubbat (Moyen Atlas-Maroc). *Notes et Mémoires de la Service Géologique du Maroc*, 30/225, 51-98.

Saavedra, J.L. 1965. La evolución de los Globigerináceos. *Boletín de la Real Sociedad Española de Historia Natural*, 63, 317-349.

Sacal, V., Debourle, A., 1957. Foraminifères d'Aquitaine 2e partie. Peneroplidae a Victoriellidae. *Mémoires de la Société Géologique de France*, 78, 1-87.

Said, R., Kenawy, A., 1956. Upper Cretaceous and Lower Tertiary foraminifera from northern Sinai, Egypt. *Micropaleontology*, 2, 105-173.

Said, R., Kerdany, M.T., 1961. The geology and micropaleontology of the Farfara Oasis, Egypt. *Micropaleontology*, 7, 317-336.

Said, R., Sabry, H., 1964, Planktonic foraminifera from the type locality of the Esna Shale in Egypt. *Micropaleontology*, 10, 375-395.

Salaj, J., 1980. *Microbiostratigraphie du Crétacé et du Paléogène de la Tunisie septentrionale et orientale (Hypostratotypes tunisiens)*. Bratislava: Geologický Ústav Dionýza Štúra, 238 p.

Salaj, J., 1983. Quelques problèmes taxinomiques concernant les foraminifères planctiques et la zonation du Sénonien supérieur d'El Kef. *Geologický Zborník*, 34, 187-212.

Salaj, J., Samuel, O., 1966. *Foraminifera der Westkarpaten-Kreide*. Bratislava: Geologický Ústav Dionýza Štúra, 291 p.

Sandidge, J.R., 1932. Additional foraminifera from the Ripley Formation in Alabama. *American Midland Naturalist*, 13, 333-377.

Scott, J.C., Watson, H.M., Copeland, C.W., Cepek, P., Hay, W.W., Masters, B.A., Worsley, T.R., 1968. *Facies changes in the Selma Group in Central and Eastern Alabama. A guidebook for the Sixth Annual Field Trip of the Alabama Geological Society, December 6-7, 1968*. Tuscaloosa: Alabama Geological Society, 69 p.

Sellier de Civrieux, J.M., 1952. Estudio de la microfauna de la seccion–tipo del Miembro Socuy de la Formacion Colon Distrito Mara, Estado Zulia. *Ministerio de Minas e Hidrocarburos Direccion de Geologia*, 2, 231-310.

Shipboard Scientific Party, 1972. Site 111. In: *Initial Reports of the Deep Sea Drilling Project, Volume 12* (Laughton, A.E. and others, Eds). Washington, D.C.: United States Government Printing Office, 12, 33-159.

Skinner, H.C., 1962. Arkadelphia foraminiferida. *Tulane Studies in Geology*, 1, 1-67.

Smith, C.C., 1978. Taxonomic comments on some Upper Cretaceous planktonic foraminiferal genera. *Journal of Foraminiferal Research*, 8, 314-318.

Smith, C. C., Pessagno, E. A. Jr., 1973. Planktonic foraminifera and stratigraphy of the Corsicana Formation (Maestrichtian), north-central Texas. *Cushman Foundation for Foraminiferal Research, Special Publications*, 13, 5-68.

Todd, R., 1970. Maestrichtian (Late Cretaceous) foraminifera from a deep–sea core off southwestern Africa. *Revista Española de Micropaleontología*, 2, 131-154.

Voorwijk, G.H., 1937. Foraminifera from the Upper Cretaceous of Habana, Cuba. *Proceedings of the Koninklijke Akademie van Wetenschappen te Amsterdam,* 40, 190-198.

Webb, P.N., 1973. Upper Cretaceous-Paleocene foraminifera from Site 208 (Lord Howe Rise, Tasman Sea), DSDP Leg 21. In: *Initial Reports of the Deep Sea Drilling Project Volume 21* (R.E. Burns and others, Eds). Washington, D.C.: United States Government Printing Office, 541-573.

Weiss, W., 1983. Heterohelicidae (seriale planktonische Foraminiferen) der tethyalen Oberkreide (Santon bis Maastricht). *Geologisches Jahrbuch*, A72, 3-93.

Wille-Janoschek, U., 1966. Stratigraphie und Tektonik der Schichten der Oberkreide und des Alttertiärs im Raume von Gosau und Abtenau (Salzburg). *Jahrbuch der Geologisches Bundesanstalt*, 109, 91-172.

Wright, C.A., Apthorpe, M., 1976. Planktonic foraminiferids from the Maastrichtian of the northwest shelf, Western Australia. *Journal of Foraminiferal Research*, 6, 22-241.

Yassini, I., 1979. Maastrichtian-lower Eocene biostratigraphy and the planktic foraminiferal biozonation of Jordan. *Revista Española de Micropaleontología*, 11, 5-57.

Zapeda, M., 1998. Planktonic foraminiferal diversity, equitability and biostratigraphy of the uppermost Campanian-Maastrichtian, ODP Leg 122, Hole 762C, Exmouth Plateau, NW Australia, eastern Indian Ocean. *Cretaceous Research*, 19, 117-152.

In: Evolutionary Classification ...   ISBN: 978-1-63321-959-5
Editors: M. Dan Georgescu and C. M. Henderson   © 2015 Nova Science Publishers, Inc.

Chapter 5

EVOLUTION OF CENTRAL PERFORATE PLATE IN THE NEW CONDENSED LINEAGE *EICHERIELLA*

*M. Dan Georgescu**

Department of Geosciences, University of Calgary,
Calgary, Alberta, Canada

ABSTRACT

Morphological and taxonomic review of the Cretaceous planktic foraminiferal species with chambers alternately added with respect to the test growth axis *Gümbelina semicostata* Cushman 1938 indicate that it belongs to a distinct lineage, which is named herein *Eicheriella*. A new kind of lineage is defined in the evolutionary classification: the condensed lineage consists only of the initiating species and is the fourth kind of lineage in this classification system after the directional, branched and iterative lineages.

Keywords: Foraminifera, planktic, Campanian, Maastrichtian, new lineage, evolutionary classification

INTRODUCTION

Development of an evolutionary classification framework for the Cretaceous planktic foraminifera with chambers alternately added with respect to the test growth axis is based on a significant increase in observation accuracy by extensive use of the Scanning Electron Microscope (SEM). The high amount of novelties in test morphology and species grouping resulted in a high number of studies focused at lineage or group of lineages level, and practically the morphology and taxonomic status of each species has to be reassessed.

One species of the group of Cretaceous planktics with chambers alternately added with respect to the test growth axis, which has a test morphology that cannot be readily assigned to

---

* E-mail: dgeorge@ucalgary.ca.

one of the known lineages, is *Gümbelina semicostata* Cushman 1938. High resolution observations on well-preserved specimens show that this species evolved from *Pseudoguembelina costulata* (Cushman 1938) in the late Campanian and became extinct in the early Maastrichtian. The evolution of the new species resulted in the formation of one new morphological feature in the evolutionary history of this group, namely a perforate central plate developed over the central portion of the test; such structure occurs symmetrically on each side of the test.

A new kind of lineage with significance in evolutionary classification is herein recognized; it is named Condensed Lineage (CL) and includes only the initiating species. *Eicheriella* is the new name given herein to the condensed lineage that includes the species originally named *G. semicostata*.

STUDIED MATERIAL

The holotype of *Gümbelina semicostata* collected form the Taylor Marl north of Lake City, Delta County, Texas and illustrated by Cushman (1938, pl. 3, fig. 6) was examined in the Cushman Collection of the National Museum of National History, Smithsonian Institution, Washington, D.C.

Most of the material used in this study comes from two Ocean Drilling Program (ODP) holes from the East Indian Ocean: ODP Hole 761B (Wombat Plateau) (Table 1) and ODP Hole 762C (Exmouth Plateau). The specimens collected from the upper Campanian-lower Maastrichtian from the two locations are well-preserved and present low magnitude recrystallization. Test fragmentation occurs frequently and the tests present broken last-formed chambers.

Additional well-preserved specimens were examined in the van Morkhoven Collection, National Museum of National History, Smithsonian Institution, Washington, D.C. These specimens were collected from the upper Campanian sediments of the Gulf of Mexico, well Eureka 67-128.

A NEW KIND OF UNIT IN EVOLUTIONARY CLASSIFICATION

In evolutionary classification the lineages are of different kinds and the occurrence of these different lineage types define an open system that contrasts to the axiomatic system of the typological classification. Georgescu (2010) recognized two lineage types: directional and branched (Figure 1:A and B respectively), which were recognized in the past by different authors, but without being conferred a role in evolutionary classification. The existence of an open system at lineage level was further demonstrated by Georgescu (2013) who described a new pattern within one lineage, namely the iterative lineage (Figure 1:C).

A new kind of lineage is herein defined, namely the condensed lineage (CL). Such a lineage consists of only the initiating species and is characterized by the rapid evolution of a new morphological feature (Figure 1:D). This kind of lineage may be associated in part with the *quantum evolution* as defined by Simpson (1944, 1961) or represent a dead-end evolutionary experiment.

Table 1. Stratigraphic distribution of selected taxa of Cretaceous planktic foraminifera with chambers alternately added with respect to the test growth axis in the upper Santonian-Maastrichtian sediments of the ODP Hole 761B (East Indian Ocean, Wombat Plateau)

ODP Leg 122, Hole 761B (samples)	Depth below sea floor (sample top)	*Ehrenbergites striata*	*Sigalia incipiens*	*Gublerina rajagopalani*	*Ehrenbergites niograndensis*	*Braunella punctulata*	*Eicheriella semicostata*	*Gublerina cuvillieri*	*Pseudotextularia varians*	*Pseudotextularia intermedia*	*Racemiguembelina powelli*	*Nederbragtina hariaensis*	*Praegublerina robusta*	*Racemiguembelina fructicosa*	*Planoglobulina acervulinoides*	Planktic foraminiferal zonation	Cretaceous stages
21-4, 145-146 cm	176.15	x		x				x	x		x	x	x	x	x	*N. hariaensis*	middle-upper Maastrichtian
21-5, 70-71 cm	176.90	x		x				x		x	x	x		x			
21-6, 70-71 cm	178.40			x				x			x	x	x	x			
22-1, 71-72 cm	180.41			x				x			x	x					
22-2, 71-72 cm	181.91	x		x				x		x	x	x					
22-3, 68-69 cm	183.38			x							x	x					
22-4, 71-72 cm	184.91	x		x				x			x	x					
22-5, 70-71 cm	186.40	x		x				x	x		x	x					
22-6, 29-30 cm	187.49	x		x				x	x		x	x	x				
23-1, 66-67 cm	189.86	x		x					x	x	x	x					
23-2, 70-71 cm	191.40	x						x	x	x	x					*A. mayaroensis*	
23-3, 71-72 cm	192.91	x		x				x	x		x						
23-4, 70-71 cm	194.40	x		x	x			x	x	x							
23-5, 70-71 cm	195.90	x		x	x	x		x	x								
24-1, 69.5-70.5 cm	199.40	x		x	x			x	x								
24-2, 72-73 cm	200.92	x		x	x	x		x	x								
24-3, 71-72 cm	202.41	x		x	x	x	x	x	x							*G. subpetaloidea*	upper Campanian
24-4, 70-71 cm	203.90	x		x		x	x	x	x								
25-1, 71-72 cm	208.91	x		x	x	x	x										
25-2, 70-71 cm	210.40	x		x			x	x								*R. calcarata* equivalent	
25-3, 71-72 cm	211.91	x		x	x	x											
25-4, 72-73 cm	213.42	x		x													
25-5, 68-69 cm	214.88	x		x													
26-1, 70-71 cm	218.40	x		x													
26-2, 71-72 cm	219.91	x															
27-1, 70-71 cm	227.90	x	x													*D. asymetrica* equivalent	upper Santonian
27-2, 70-71 cm	229.40	x	x														
27-3, 70-71 cm	230.90	x															
28-core catcher	236.90	barren														not zoned	

PALEONTOLOGICAL DESCRIPTION

Evolutionary classification units are after Georgescu (2010, 2011). Abbreviations: CL - condensed lineage, IS - initiating species. Only the reports of detached specimens are included in the synonymy list.

Condensed Lineage: *Eicheriella* - new

Remarks. The description, age and geographic distribution of *Eicheriella* correspond to that of the included species.

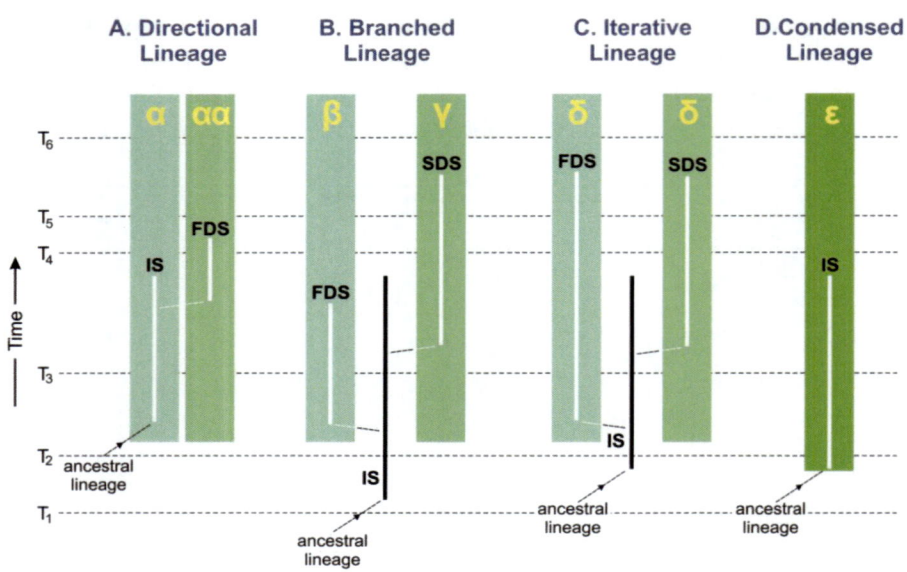

Figure 1. Kinds of lineages in evolutionary classification after Georgescu (2013) with modifications. Species kind abbreviations: IS-initiating species, FDS-first descendant species, SDS-second descendant species. Morphological features: α, β, γ, δ and ε. Note that in the case of directional lineage (A) one feature (α) is accentuated as the evolution progresses; two distinct features (β and γ) evolve independently from the same IS in the case of a branched lineage (B); in the case of the iterative lineage (C) one feature (δ) evolved independently and successively from the IS; in the case of one condensed lineage (D) the newly evolved morphological feature (ε) occurs only in the IS.

IS: *Eicheriella semicostata* (Cushman 1938)
(Figure 2:1-11)

Gümbelina semicostata CUSHMAN 1938, p. 16, pl. 3, fig. 6.

Gümbelina semicostata Cushman. Cushman 1946, p. 107, pl. 46, figs 1-5. Kikoïne 1948, pl. 1, fig. 6.

Ventilabrella (?) sp. Sacal and Debourle 1957, pl. 3, fig. 7.

Gümbelina postsemicostata VASSILENKO 1961, p. 206, pl. 41, figs 8-9.

Gublerina semicostata (Cushman). Barr 1968, pl. 1, fig. 5.

Sigalia bejaouensis SALAJ and MAAMOURI 1970, p. 75, fig. 5.

Pseudoguembelina postsemicostata (Vassilenko). Aliyulla 1977, pl. 1, fig. 7.

Heterohelix semicostata (Cushman). Pessagno 1967, p. 263, pl. 98, fig. 21. Masters 1977, p. 352, pl. 2, figs 4-5, pl. 3, figs 1-2. Nederbragt 1991, p. 348, pl. 4, figs 2, 4. Thompson 1991, p. 36, pl. 1, fig. 6. Gawor-Biedowa 1992, p. 70, pl. 10, figs 14-15. Quilty 1992, p. 379, pl. 1, fig. 9. Howe and others 2003, pl. 7, figs 22-23. Campbell and others 2004, fig. 13: Q-R.

Species included. IS: *E. semicostata* (Cushman 1938).

Condensed lineage name derivation. Condensed lineage name is after D.L. Eicher (University of Colorado) for his outstanding studies in Cretaceous planktic foraminifera; to his name the suffix *-iella* is added.

Diagnosis (common at species and lineage levels). A condensed lineage with chambers added alternately with respect to the test growth axis throughout the ontogeny, pycnocostate ornamentation and perforate central plate over the central suture on both test sides.

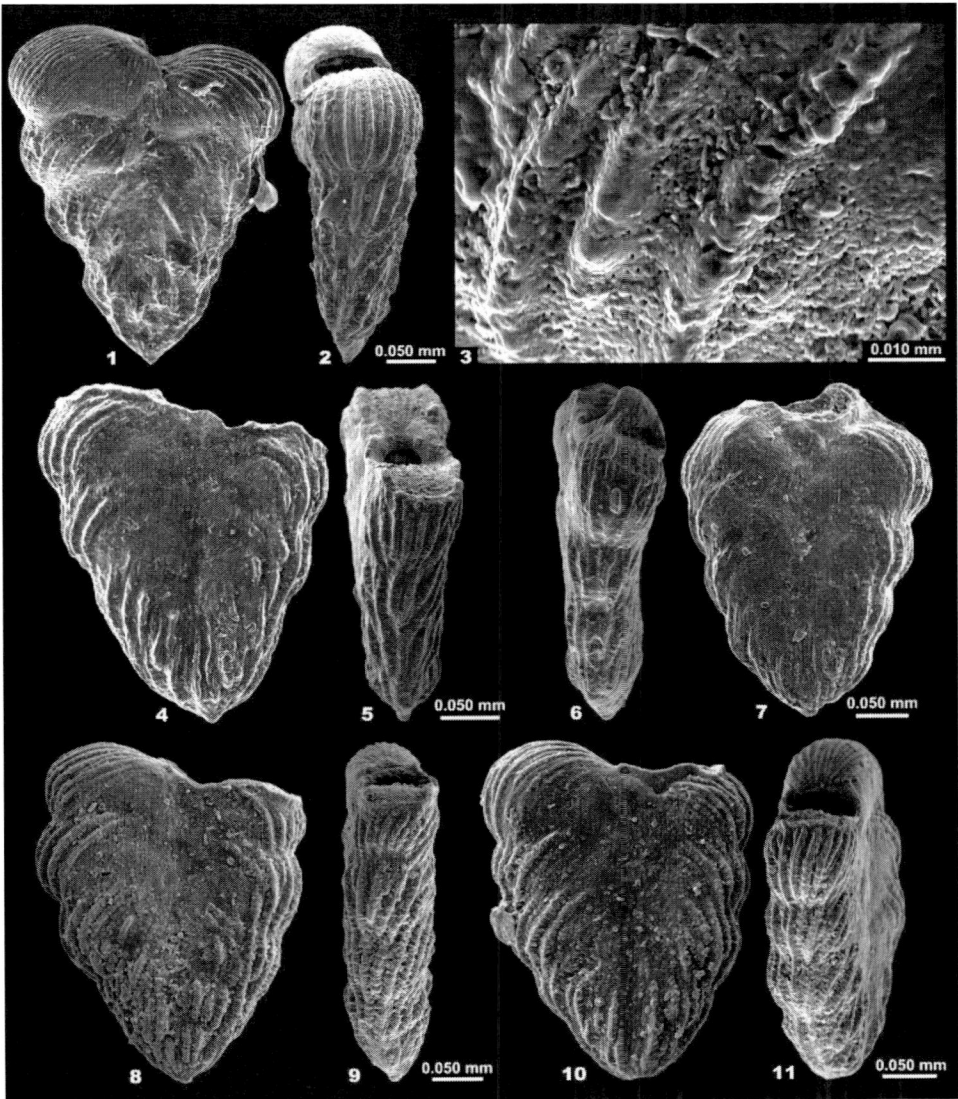

Figure 2. Specimens of *Eicheriella semicostata* from the Gulf of Mexico (1-7) and East Indian Ocean (Exmouth Plateau) (8-11). **1-3** Specimen from the upper Campanian sediments of the well Eureka 67-128, Gulf of Mexico; specimen from the van Morkhoven Collection. **4-5** Specimen from the upper Campanian sediments of the well Eureka 67-128, Gulf of Mexico; specimen from the van Morkhoven Collection. **6-7** Specimen from the upper Campanian sediments of the well Eureka 67-128, Gulf of Mexico; specimen from the van Morkhoven Collection. **8-9** Specimen from the upper Campanian sediments of Exmouth Plateau; Sample 122-761B-25-2, 70-71 cm. **10-11** Specimen from the upper Campanian sediments of Exmouth Plateau; Sample 122-761B-25-2, 70-71 cm.

Description (common at species and lineage levels). Test consists of the proloculus followed by 11-16 chambers alternately added with respect to the test growth axis. Chambers of the adult stage are subrectangular in shape, overlap at various rate and have a gradual size increase. Sutures are often indistinct in the earlier portion of the test and distinct and depressed, slightly curved between the last-formed chambers of the test. The central suture is not visible due to the development of one perforate central plate over the central portion of the test and on both test sides; the pores over this structure are circular in shape and with a diameter range of 0.0002-0.0005 mm. Test is compressed in edge view and presents nearly parallel sides except for the earliest portion of the test where the chamber size increase was faster; periphery is rounded and simple, without peripheral structures. Aperture is a low arch at the base of the last-formed chamber; aperture is bordered by symmetrically developed structures, which are often obscured by the perforate central plate. Chamber surface is ornamented longitudinal pycnocostae, which are more prominent over the peripheral regions; the pycnocostae present a thickness of 0.0060-0.0144 mm. Thinner ornamentation elements occur occasionally over the last-formed chambers of the test. Pores are situated in the space between the pycnocostae and present a wide variability in outline: circular, elliptical and irregular. Pore diameter or maximum dimension is 0.0006-0.0012 mm.

Remarks (common at species and lineage levels). *Eicheriella semicostata* is characterized among the other taxa of Cretaceous planktics with chambers alternately added with respect to the test growth axis by the occurrence of one perforate central plate on each test side. This condensed lineage evolved probably from *Pseudoguembelina costulata* (Cushman 1938) through the disappearance of the supplementary apertures covered by lid-like wall extensions, which occur in the last-formed chambers of the test and increase in pycnocostae thickness from 0.0054-0.0118 mm in *P. costulata* to 0.0060-0.0144 mm in *E. semicostata*. One morphological feature in *E. semicostata* requires further study and this is the periapertural structure; they were observed in good state of preservation only in one specimen, which is herein figured in Figure 2:1. The periapertural structures in *E. semicostata* appear oriented towards and are attached forward to the penultimate chamber rather than backwards as in *P. costulata*; as these structures in the descendant species appear much reduced in size when compared to the corresponding structures in the ancestor, we cannot rule out that they were atrophied during evolution of *E. semicostata*.

Age (common at species and lineage levels). Late Campanian-early Maastrichtian.

Geographic distribution (common at species and lineage levels). Cosmopolitan.

CONCLUSION

The taxonomic and morphological revisions of the Cretaceous planktic foraminiferal species with chambers alternately added with respect to the test growth axis *Gümbelina semicostata* Cushman 1938 show that it evolved in the late Campanian from *Pseudoguembelina costulata* (Cushman 1938). Despite the morphological resemblances between the two species, which are given by the common ancestry, and overlapping stratigraphic ranges there are also significant morphological differences between them. Based on the distinct evolutionary trend as supported by the morphological distinctiveness given especially by the evolution of the perforate central plate over the central suture the species *G. semicostata* is assigned to a new lineage: *Eicheriella*. *Eicheriella* is a condensed lineage (CL) that consists of only one species and this new kind of lineage is herein described.

REFERENCES

Aliyulla, K., 1977. *Upper Cretaceous and foraminiferal development in the Lesser Caucasus (Azerbaijan).* Baku: Akademia Nauk Azerbaydzhanskoy SSR, Institut Geologia im. Akad. I. M. Gubkina. Izdatelstvo "Elm", 229 p. [in Russian].

Barr, F. T., 1968. Upper Cretaceous stratigraphy of Jabal al Akhdar, northern Cyrenaica. In: *Geology and Archaeology of northern Cyrenaica, Libya* (F. T. Barr, Ed.). Petroleum Exploration Society of Libya, Tenth Annual Field Conference, 131-146.

Campbell, R. J., Howe, R. W., Rexilius, J. P., 2004. Middle Campanian–lowermost Maastrichtian nannofossil and foraminiferal biostratigraphy of the northwestern Australian margin. *Cretaceous Research*, 25, 827-864.

Cushman, J. A., 1938. Cretaceous species of *Gümbelina* and related genera. *Contributions from the Cushman Laboratory for Foraminiferal Research*, 14, 2-28.

Cushman, J. A., 1946. Upper Cretaceous foraminifera of the Gulf coastal region of the United States and adjacent areas. *United States Geological Survey Professional Paper*, 206, 1-241.

Gawor-Biedowa, E., 1992. Campanian and Maastrichtian foraminifera from the Lublin Upland, Eastern Poland. *Palaeontologica Polonica*, 52, 1-187.

Georgescu, M. D., 2010. Origin, taxonomic revision and evolutionary classification of the late Coniacian-early Campanian (Late Cretaceous) planktic foraminifera with multichamber growth in the adult stage. *Revista Española de Micropaleontología*, 42, 59-118.

Georgescu, M. D., 2011. Iterative evolution, taxonomic revision and evolutionary classification of the praeglobotruncanid planktic foraminifera, Cretaceous (late Albian-Santonian). *Revista Española de Micropaleontología*, 43, 173-207. [published in 2012].

Georgescu, M. D., 2013. Revised evolutionary systematics of the Cretaceous planktic foraminifera described by C.G. Ehrenberg. *Micropaleontology*, 59, 1-49.

Howe, R. W., Campbell, R. J., Rexilius, J. P., 2003. Integrated uppermost Campanian-Maastrichtian calcareous nannofossil and foraminiferal biostratigraphic zonation of the northwestern margin of Australia. *Journal of Micropaleontology*, 22, 29-62.

Kikoïne, J., 1948. Les Heterohelicidae du Crétacé supérieur pyrénéen. *Bulletin de la Société Géologique de France*, 18, 15-35.

Masters, B. A., 1977. Mesozoic planktonic foraminifera. A world-wide review and analysis. In: *Oceanic Micropaleontology* (A.T.S. Ramsay, Ed.). London-New York-San Francisco: Academic Press, 1, 301-731.

Nederbragt, A. J., 1989. Chamber proliferation in the Cretaceous planktonic foraminifera Heterohelicidae. *Journal of Foraminiferal Research*, 19, 105-114.

Nederbragt, A. J., 1991. Late Cretaceous biostratigraphy and development of Heterohelicidae (planktic foraminifera). *Micropaleontology*, 37, 329-372.

Pessagno, E. A. Jr., 1967. Upper Cretaceous planktonic foraminifera from the Western Gulf coastal plain. *Palaeontographica Americana*, 5(37), 243-445.

Quilty, P. G., 1992. Upper Cretaceous planktonic foraminifers and biostratigraphy, Leg 120, southern Kerguelen Plateau. In: *Proceedings of the Ocean Drilling Program, Scientific Results*, Volume 120 (S. W.Jr., Wise, Eds). College Station: Ocean Drilling Program, 371-392.

Sacal, V., Debourle, A., 1957. Foraminifères d'Aquitaine 2e partie. Peneroplidae a Victoriellidae. *Mémoires de la Société Géologique de France*, 78, 1-87.

Salaj, J., Maamouri, A. L., 1970. Remarques microbiostratigraphiques sur le Sénonien supérieur de l'Anticlinal de l'Oued Bazina (Région de Béja, Tunisie septentrionale). *Notes Service Géologique Tunisie*, 32, 65-78. [published in 1971].

Simpson, G. G., 1944. *Tempo and Mode in Evolution*. New York: Columbia University Press, 237 p.

Simpson, G. G., 1961. *Principles of Animal Taxonomy*. New York and London: Columbia University Press, 247 p.

Thompson, L. B., 1991. Late Santonian to early Maastrichtian planktonic foraminiferal biostratigraphy and zonation of northeast Texas. *Micropaleontology, Special Publication*, 5, 9-66.

Vassilenko, V. P., 1961. Upper Cretaceous foraminifera of the Mangyshlak Peninsula (descriptions, phylogenetical schemes for some groups and stratigraphic analysis). *Trudy VNIGRI*, 171, 1-487 [in Russian].

In: Evolutionary Classification ...
Editors: M. Dan Georgescu and C. M. Henderson

ISBN: 978-1-63321-959-5
© 2015 Nova Science Publishers, Inc.

Chapter 6

EVOLUTION AND EVOLUTIONARY CLASSIFICATION OF THE LATE CAMPANIAN-MAASTRICHTIAN PLANKTIC FORAMINIFERA THAT EVOLVED MULTIPLANE CHAMBER PROLIFERATION (*PSEUDOTEXTULARIA* AND *RACEMIGUEMBELINA*)

M. Dan Georgescu[*]

Department of Geosciences, University of Calgary,
Calgary, Alberta, Canada

ABSTRACT

The taxonomic revision of the late Campanian-Maastrichtian planktic foraminifera with the chambers alternately added with respect to the test growth axis that evolved transversally elongate chambers and multiplane chamber proliferation in the adult stage reveals the occurrence of two directional lineages: *Pseudotextularia* Rzehak 1891 and *Racemiguembelina* Montanaro Gallitelli 1957; both were originally described as genera in typological classification and are herein revised as units with significance in evolutionary classification. *Pseudotextularia* is reviewed as directional lineage consisting of the initiating species *P. elegans* (Rzehak 1891) and its first descendant *P. varians* Rzehak 1895; this directional lineage evolved apparently from *Mihaia reussi* (Cushman 1938). *Racemiguembelina* Montanaro Gallitelli 1957 is revised as directional lineage consisting of the initiating species *R. intermedia* (de Klasz 1953), first descendant species *R. powelli* Smith and Pessagno 1973 and second descendant species *R. fructicosa* (Egger 1899); the directional lineage *Racemiguembelina* evolved in the late Campanian from *P. elegans*.

Keywords: Foraminifera, planktic, late Campanian, Maastrichtian, evolutionary classification

---

[*] Corresponding author: M. Dan Georgescu. Department of Geosciences, University of Calgary, 2500 University Drive NW, Calgary, Alberta T2N 1N4, Canada. E-mail: dgeorge@ucalgary.ca.

INTRODUCTION

The planktic foraminifera with chambers alternately added with respect to the test growth axis, which evolved during the late Albian, evolved chamber proliferation in the adult stage multiple times throughout the lower Turonian-upper Campanian stratigraphic interval. Whenever such an evolutionary process happened, the adult proliferating stage presents two characteristics: it begins with one biaperturate chamber known as progressive chamber and chamber proliferation is in the same plane in which the chambers are alternately added with respect to the test growth axis in the juvenile stage. A new architecture of the adult stage with chamber proliferation was developed during the late Maastrichtian in two lineages that evolved during the late Campanian. In this case the chamber proliferation was described either as tridimensional or in a plane perpendicular to that of chamber addition in the juvenile stage. In this study it is shown that the chamber proliferation is characterized by the absence of the progressive biaperturate chamber and multiplane chamber proliferation in the adult stage. The new term multiplane is herein proposed to replace those previously used in describing this kind of chamber proliferation.

The multiplane chamber proliferation can be easily recognized by the general conical aspect of the tests that developed it. It was probably this morphological distinctiveness that led to the report of this type of chamber proliferation by Rzehak (1895) a short time after the first report of the chamber transversal elongation (Rzehak 1891) (Figure 1). The possible taxonomic diversity among the taxa displaying this kind of chamber proliferation was further explored by Egger (1899), White (1929), Glaessner (1936) and Voorwijk (1937). Montanaro Gallitelli (1957) described the typological unit of genus rank *Racemiguembelina* to include the species with this kind. The earliest phylogenetic interpretations by Aliyulla (1965) and Pessagno (1967) followed and they focused on the genus level but the subsequent studies on the representatives of this group referred only at the test morphology (Masters 1977; Weiss 1983).

Nederbragt (1991) inferred the first ancestor-descendant relationships between the representatives of the Cretaceous planktic foraminifera with chambers alternately added with respect to the test growth axis at species level.

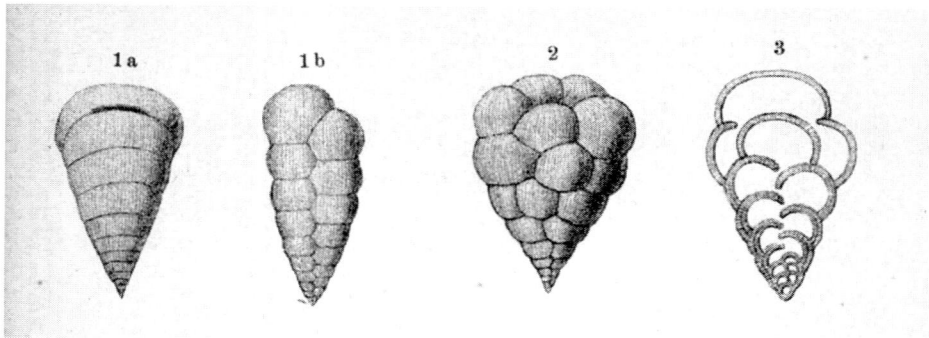

Figure 1. Original illustrations of specimens of *Pseudotextularia varians* by Rzehak (1895, pl. 7, figures 1-3). Note the adult stage with chamber proliferation in figure 3; this specimen was omitted in the re-illustration by Glaessner (1936, text-Figure 1) as *P. elegans*.

Subsequent studies, such as that by Georgescu (2010), confirmed at least in part the validity of the newly inferred evolutionary relationships and at the same time revealed the many inconsistencies in generic assignments of the various species of the group. Despite its inconsistencies the evolutionary relationships inferred by Nederbragt (1991) represented one of the ideas that led to the beginnings of development of the evolutionary classification in the representatives of this group.

This taxonomic revision based on scanning electron microscope (SEM) high-resolution morphologic observations on tests with chamber transversal elongation and multiplane proliferation in the adult stage reveals the occurrence of an evolutionary continuum between *Pseudotextularia* and *Racemiguembelina*; these two genera are herein defined as directional lineages in evolutionary classification, which have the chamber ornamentation consisting of longitudinal pycnocostae.

MATERIAL STUDIED

The material used in this study was collected from a variety of localities worldwide of which most are wells drilled under the auspices of the Deep Sea Drilling Project (DSDP)/ Ocean Drilling Program (ODP): DSDP Hole 111A (Orphan Knoll, North Atlantic Ocean), DSDP Site 152 (Nicaragua Rise, Caribbean Region), DSDP Site 305 (Shatsky Rise, Central Pacific Ocean), DSDP Site 356 (São Paulo Plateau, South Atlantic Ocean), DSDP Site 357 (Rio Grande Rise, South Atlantic Ocean), DSDP Site 384 (J-Anomaly Ridge, North Atlantic Ocean), ODP Hole 761B (Wombat Plateau, East Indian Ocean), ODP Hole 763B (Exmouth Plateau, East Indian Ocean) and ODP Holes 1050C and 1052E (Blake Nose, North Atlantic Ocean). The sediments from which relevant foraminiferal tests were collected are of late Campanian-Maastrichtian age. The representatives of the directional lineages *Pseudotextularia* and *Racemiguembelina* at all these locations were collected from white and yellow deep water nannofossil chalks accumulated in outer shelf and deep oceanic conditions.

Specimens from the early evolution of the two lineages (late Campanian-early Maastrichtian) were studied in the McGugan Collection (University of Calgary). The specimens were collected from the cropping out Ripley Formation and Prairie Bluff Chalk of northern Alabama (US). The material was collected by Dr Alan McGugan during a field trip in 1968 and the lithostratigraphic framework of the sediments adopted herein is that of Liu (2007).

Specimens from the Kemp Clay were studied in a number of samples from the wells Mullinax-1 and Mullinax-3 from Texas (US); the samples were provided by Dr Sigal Abramovich (Ben Gurion University of the Negev, Israel). One sample from the Kemp Clay collected from the surroundings of the Limestone City of Texas (US), which yielded the earliest species of the directional lineage *Pseudotextularia*, was studied at the National Museum of Natural History, Smithsonian Institution, Washington, D.C.

One topotype of *Pseudotextularia elegans* was examined in the Loeblich and Tappan Topotype Collection from the National Museum of Natural History, Smithsonian Institution, Washington, D.C.; the specimen is herein illustrated (Plate 1, Figures 1-2). The specimen was identified by V. Pokorný during the curation of the original material of A. Rzehak according to the text on the slide.

SYSTEMATIC DESCRIPTIONS

Evolutionary classification units at lineage level are after Georgescu (2010, 2013). Species types within a lineage are after Georgescu (2011): IS-initiating species, FDS-first descendant species and SDS-second descendant species.

Directional Lineage: *Pseudotextularia* Rzehak 1891 - emended

Pseudotextularia RZEHAK 1891, p. 4.
Pseudotextularia Rzehak 1886. Cushman 1925, p. 133.
Pseudotextularia Rzehak 1886. Cushman 1927a, p. 157.
Pseudotextularia Rzehak 1886. Cushman 1927b, p. 59.
Pseudotextularia Rzehak 1886. White 1929, p. 40.
Pseudotextularia Rzehak 1886. Cushman 1933, p. 210.
Pseudotextularia Rzehak 1886. Galloway 1933, p. 347.
Pseudotextularia Rzehak 1891. Glaessner 1936, p. 99.
Pseudotextularia Rzehak 1886. Cushman 1938, p. 21.
Pseudotextularia Rzehak 1886. Cushman 1950, p. 256.
Pseudotextularia Rzehak 1891. Montanaro Gallitelli 1957, p. 138.
Pseudotextularia Rzehak 1891. Loeblich and Tappan 1964, p. C656.
Pseudotextularia Rzehak 1891. Saavedra 1965, p. 344.
Pseudotextularia Rzehak 1891. Wille-Janoschek 1966, p. 118.
Pseudotextularia Rzehak 1891. Brown 1969, p. 43.
Pseudotextularia Rzehak 1891. Govindan 1972, p. 170.
Pseudotextularia Rzehak 1891. Masters 1976, p. 320.
Pseudotextularia Rzehak 1891. Aliyulla 1977, p. 201.
Pseudotextularia Rzehak 1891. Masters 1977, p. 378.
Pseudotextularia Rzehak 1891. Smith 1978, p. 314.
Pseudotextularia Rzehak 1891. Weiss 1983, p. 59.
Pseudotextularia Rzehak 1891. Caron 1985, p. 24.
Pseudotextularia Rzehak 1891. Aliyulla in Ali-zade and others 1988, p. 129.
Pseudotextularia Rzehak 1891. Loeblich and Tappan 1987, p. 455.
Pseudotextularia Rzehak 1891. Nederbragt 1989a, p. 113.
Pseudotextularia Rzehak 1891. Nederbragt 1991, p. 362.

Species included. IS: *P. elegans* (Rzehak 1891) and FDS: *P. varians* Rzehak 1895.

Diagnosis. A directional lineage with pycnocostate ornamentation with well-developed chamber transversal elongation throughout and in the FDS with the main aperture bordered by symmetrically developed leptoflanges and occasional adult stage with chamber multiplane proliferation.

Description. The test consists of proloculus followed by chambers alternately added with respect to the test growth axis in the IS and with an adult stage with multiplane chamber proliferation that occurs occasionally in the FDS. Chambers present a gradual size increase in the portion of the test with alternate chamber addition.

The proliferating adult stage lacks a progressive chamber and consists of three chamber sets: there are two chambers in the first set, each of them on each side of the test and three chambers in each of the subsequent two chamber sets. Sutures are distinct and depressed, straight to slightly curved between all the chambers of the test. Periphery is broadly rounded and simple, without peripheral structures. Aperture in the stage with alternately added chambers has an arch shape and is situated at the base of the last-formed chamber. In this portion of the test the aperture is bordered by two symmetrical metaflanges in the IS and leptoflanges in the FDS with one imperforate lip between them; relict leptoflanges occur along the central suture in the portion of the last-formed chambers in the FDS resulting in false supplementary apertures. There is one aperture for each of the chambers of the adult stage with chamber proliferation; each chamber of this stage is protected by Y-shaped bridges, which are extensions of the test wall. Chambers are ornamented with longitudinal pycnocostae with the observed thickness of 0.0068-0.0148 mm in the IS and 0.0063-0.0135 mm in the FDS. Test wall is calcitic, hyaline, simple and perforate; pores have circular to elliptical outline and diameters of 0.0010-0.0029 mm in the IS and 0.0022-0.0043 mm in the FDS.

Remarks. The *Pseudotextularia* directional lineage evolved from leptocostate *Mihaia reussi* (Cushman 1938) during the late Campanian mainly through the development of chamber transversal elongation and pycnocostate ornamentation. There is a significant increase in pore size with the initiation of the directional lineage *Pseudotextularia* from 0.0006-0.0012 mm in *M. reussi* to 0.0010-0.0029 mm in *P. elegans*.

Age. Late Campanian-Maastrichtian.

Geographic distribution. Cosmopolitan.

IS: *Pseudotextularia elegans* (Rzehak 1891)
(Figure 2:1-12)

Cuneolina elegans RZEHAK 1891, p. 4.

Textularia biarritzensis HALKYARD 1917, p. 60, pl. 2, fig. 6.

Pseudotextularia elegans (Rzehak). Glaessner 1936, p. 99, pl. 1, figs 1-2. Bettenstaedt and Wicher 1955, pl. 1, fig. 6 (right only). Montanaro Gallitelli 1957, pl. 33, fig. 6. Seiglie 1958, p. 55, pl. 1, figs 1, 3. Olsson 1960, p. 28, pl. 4, figs 9-10. Said and Kerdany 1961, p. 332, pl. 2, fig. 9. Hiltermann and Koch 1962, p. 337, pl. 46, fig. 11 (only). Skinner 1962, p. 40, pl. 5, fig. 17. Kavary in Kavary and Frizzell 1963, p. 64, pl. 13, figs 9-10. Said and Sabry 1964, p. 392, pl. 3, fig. 29. Perlmutter and Todd 1965, p. 114, pl. 2, fig. 17. Lehmann 1966, p. 316, pl. 2, fig. 10. Salaj and Samuel 1966, p. 232, pl. 37, fig. 11. Wille-Janoschek 1966, p. 120, pl. 8, fig. 10. Barr 1968, pl. 1, fig.10. Sliter 1968, p. 98, pl. 14, figs 13-15. Bate and Bayliss 1969, pl. 4, fig. 3. Dupeuble 1969, p. 158, pl. 4, fig. 13. Funnell and others 1969, p. 23, pl. 1, figs 9-10, text-figure 5. Neagu 1970, p. 61, pl. 14, fig. 1. Sturm in Faupl and others 1970, p. 113, pl. 8, fig. 2. Todd 1970, p. 151, pl. 5, fig. 5. Govindan 1972, p. 170, pl. 1, figs 5-6, pl. 2, figs 6-7. Hanzlikova 1969, p. 43, pl. 8, fig. 12 (only). Hanzliková 1972, p. 95, pl. 24, figs 8, 10. Pessagno and Longoria in Shipboard Scientific Party 1972, pl. 10, fig. 3. Saito and Van Donk 1974, p. 165, pl. 1, figs 11-13. Darmoian 1975, p. 199, pl. 3, figs 16-17. Wright and Apthorpe 1976, pl. 1, fig. 6 (only). Aliyulla 1977, pl. 1, fig. 8. Linares-Rodríguez 1977, pl. 48, fig. 3. Masters 1977, p. 383, pl. 6, figs 3-4. Salaj 1980, pl. 16, fig. 9. Nash 1981, p. 72, pl. 1, figs 1-7. Abdel-Kireem 1986, p. 226, pl. 2, figs 13-16, pl. 3, figs 12-13. Jansen and Kroon 1987, p. 564, pl. 8, figs 1-2. Nederbragt 1989b, p. 204, pl. 8, figs 5-6. Huber 1990, p. 503, pl. 1, fig. 16. D'Hondt and Keller 1991, pl. 2, fig. 4. Malmgren 1991, pl. 1, fig. 11. Nederbragt 1991, p. 364, pl. 10, figs 1-2. Georgescu 1995, p. 405, pl. 3, figs 1-3. Mancini and others 1996, fig. 4: 8. Georgescu 1997, fig. 9: 1. Zapeda 1998, p. 138, fig. 11: 3 (only). Arz and Molina 2001, pl. 2, figs 10-11. Abdelghany 2003, fig. 8: 15. Howe and others 2003, pl. 8, figs 3-4. Alencáster and Omaña 2006, fig. 7: 8. Omaña 2006, fig. 4: 5. Mukhopadhyay 2012, fig. 6: 1-16. Pérez-Rodriguez and others 2012, fig. 7: F.

  Gümbelina striata (Ehrenberg). Voorwijk 1937, p. 194, figs 9-10.

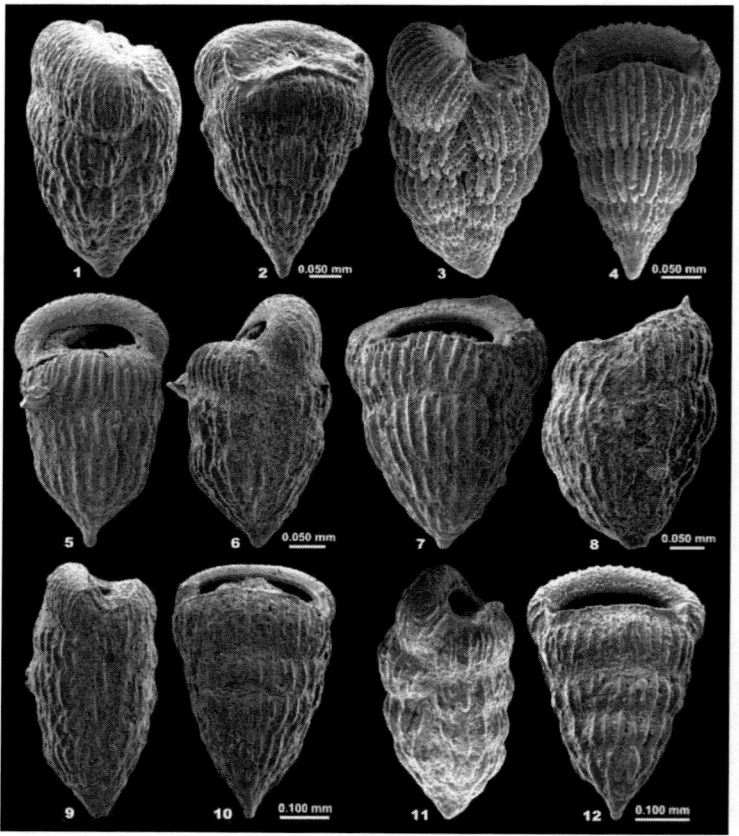

Figure 2. Specimens of *Pseudotextularia elegans* (Rzehak 1891). 1-2 Topotype from Bruderndorf, Niederöstereich; from the Loeblich and Tappan Topotype Collection, National Museum of Natural History, Smithsonian Institution, Washington, D.C. (USNM 480989). 3-4 Hypotype from the Maastrichtian Kemp Clay from the surroundings of the Limestone City, Texas; Sample Huber, National Museum of Natural History, Smithsonian Institution, Washington, D.C. 5-6 Hypotype from the upper Maastrichtian sediments of Blake Nose, North Atlantic Ocean (Sample 171B-1050C-11-1, 127-130 cm). 7-8 Hypotype from the upper Maastrichtian sediments of Blake Nose, North Atlantic Ocean (Sample 171B-1050C-11-2, 123-126 cm). 9-10 Hypotype from the upper Maastrichtian sediments of Blake Nose, North Atlantic Ocean (Sample 171B-1050C-11-1, 127-130 cm). 11-12 Hypotype from the upper Maastrichtian sediments of Blake Nose, North Atlantic Ocean (Sample 171B-1050C-11-1, 75-78 cm).

  Gümbelina striata deformis KIKOÏNE 1948, p. 20, pl. 1, fig. 8.
  Gümbelina plummerae Loetterle. Sellier de Civrieux 1952, p. 270, pl. 6, fig. 11. Said and Kenawy 1956, p. 139, pl. 3, fig. 33.

Gümbelina deformis Kikoïne. Sacal and Debourle 1957, p. 12, pl. 3, fig. 3.

Pseudotextularia bronnimanni SEIGLIE 1958, p. 57, pl. 1, figures 5-8.

Pseudotextularia pecki KAVARY in Kavary and Frizzell 1963, p. 63, pl. 13, fig. 4.

Pseudotextularia deformis (Kikoïne). Pessagno 1967, p. 269, pl. 90, fig. 16, pl. 92, figs 19-21. Smith and Pessagno 1973, p. 29, pl. 9, figs 1-3, pl. 10, figs 7-9. Webb 1973, p. 552, pl. 1, figs 5-8. Linares-Rodríguez 1977, pl. 48, fig. 4. Yasini 1979, p. 23, pl. 2, figs 15-17. Salaj 1983, pl. 1, fig. 36. Weiss 1983, p. 60, pl. 8, figs 1-2. Abdel-Kireem 1986, p. 226, pl. 2, figs 10-12, pl. 3, fig. 11. Keller 1988, pl. 2, fig. 9. Ayyad and others 1996, fig. 7: a. Luciani 1997, fig. 3: 13. Zapeda 1998, p. 138, fig. 11: 1-2. Abramovich and others 2002, pl. 1, figs 8-9. Keller and others 2002, pl. 2, fig. 8. Abramovich and others 2003, pl. 4, fig. 1. Obaidalla 2005, pl. 1, fig. 7.

Pseudotextularia bronnimanni Seiglie. Aliyulla 1977, pl. 2, fig. 1. Aliyulla in Ali-zade and others 1988, p. 129, pl. 25, fig. 3.

Diagnosis. *Pseudotextularia* with chambers alternately added with respect to the test growth axis throughout the ontogenetic development and main aperture bordered by two symmetrically developed narrow metaflanges and one imperforate lip between the metaflanges.

Description. The test consists of the proloculus followed by 14-16 chambers alternately added with respect to the test growth axis; chambers present a gradual size increase. Earlier chambers are subglobular in shape and those of the adult stage present a well-developed transversal elongation and a rapid rate of thickness increase; specimens with a distinct decrease in width in the last-formed chambers occur frequently. Sutures are distinct and depressed, straight to slightly curved throughout; central suture is almost straight and often hardly visible. Test is symmetrical in edge view with a broad and simple periphery, without peripheral structures. Aperture is a low or rarely medium high arch at the base of the last-formed chamber; it is bordered by two narrow metaflanges one on each side of the test and an imperforate lip between them. Chamber surface is ornamented with longitudinally arranged pycnocostae; the pycnocostae have a thickness of 0.0068-0.0148 mm but thinner ornamentation elements occur occasionally over the last-formed chambers of the test; one pustulose area consisting of scattered dome-like pustules occurs in the anterior portion of the chambers. Test wall is calcitic, hyaline, simple and perforate; pores are situated in the space between the pycnocostae and present a circular or elliptical outline with a diameter or maximum dimension of 0.0010-0.0029 mm.

Remarks. The general geologic setting at the type locality was discussed in detail by Gohrbandt (1967) and the conclusions are completely accepted herein. One topotype specimen from the Loeblich and Tappan Topotype Collection (National Museum of Natural History, Smithsonian Institution, Washington, D.C.) and illustrated herein (Figure 2:1-2) confirm the validity of the neotype selected by Nash (1981). *Pseudotextularia elegans* differs from *Mihaia reussi* (Cushman 1938) and *Planoglobulina sphaeralis* Georgescu 2014 mainly by chamber transversal elongation; in addition *M. reussi* is ornamented with leptocostae rather than pycnocostae as in *P. elegans*.

Age. Late Campanian-Maastrichtian.

Geographic distribution. Cosmopolitan.

FDS: *Pseudotextularia varians* Rzehak 1895
(Figures 3:1-10, 4:1-12)

Pseudotextularia varians RZEHAK 1895, p. 217, pl. 7, figs 1-3.
Pseudotextularia varians textulariformis WHITE 1929, p. 41, pl. 4, fig. 17.
Gümbelina plummerae Loetterle. Van Wessem 1943, p. 45, pl. 1, figs 37-38.
Pseudotextularia varians Rzehak. Djafarov and others 1951, p. 110, pl. 16, figs 1-2. (?)
Hamilton 1953, p. 235, pl. 30, fig. 9. Vassilenko 1961, p. 207, pl. 41, fig. 10. Salaj and
Samuel 1966, p. 233, pl. 37, fig. 13.

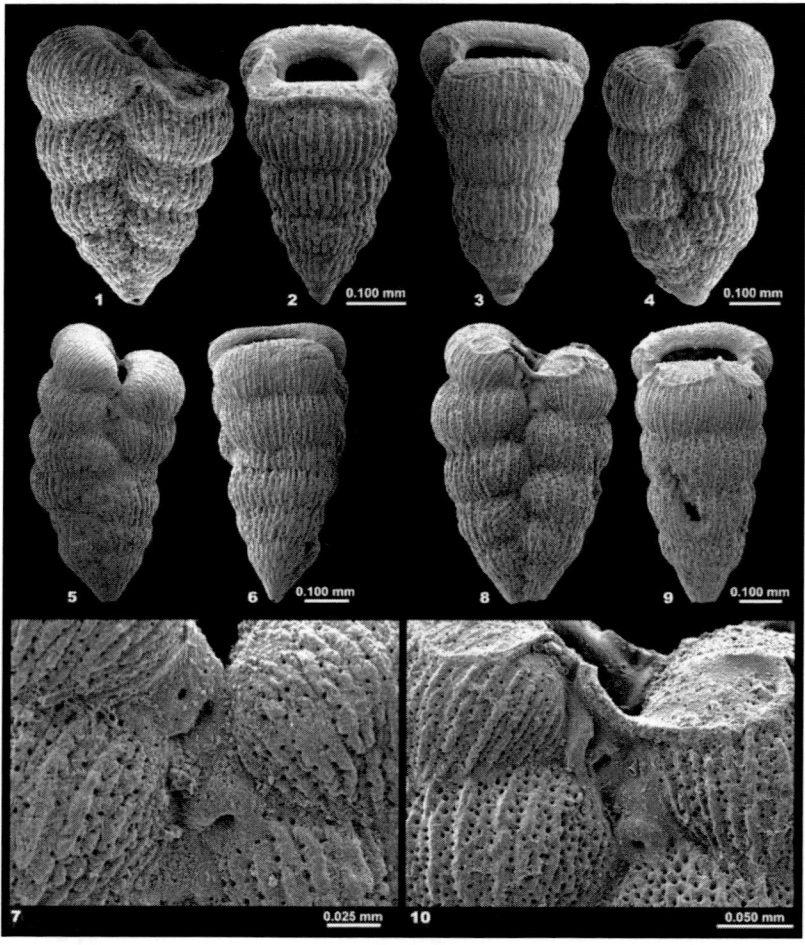

Figure 3. Specimens of *Pseudotextularia varians* Rzehak 1895 with chambers alternately added with respect to the test axis throughout. 1-2 Hypotype from the upper Maastrichtian sediments of the Wombat Plateau, eastern Indian Ocean (Sample 122-761B-24-1, 25-26 cm). 3-4 Hypotype from the upper Maastrichtian sediments of the Wombat Plateau, eastern Indian Ocean (Sample 122-761B-24-1, 111-112 cm). 5-7 Hypotype from the upper Maastrichtian sediments of the Wombat Plateau, eastern Indian Ocean (Sample 122-761B-24-1, 89-90 cm). 8-10 Hypotype from the upper Maastrichtian sediments of the Wombat Plateau, eastern Indian Ocean (Sample 122-761B-24-1, 25-26 cm).

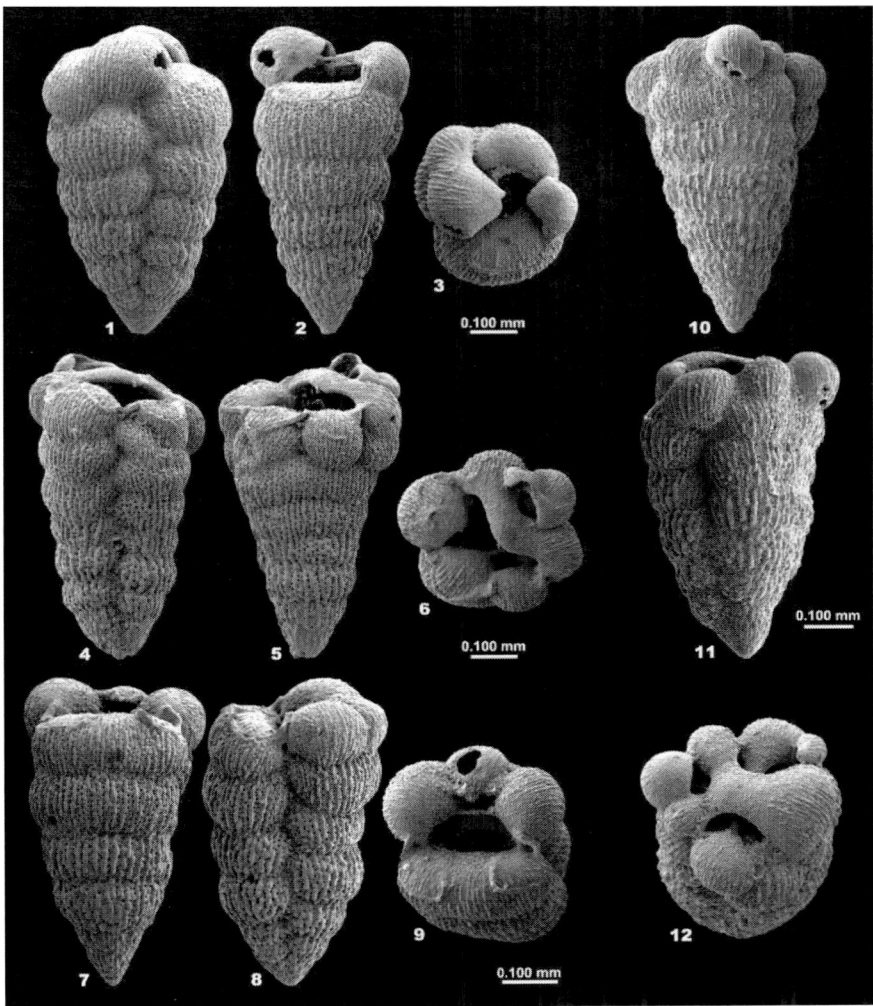

Figure 4. Specimens of *Pseudotextularia varians* Rzehak 1895 with chambers alternately added with respect to the test axis in the early stage and multiplane chamber proliferation in the adult stage. 1-3 Hypotype from the upper Maastrichtian sediments of the Wombat Plateau, eastern Indian Ocean (Sample 122-761B-24-1, 89-90 cm). 4-6 Hypotype from the upper Maastrichtian sediments of the Wombat Plateau, eastern Indian Ocean (Sample 122-761B-22-3, 68-69 cm). 7-9 Hypotype from the upper Maastrichtian sediments of the Wombat Plateau, eastern Indian Ocean (Sample 122-761B-24-1, 25-26 cm). 10-12 Hypotype from the upper Maastrichtian sediments of the Wombat Plateau, eastern Indian Ocean (Sample 122-761B-24-1, 111-112 cm).

Pseudotextularia elongata SEIGLIE 1958, p. 58, pl. 1, figs 2, 4, pl. 2, figs 1-2, 4, 6, pl. 3, fig. 1.

Pseudotextularia elegans (Rzehak). Pringgoprawiro 1965, p. 37, pl. 6, fig. 3 (only). Wright and Apthorpe 1976, pl. 1, fig. 4. Butt 1981, pl. 18, fig. P. Busulini and others 1984, fig. 5: g-h. Villars 1988, pl. 2, fig. 7. Korchagin 2011, pl. 1, figs 1-2.

Racemiguembelina textulariformis (White). Salaj and Samuel 1966, p. 233, pl. 37, fig. 12. Salaj 1980, pl. 16, fig. 8. Salaj 1983, pl. 1, fig. 37.

Pseudotextularia elongata Seiglie. Thompson 1991, p. 38, pl. 1, fig. 10.

Diagnosis. *Pseudotextularia* with chambers alternately added with respect to the test growth at least in the early stage and occasional adult stage with multiplane chamber proliferation; aperture is bordered by two symmetrically developed leptoflanges.

Description. The test consists of the proloculus followed by 16-24 chambers alternately added with respect to the test growth axis at least in the juvenile stage; an adult stage with multiplane chamber proliferation occurs occasionally. The adult stage consists of two or maximum three chamber-sets but lacks the progressive chamber; the first chamber set of the adult stage with chamber proliferation consists of two chambers and the remaining two of three chambers each. Chambers overlap at various rates throughout the ontogenetic development. The sutures are distinct and depressed, straight to curved throughout; the central sutures are distinct and with indented aspect. At least the earlier portion of the test is symmetrical in edge view but symmetry is lost in the adult stage with chamber proliferation; periphery is broadly rounded and simple, without peripheral structures. Aperture in the test portion with chambers alternately added with respect to the test growth axis has the shape of a low or rarely medium high arch and is situated at the base of the last-formed chamber; the aperture in this stage is bordered by leptoflanges, which are connected by an imperforate lip, and relict leptoflanges occur frequently in the anterior portion of the test along the central structure. The apertures of the chambers of the adult stage with multiplane proliferation are situated beneath bridges, which are often Y-shaped.

Chamber surface is ornamented with longitudinally oriented pycnocostae with a measured thickness of 0.0063-0.0135 mm. A pustulose periapertural area consisting of scattered and small-sized dome-like pustules occurs in the chamber anterior portion. Test wall is calcitic, hyaline, simple and perforate; pores are simple, with a circular and elliptical outline and diameter or maximum dimension of 0.0022-0.0043 mm.

Remarks. *Pseudotextularia varians* differs from its ancestor *P. elegans* mainly by having a test consisting of more numerous chambers (16-24 rather than 14-16) and development of an adult stage with multiplane chamber proliferation.

Age. Late Maastrichtian.

Geographic distribution. Cosmopolitan.

Directional Lineage: *Racemiguembelina* Montanaro Gallitelli 1957

Racemiguembelina MONTANARO GALLITELLI 1957, p. 142.

Pseudotextularia Rzehak 1891. Loeblich and Tappan 1964, p. C656.

Racemiguembelina Montanaro Gallitelli 1957. Saavedra 1965, p. 344.

Racemiguembelina Montanaro Gallitelli 1957. Pessagno 1967, p. 270.

Racemiguembelina Montanaro Gallitelli 1957. Smith and Pessagno 1973, p. 32.

Racemiguembelina Montanaro Gallitelli 1957. Aliyulla 1977, p. 202.

Racemiguembelina Montanaro Gallitelli 1957. Smith 1978, p. 315.

Racemiguembelina Montanaro Gallitelli 1957. Weiss 1983, p. 63.

Racemiguembelina Montanaro Gallitelli 1957. Caron 1985, p. 24.

Racemiguembelina Montanaro Gallitelli 1957. Aliyulla in Ali-zade and others 1988, p.130.

Racemiguembelina Montanaro Gallitelli 1957. Loeblich and Tappan 1987, p. 455.

Racemiguembelina Montanaro Gallitelli 1957. Nederbragt 1989a, p. 113.

Racemiguembelina Montanaro Gallitelli 1957. Nederbragt 1991, p. 366.

Species included. IS: *R. intermedia* (de Klasz 1953), FDS: *R. powelli* Smith and Pessagno 1973 and SDS: *R. fructicosa* (Egger 1899).

Diagnosis. A directional lineage with pycnocostate ornamentation and with multiplane chamber proliferation in the adult stage throughout.

Description. The test consists of the proloculus followed by a stage in which the chambers are alternately added with respect to the test growth axis and an adult stage with multiplane chamber proliferation; the adult stage consists of one or maximum two chamber sets in the IS, up to four in the FDS and more than four chamber sets in the SDS. Sutures are distinct and depressed and straight to slightly curved throughout. Periphery is broadly rounded and simple, without peripheral structures. Aperture is in the shape of an arch at the base of the last-formed chamber and each chamber of the adult stage has one aperture. The apertures in the adult stage are protected by bridges that represent extensions of the test wall and connect chambers across the test cone-like internal cavity. Chamber surface is ornamented with pycnocostae with a thickness of 0.0066-0.0133 mm in the IS, 0.0050-0.0149 mm in the FDS and 0.0054-0.0206 mm and SDS. Pores have a circular, elliptical or irregular outline, are situated in the space between the pycnocostae and diameter or maximum dimension of 0.0012-0.0053 mm in the IS, 0.0018-0.0053 mm in the FDS and 0.0019-0.0057 mm in the SDS.

Remarks. *Racemiguembelina* differs from *Pseudotextularia* mainly by the occurrence of the adult stage with multiplane chamber proliferation throughout the directional lineage, rather than occasionally occurring in the FDS.

Age. Late Campanian-Maastrichtian.

Geographic distribution. Cosmopolitan.

IS: *Racemiguembelina intermedia* (de Klasz 1953)
(Figure 5:1-9)

Pseudotextularia textulariformis White. Kikoïne 1948, p. 23, pl. 2, fig. 5.

Pseudotextularia intermedia DE KLASZ 1953, p. 231, pl. 5, fig. 2.

Pseudotextularia intermedia de Klasz. Pessagno 1967, p. 269, pl. 86, fig. 11. Pessagno and Longoria in Shipboard Scientific Party 1972, pl. 10, fig. 2. Linares-Rodríguez 1977, pl. 48, fig. 5. Nederbragt 1989a, pl. 4, fig. 1. Nederbragt 1989b, p. 202, pl. 8, figs 1-2. Nederbragt 1991, p. 364, pl. 10, fig. 3. Abramovich and others 2002, pl. 1, fig. 14. Alencáster and Omaña 2006, fig. 7: 9.

Pseudotextularia deformis (Kikoïne). Peryt 1980, p. 44, pl. 6, fig. 6.

Racemiguembelina intermedia (de Klasz). Weiss 1983, p. 65, pl. 8, figs 8-10, pl. 9, figs 1-2. Jansen and Kroon 1987, p. 565, pl. 8, fig. 12. Keller and others 2002, pl. 3, fig. 1. Abramovich and others 2003, pl. pl. 4, figs 3-4.

Racemiguembelina fructicosa (Egger). Quilty 1992, p. 380, pl. 1, fig. 12. Robaszynski 1998, pl. 18, fig. 3.

Diagnosis. *Racemiguembelina* with incipient chamber proliferation in the adult stage, which consists of one or rarely two chamber sets.

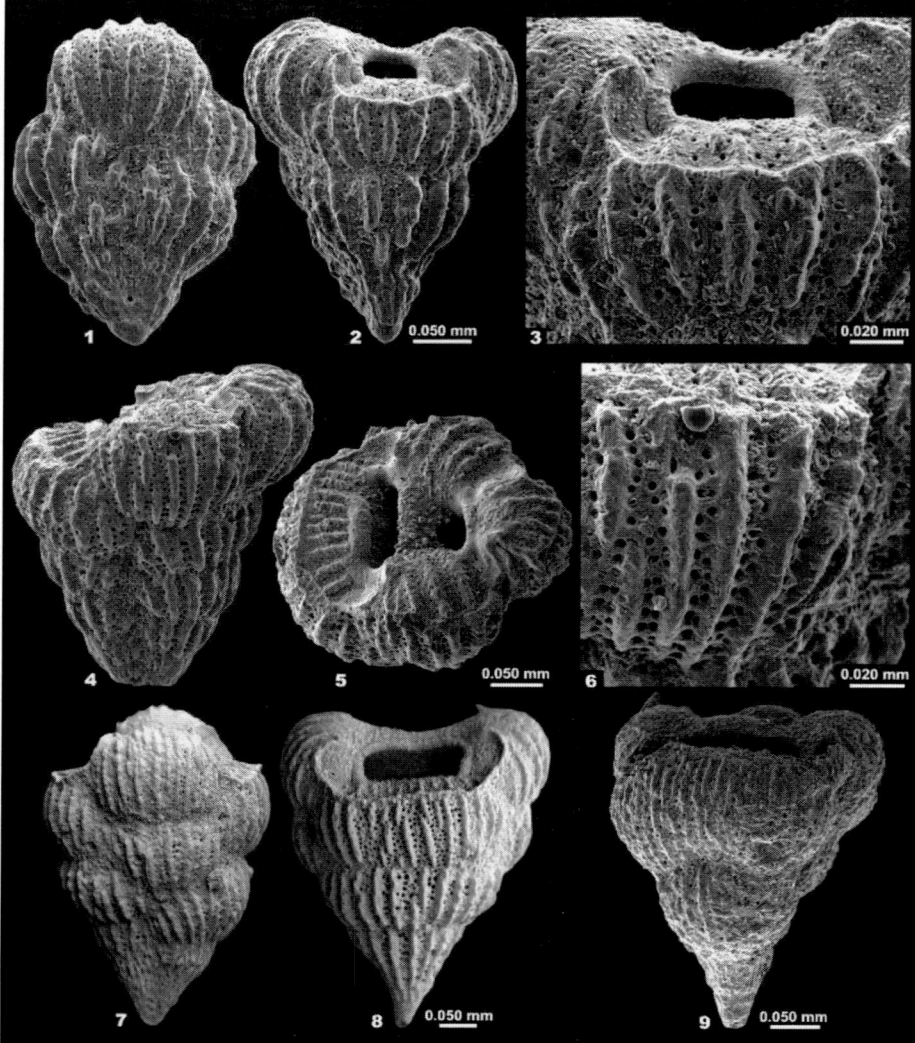

Figure 5. Specimens of *Racemiguembelina intermedia* (de Klasz 1953). 1-3 Hypotype from the upper Maastrichtian sediments of the Blake Nose, North Atlantic Ocean (Sample 171B-1050C-11-core catcher). 4-6 Hypotype from the upper Maastrichtian sediments of the Blake Nose, North Atlantic Ocean (Sample 171B-1050C-11-core catcher). 7-8 Hypotype from the upper Maastrichtian sediments of the Blake Nose, North Atlantic Ocean (Sample 171B-1050C-11-2, 123-126 cm). 9 Hypotype from the São Paulo Plateau, South Atlantic Ocean (Sample 39-356-30-4, 42-56 cm).

Description. The test consists of the proloculus followed by an early stage consisting of 9-13 chambers alternately added with respect to the test growth axis and adult stage with chamber proliferation. The adult stage consists in most of the specimens of one chamber set consisting of two chambers symmetrically added on each side of the early stage and a second set also consisting of two chambers; there is no progressive chamber. Chambers of the early stage presents a gradual size increase and overlap at various rates. Sutures are distinct and depressed, straight to slightly curved throughout. Periphery is broadly rounded and simple, without peripheral structures. Aperture in the early stage has the shape of an arch and is situated at the base of the last-formed chamber; the aperture is bordered by one weakly prominent imperforate lip. The chambers of the adult proliferating stage are protected by one transversal bridge, which is rarely preserved. Chamber surface is ornamented with longitudinally arranged pycnocostae with a thickness of 0.0066-0.0133 mm; one periapertural pustulose area consisting of scattered dome-like pustules occurs in the anterior portion of the chambers. Test wall is calcitic, hyaline, simple and perforate; pores are circular to elliptical in shape and have a diameter or maximum dimension of 0.0012-0.0053 mm.

Remarks. *Racemiguembelina intermedia* differs from *P. elegans* mainly by the occurrence of an adult stage with multiplane chamber proliferation rather than having the chambers alternately added with respect to the test growth axis throughout the ontogenetic development. The similarities in the test architecture of *P. elegans* and the early stage of *R. intermedia* and ornamentation consisting of longitudinally arranged pycnocostae indicate that the two are connected by an ancestor-descendant relationship in which *P. elegans* in the ancestor and *R. intermedia* its descendant. This species differs from *P. varians* mainly by having shorter juvenile stage with chambers having a more rapid thickness increase and by lacking symmetrical developed leptoflanges bordering the aperture.

Age. Late Campanian-Maastrichtian.

Geographic distribution. Cosmopolitan.

FDS: *Racemiguembelina powelli* Smith and Pessagno 1973
(Figure 6:1-4)

Pseudotextularia elegans varians Rzehak. Glaessner 1936, p. 101, pl. 1, fig. 5.

Pseudotextularia varians Rzehak. Voorwijk 1937, p. 194, pl. 1, figs 14-15. Wille-Janoschek 1966, p. 121, pl. 8, fig. 9. Hofker 1978, pl. 1, figs 10-12.

Pseudotextularia elegans (Rzehak). Hofker 1956, pl. 9, fig. 78.

Pseudotextularia fructicosa (Egger). Montanaro Gallitelli 1957, pl. 32, figs 14-15. Govindan 1972, p. 171, pl. 1, figs 7-8, pl. 2, figs 8-9.

Racemiguembelina fructicosa (Egger). Sliter 1968, p. 98, pl. 14, fig. 16. Pessagno and Brown 1969, pl. 1, figs 1-4. Neagu 1970, p. 61, pl. 14, fig. 3. Todd 1970, p. 152, pl. 5, fig. 7. Linares-Rodríguez 1977, pl. 48, fig. 6. Yasini 1979, p. 23, pl. 2, figs 12-13. Salaj 1983, pl. 1, fig. 42. Obaidalla 2005, pl. 1, fig. 4.

Racemiguembelina sp. Pessagno and Longoria in Shipboard Scientific Party 1972, pl. 10, fig. 1.

Racemiguembelina powelli SMITH and PESSAGNO 1973, p. 35, pl. 1, figs 4-12.

Racemiguembelina powelli Smith and Pessagno. Aliyulla 1977, pl. 2, fig. 221. Peryt 1980, p. 45, pl. 6, fig. 7. Weiss 1983, p. 65, pl. 9, figs 3-4. Abdel-Kireem 1986, p. 228, pl. 2, figs 5-6. Peryt 1988, pl. 2, fig. 3. Nederbragt 1989b, p. 204, pl. 8, figs 5-6. D'Hondt and Keller 1991, pl. 2, fig. 3. Nederbragt 1991, p. 368, pl. 11, fig. 1. Ayyad and others 1996, fig. 7: d-e. Arenillas and others 2000, pl. 1, fig. 2. Abramovich and others 2002, pl. 1, fig. 15. Keller and others 2002, pl. 3, fig. 3. Abramovich and others 2003, pl. 4, figs 5-6. Chacón and others 2004, fig. 4: M. Pérez-Rodriguez and others 2012, fig 7: H.

Racemiguembelina varians fructicosa (Egger). Salaj 1980, pl. 16, fig. 7. Salaj 1983, pl. 1, figs 38-39.

Diagnosis. *Racemiguembelina* with adult stage with multiplane chamber proliferation consisting of three to four chamber sets.

Description. The test consists of the proloculus followed by a juvenile stage with chambers alternately added with respect to the test growth axis; the adult stage with multiplane chamber growth consists of three to four chamber sets. Chambers of the juvenile stage present a gradual size increase and overlap at various rates. The adult stage lacks a progressive chamber. Sutures are distinct and depressed, straight to slightly curved throughout. Periphery is broadly rounded and simple, without peripheral structures. Aperture in the early stage is an arch situated at the base of the last-formed chamber. Each chamber of the adult stage with chamber proliferation has one aperture, which is connected across the test central cavity by bridges that represent extensions of the test wall. Chamber surface is ornamented with longitudinal leptocostae with a thickness of 0.0050-0.0149 mm; a pustulose periapertural area consisting of scattered dome-like pustules occur in the chamber anterior part. Test wall is calcitic, hyaline, simple and perforate; pores have a circular, elliptical or irregular outlines and a diameter or maximum dimension respectively of 0.0018-0.0053 mm.

Remarks. *Racemiguembelina powelli* differs from its ancestor *R. intermedia* mainly by having the adult stage consisting of up to four chamber sets rather than one, rarely two. There are frequent tests with intermediate morphological features between the two species, and such tests occur in all of the studies sites.

Age. Maastrichtian.

Geographic distribution. Cosmopolitan.

SDS: *Racemiguembelina fructicosa* (Egger 1899)
(Figure 6:5-8)

Gümbelina fructicosa EGGER 1899, p. 35, pl. 14, figs 8-9.

Pseudotextularia varians Rzehak. Cushman 1926, p. 17, pl. 2, fig. 4. Cushman 1927a, p. 157, pl. 27, fig. 2. White 1929, p. 40, pl. 4, fig. 15. Cushman 1938, p. 21, pl. 4, figs 1-2, 4. Keller 1939, pl. 2, fig. 2. Cushman 1946, p. 110, pl. 47, figs 4-9. Cita 1948, p. 125, pl. 2, fig. 7. Kikoïne 1948, p. 23, pl. 2, fig. 4. Cushman 1949, p. 8, pl. 3, fig. 26. Noth 1951, p. 62, pl. 7, figs 20-21. Itzhaki 1952, p. 188, figs 1-8. Bukowy and Geroch 1957, pl. 28, figs 12-13. Sacal and Debourle 1957, p. 13, pl. 3, fig. 10.

Pseudotextularia varians mendezensis WHITE 1929, p. 41, pl. 4, fig. 16.

Pseudotextularia elegans varians Rzehak. Glaessner 1936, p. 101, pl. 1, figs 3-4.

Pseudotextularia varians mendezensis White. Voorwijk 1937, p. 194, figs 18, 24.

Pseudotextularia fructicosa (Egger). Seiglie 1958, p. 56, pl. 2, figs 3, 5. Hanzliková 1972, p. 96, pl. 24, fig. 9.

Pseudotextularia elegans (Rzehak). Witwika 1958, p. 195, figs 6-7. Pringgoprawiro 1965, p. 37, pl. 6, fig. 3 (only).

Racemiguembelina fructicosa (Egger). Eternod Olvera 1959, p. 78, pl. 2, figs 5-7, 11. Said and Kerdany 1961, p. 334, pl. 2, fig. 17. Skinner 1962, p. 41, pl. 5, fig. 14. Pessagno 1967, p. 270, pl. 90, figs 14-15. Ansary and Tewfik 1968, p. 43, pl. 3, fig. 16. Bate and Bayliss 1969, pl. 4, fig. 1. Cita 1969, pl. 4, fig. 6. Dupeuble 1969, p. 158, pl. 4, fig. 12. Funnell and others 1969, p. 25, pl. 2, figs 3-4, text-Figure 8. Hanzliková 1969, p. 44, pl. 8, fig. 14. Cita and Gartner 1971, pl. 3, fig. 2. Smith and Pessagno 1973, p. 33, pl. 12, figs 1-8.

Saito and Van Donk 1974, p. 165, pl. 1, figs 3-5. Linares-Rodríguez 1977, pl. 48, fig. 7. Butt 1981, pl. 19, figs H, J. Weiss 1983, p. 64, pl. 9, fig. 5. Premoli Silva and McNulty 1984, pl. 3, fig. 11.

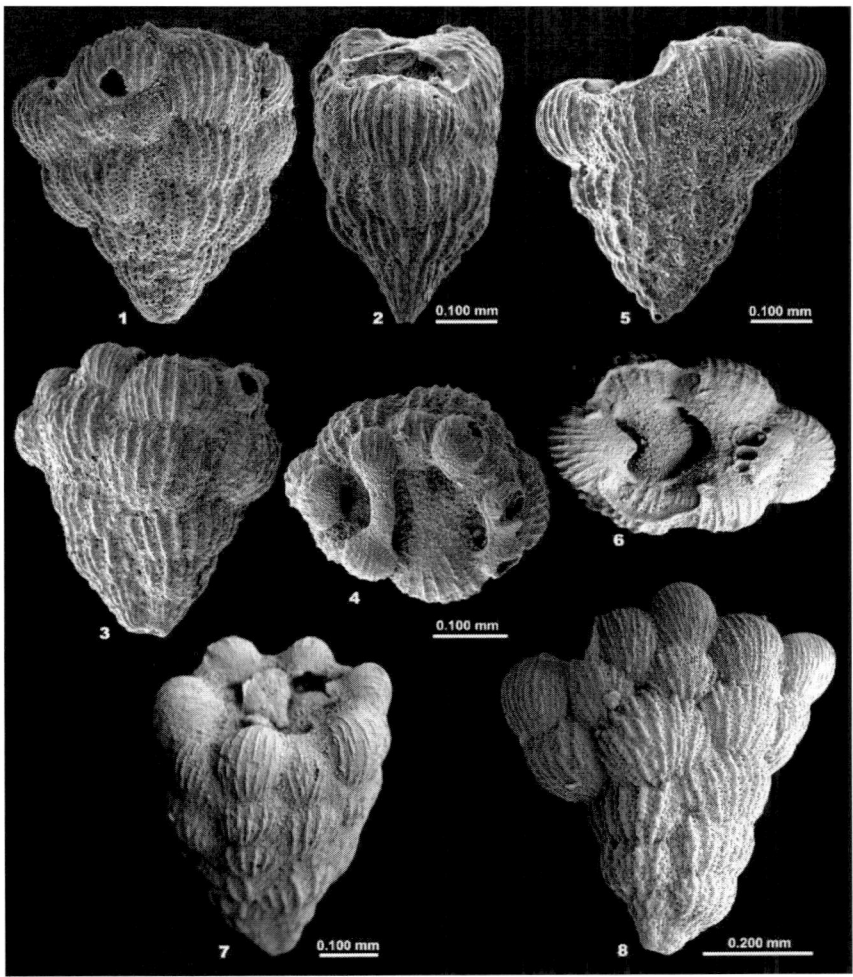

Figure 6. Specimens of *Racemiguembelina powelli* Smith and Pessagno 1973 (1-4) and *R. fructicosa* (Egger 1899) (5-8). 1-2 Hypotype from the upper Maastrichtian sediments of the Wombat Plateau, eastern Indian Ocean (Sample 122-761B-24-1, 145-146 cm). 3-4 Hypotype from the São Paulo Plateau, South Atlantic Ocean (Sample 39-356-31-5, 70-84 cm). 5-6 Hypotype from the upper Maastrichtian sediments of the Blake Nose, North Atlantic Ocean (Sample 171B-1050C-13-4, 110-114 cm). 7 Hypotype from the upper Maastrichtian sediments of Blake Nose, North Atlantic Ocean (Sample 171B-1050C-11-2, 70-74 cm). 8 Hypotype from the upper Maastrichtian sediments of the *J*-Anomaly Ridge, North Atlantic Ocean (Sample 43-384-13-6, 70-72 cm).

Caron 1985, p. 67, fig. 24: 22-23. Abdel-Kireem 1986, p. 226, pl. 2, fig. 7. Jansen and Kroon 1987, p. 564, pl. 8, fig. 13. Aliyulla in Ali-zade and others 1988, p. 130, pl. 25, figs 4-6. Keller 1988, pl. 1, fig. 15. Nederbragt 1989a, pl. 2, figs 2-3, pl. 4, figs 2-5. Nederbragt 1989b, p. 204, pl. 8, fig. 4. D'Hondt and Keller 1991, pl. 2, fig. 2. Malmgren 1991, pl. 1, fig. 13. Nederbragt 1991, p. 366, pl. 10, fig. 5. Georgescu 1995, p. 405, pl. 3, figs 4-5. Georgescu

1997, fig. 9: 3. D'Hondt and Zachos 1998, fig. 1 (upper row, left). Zapeda 1998, p. 139, fig. 10: 2.

Abramovich and others 2002, pl. 1, fig. 13. Keller and others 2002, pl. 3, fig. 2. Abramovich and others 2003, pl. 4, figs 7-8. Pérez-Rodriguez and others 2012, fig. 7: G.

Planoglobulina fructicosa (Egger). Pandey 1980, p. 64, pl. 3, figs 13-14.

Racemiguembelina pomeroli SALAJ 1983, p. 209, pl. 1, fig. 41, pl. 8, figs 11-12.

Diagnosis. *Racemiguembelina* with the adult stage with multiplane chamber proliferation consisting of more than four chamber sets.

Description. The test consists of the proloculus followed by a juvenile stage consisting of chambers alternately added with respect to the test growth axis and adult stage with multiplane chamber proliferation. Chambers of the early stage present a gradual size increase and overlap at various rates. Periphery is broadly rounded and continuous conferring the test a conical appearance; there are no peripheral structures. Chambers of the adult stage are connected by bridges across the test cone-like central cavity; bridges are extensions of the test wall. Narrow and prominent imperforate periapertural lips occur between the connected chambers and connective ridges. Chamber surface is ornamented with longitudinal leptocostae with a thickness of 0.0054-0.0206 mm. Test wall is calcitic, hyaline, simple and perforate; pores are circular, elliptical or with irregular outline, presenting diameters or maximum dimensions of 0.0019-0.0057 mm.

Remarks. *Racemiguembelina fructicosa* differs from its ancestor *R. powelli* by having the adult stage with multiplane chamber proliferation with more than four chamber sets and thicker pycnocostae (0.0054-0.0206 mm rather than 0.0050-0.0149 mm). Specimens with intermediate morphological features between *R. fructicosa* and *R. powelli* occur frequently throughout the late Maastrichtian.

Age. Late Maastrichtian.

Geographic distribution. Cosmopolitan.

CONCLUSION

Evolution of transversally elongate chambers in the foraminiferal tests with chambers alternately added with respect to the test growth axis occurred twice in the group history, namely during the Santonian and late Campanian.

The earliest evolution led to the development of tests with leptocostate ornamentation and they are included in the directional lineage *Bronnimannella* Montanaro Gallitelli 1956. The second evolution happened during the late Campanian with the initiation of the directional lineage *Pseudotextularia* Rzehak 1891.

Pseudotextularia is revised as a directional lineage in evolutionary classification and consists of the initiating species *P. elegans* (Rzehak 1891) and first descendant species *P. varians* Rzehak 1895. The former presents the chambers alternately added with respect to the test growth axis throughout, whereas an adult stage with multiplane chamber proliferation occasionally occurs in the latter. The tests included in this lineage are ornamented with pycnocostae and there is a pore size increase from 0.0010-0.0029 mm in the IS and 0.0022-0.0043 mm in the FDS.

The study of the chamber arrangement in the adult stage with multiplane proliferation shows that there is no progressive biaperturate chamber. *Pseudotextularia elegans* evolved from the globular-chambered *Mihaia reussi* (Cushman 1938).

Racemiguembelina is a directional lineage that evolved in the late Campanian from *P. elegans*, and includes three species: *R. intermedia* (de Klasz 1953) as initiating species, *R. powelli* Smith and Pessagno 1973 as first descendant species and the second descendant species *R. fructicosa* (Egger 1899). An increase of the number of chambers in the adult stage with multiplane chamber proliferation occurs throughout the directional lineage and as a result the tests of the FDS and SDS present a conical appearance. The chamber surface is ornamented with longitudinal pycnocostae that show an increase in observed thickness from 0.0066-0.0133 mm in the IS to 0.0050-0.0149 mm in the FDS and 0.0054-0.0206 mm and SDS. There is a gradual increase in the observed pore size from 0.0012-0.0053 mm in the IS to 0.0018-0.0053 mm in the FDS and 0.0019-0.0057 mm in the SDS.

The two directional lineages became extinct at the Cretaceous/Paleogene boundary.

REFERENCES

Abdel-Kireem, M. R., 1986. Planktonic foraminifera and stratigraphy of the Tanjero Formation (Maastrichtian), northeastern Iraq. *Micropaleontology*, 32, 215-231.

Abdelghany, O., 2003. Late-Campanian-Maastrichtian foraminifera from the Simsima Formation on the western side of the Northern Oman Mountains. *Cretaceous Research*, 24, 391-405.

Abramovich, S., Keller, G., Adatte, T., Stinnesbeck, W., Hottinger, L., Stueben, D., Berner, Z., Ramanivosoa, B., Randiriamanantenasoa, A., 2002. Age and paleoenvironment of the Maastrichtian to Paleocene of the Mahajanga Basin, Madagascar: a multidisciplinary approach. *Marine Micropaleontology*, 47, 17-70.

Abramovich, S., Keller, G., Stüben, D., Berner, Z., 2003. Characterization of late Campanian and Maastrichtian planktonic foraminiferal depth habitats and vital activities based on stable isotopes. *Palaeogeography, Palaeoclimatology, Palaeoecology*, 202, 1-29.

Alencáster, G., Omaña, L., 2006. Maastrichtian inoceramid bivalves from Central Chiapas, southeastern México. *Journal of Paleontology*, 80, 946-957.

Ali-zade, A., Aliev, G. A., Aliev, M. M., Aliyulla, K., Khalilov, A. G., 1988. *Cretaceous fauna of Azerbaijan*. Baku: Akademia Nauk Azerbaydzhanskoy SSR, Institut Geologia im. Akad. I. M. Gubkina. Izdatelstvo "Elm", 447 p. [in Russian].

Aliyulla, K. 1965. On the state of the knowledge of the family Heterohelicidae and the way of its subsequent study. *Voprosii Mikropaleontologii*, 9, 215-228. [in Russian].

Aliyulla, K., 1977. *Upper Cretaceous and foraminiferal development in the Lesser Caucasus (Azerbaijan)*. Baku: Akademia Nauk Azerbaydzhanskoy SSR, Institut Geologia im. Akad. I. M. Gubkina. Izdatelstvo "Elm", 229 p. [in Russian].

Ansary, S. E., Tewfik, N. M., 1968. Planktonic foraminifera and some benthonic species from the subsurface Upper Cretaceous of Ezz El Orban area, Gulf of Suez. *Journal of Geology of the United Arab Republic*, 10, 37-76.

Arenillas, I., Arz, J. A., Molina, E., Dupuis, C., 2000. The Cretaceous/Paleogene (K/P) boundary at Aïn Settara: Tunisia: sudden catastrophic mass extinction in Planktic foraminifera. *Journal of Foraminiferal Research*, 30, 202-218.

Arz, J. A., Molina. E., 2001. Planktic foraminiferal quantitative analysis across the Campanian/Maastrichtian boundary at Tercis (Landes, France). In: *The Campanian-Maastrichtian Boundary. Characterisation and Correlation from Tercis-les-Bains (Landes, SW France) to Europe and Other Continents* (G. S. Odin, Ed.). Elsevier, IUGS Special Publication, 36, 338-348.

Ayyad, S. N., Abed, M. M., Abu Zied, R. H., 1996. Biostratigraphy and correlation of Cretaceous rocks in Gebel Arif El–Naga, northeastern Sinai, Egypt, based on planktonic foraminifera. *Cretaceous Research*, 17, 263-291.

Barr, F. T., 1968. Upper Cretaceous stratigraphy of Jabal al Akhdar, northern Cyrenaica. In: *Geology and Archaeology of northern Cyrenaica,* Libya (F. T. Barr, Ed.). Petroleum Exploration Society of Libya, Tenth Annual Field Conference, 131-146.

Bate, R. H., Bayliss, D. D., 1969. An outline account of the Cretaceous and Tertiary foraminifera and of the Cretaceous ostracods of Tanzania. *Proceedings of the 3rd African Micropaleontology Colloquium,* Cairo, 113-164.

Bettenstaedt, F., Wicher, C. A., 1955. Stratigraphic correlation of Upper Cretaceous and Lower Cretaceous in the Tethys and Boreal by the aid of microfossils. *Proceedings of the Fourth World Petroleum Congress, Geology and Geophysics*, I/D, 493-515.

Brown, N. K., Jr., 1969. Heterohelicidae Cushman, 1927, amended, a Cretaceous planktonic foraminiferal family. In: *Proceedings of the First International Conference on Planktonic Microfossils,* Geneva 1967 (P. Brönnimann, P. and H. H. Renz, Eds). Leiden: E.J. Brill, 2, 21-67.

Bukowy, S., Geroch, S., 1957. O wieku zlepieńców egzotykowych w Kruhelu Wielkim. *Rocznik Polskiego Towarzystwa Geologicznego*, 26, 297-329.

Busulini, A., Dieni, I., Massari, F., Pejović, D., Wiedmann, J., 1984. Nouvelles Données sur le Crétacé Supérieur de la Sardigne Orientale. *Cretaceous Research*, 5, 243-258.

Butt, A., 1981. Depositional environments of the Upper Cretaceous rocks in the northern part of the Eastern Alps. *Cushman Foundation for Foraminiferal Research Special Publication*, 20, 1-121.

Caron, M., 1985. Cretaceous planktic foraminifera. In: *Plankton Stratigraphy* (H. M. Bolli, J. B. Saunders, K. Perch-Nielsen, Eds.). Cambridge: Cambridge University Press, 17-86.

Chacón, B., Martín-Chivelet, J., Gräfe, K.-U., 2004. Latest Santonian to latest Maastrichtian planktic foraminifera and biostratigraphy of the hemipelagic successions of the Prebetic Zone (Murcia and Alicante provinces, south-east Spain). *Cretaceous Research*, 25, 585-601.

Cita, M. B., 1948. Ricerche stratigrafiche e micropaleontologiche sul Cretarcico e sull'Eocene di Tignale (Lago di Garda). II. Paleontologia. *Rivista Italiana di Paleontologia e Stratgrafia*, 54, 117-143.

Cita, M. B., 1969. Observations sur quelques aspects paléoécologiques de sondages subocéaniques effectués dans l'Atlantique Nord. *Revue de Micropaléontologie*, 12, 187-201.

Cita, M. B., Gartner, S., Jr., 1971. Deep sea Upper Cretaceous from the western North Atlantic. In: *Proceedings of the II Planktonic Conference,* Roma 1967 (A. Farinacci, Ed.). Roma: Edizioni Tecnoscienza, 1, 287-319.

Cushman, J. A., 1925. The genera *Pseudotextularia* and *Gümbelina*. *Journal of the Washington Academy of Sciences*, 15, 133-134.

Cushman, J. A., 1926. Some foraminifera from the Mendez Shale of eastern Mexico. *Contributions from the Cushman Laboratory for Foraminiferal Research*, 2, 16-28.

Cushman, J. A., 1927a. Some characteristic Mexican fossil foraminifera. *Journal of Paleontology*, 1, 147-172.

Cushman, J. A., 1927b. An outline of a re-classification of the foraminifera. *Contributions from the Cushman Laboratory for Foraminiferal Research*, 3, 1-105.

Cushman, J. A., 1933. *Foraminifera, their classification and economic use*; 2nd edition. Cambridge, Massachusetts: Harvard University Press, 349 p.

Cushman, J. A., 1938. Cretaceous species of *Gümbelina* and related genera. *Contributions from the Cushman Laboratory for Foraminiferal Research*, 14, 2-28.

Cushman, J. A., 1946. Upper Cretaceous foraminifera of the Gulf coastal region of the United States and adjacent areas. *United States Geological Survey Professional Paper*, 206, 1-241.

Cushman, J. A., 1949. The foraminiferal fauna of the Upper Cretaceous Arkadelphia Marl in Arkansas. *United States Geological Survey Professional Paper*, 221A, 1-17.

Cushman, J. A., 1950. *Foraminifera, their classification and economic use*; 4th edition. Cambridge, Massachusetts: Harvard University Press, 605 p.

Darmoian, S. A., 1975. Planktonic foraminifera from the Upper Cretaceous of southeastern Iraq: Biostratigraphy and systematics of the Heterohelicidae. *Micropaleontology*, 21, 185-214.

D'Hondt, S., Keller, G., 1991. Some patterns of planktic foraminiferal assemblage turnover at the Cretaceous-Tertiary boundary. *Marine Micropaleontology*, 17, 77-118.

D'Hondt, S., Zachos, J. C., 1998. Cretaceous foraminifera and the evolutionary history of planktic photosymbiosis. *Paleobiology*, 24, 512-523.

Djafarov, D. I., Agalarova, D. A., Khalilov, D. M., 1951. *Dictionary of microfauna of the Cretaceous deposits of Azerbaijan*. Baku: Gosudarstvenoe Nauchno-technicheskoe Izdatelstvo Neftianoi i Gorno-toplivnoi Lineraturyi Azerbaijanskoe Otdelenie, 128 p. [in Russian].

Dupeuble, P. A., 1969. Foraminifères planctoniques (Globotruncanidae et Heterohelicidae) du Maastrichtien supérieur en Aquitaine Occidentale. In: *Proceedings of the First International Conference on Planktonic Microfossils,* Geneva 1967 (P. Brönnimann, P. and H. H. Renz, Eds). Leiden: E.J. Brill, 2: 153-161.

Egger, J. G., 1899. Foraminiferen und Ostrakoden aus den Kreidemergeln der Oberbayerischen Alpen. *Abhandlungen der Mathematisch-Physikalischen Klasse der Königlich Bayerischen Akademie der Wissenschaften*, 21, 3-230. [published in 1902].

Eternod Olvera, Y., 1959. Foraminiferos del Cretacico superior de la Cuenca de Tampico-Tuxpan, Mexico. *Boletin de la Asociación Mexicana de Geólogos Petroleros*, 11, 63-134.

Faupl, P., Grün, W., Lauer, G., Maurer, R., Papp, A., Schnabel, W., Sturm, M., 1970. Zur Typisierung des Sieveringer Schichten im Flysch des Wienerwaldes. *Jahrbuch der Geologisches Bundesanstalt*, 113, 73-158.

Funnell, B. M., Friend, J. K., Ramsay, T. S., 1969. Upper Maastrichtian planktonic foraminifera from Galicia Bank, west of Spain. *Palaeontology*, 12, 19-41.

Galloway, J. J., 1933. *A Manual of Foraminifera*. Bloomington, Indiana: The Principia Press, 483 p.

Georgescu, M. D., 1995. Upper Cretaceous Heterohelicidae in the Romanian Western Black Sea offshore. *Revista Española de Micropaleontología*, 27, 91-106.

Georgescu, M. D., 1997. Upper Jurassic-Cretaceous planktonic biofacies succession and the evolution of the western Black Sea Basin. In: *Regional and petroleum geology of the Black Sea and surrounding region* (A. G. Robinson, Ed.). *The American Association of Petroleum Geologists Memoir*, 68, 169-182.

Georgescu, M. D., 2010. Origin, taxonomic revision and evolutionary classification of the late Coniacian-early Campanian (Late Cretaceous) planktic foraminifera with multichamber growth in the adult stage. *Revista Española de Micropaleontología*, 42, 59-118.

Georgescu, M. D., 2011. Iterative evolution, taxonomic revision and evolutionary classification of the praeglobotruncanid planktic foraminifera, Cretaceous (late Albian-Santonian). *Revista Española de Micropaleontología*, 43, 173-207. [published in 2012].

Georgescu, M. D., 2013. Revised evolutionary systematics of the Cretaceous planktic foraminifera described by C. G. Ehrenberg. *Micropaleontology*, 59, 1-49.

Glaessner, M. F., 1936. Die Foraminiferengattungen *Pseudotextularia* und *Amphimorphina*. *Problemyi Paleontologhyi*, 1, 97-130.

Gohrbandt, K. H. A., 1967. The geologic age of the type locality of *Pseudotextularia elegans* (Rzehak). *Micropaleontology*, 13, 68-74.

Govindan, A., 1972. Upper Cretaceous planktonic foraminifera from the Pondicherry are, south India. *Micropaleontology*, 18, 160-193.

Halkyard, E., 1917. The fossil foraminifera of the Blue Marl of Cote des Basques, Biarritz. *Memoirs and Proceedings of the Manchester Literary and Philosophical Society*, 62/6, 1-145. [published in 1918].

Hamilton, E. L., 1953. Upper Cretaceous, Tertiary, and Recent Planktonic Foraminifera from Mid-Pacific flat-topped seamounts. *Journal of Paleontology*, 27, 204-237.

Hanzlikova, E., 1969. The foraminifera of the Frýdek Formation (Senonian). *Sborník Geologických Věd, Paleontologie*, 11, 7-79.

Hanzlikova, E., 1972. Carpathian Upper Cretaceous Foraminiferida of Moravia (Turonian-Maastrichtian). *Rozpravy Ústředního Ústavu Geologického*, 39, 1-160.

Hiltermann, H., Koch, W., 1962. Oberkreide des nördlich Mitteleuropa. In: *Leitfossilien der Mikropaläontologie. Ein Abriss herausgegeben von einem Arbeitskreis deutscher Mikropaläontologen* (H. Bartenstein and others, Eds). Berlin-Nikolassee: Gebrüder Borntraeger, 229-338.

Hofker, J., 1956. Die *Pseudotextularia*-Zone der Bohrung Maasbüll I und ihre Foraminiferen-Fauna. *Paläontologische Zeitschrift*, 30, 59-79.

Hofker, J., 1978. Analysis of a large succession of samples through the upper Maastrichtian and the lower Tertiary of drill hole 47.2, Shatsky Rise, Pacific, Deep Sea Drilling Project. *Journal of Foraminiferal Research*, 8, 46-75.

Howe, R. W., Campbell, R. J., Rexilius, J. P., 2003. Integrated uppermost Campanian-Maastrichtian calcareous nannofossil and foraminiferal biostratigraphic zonation of the northwestern margin of Australia. *Journal of Micropaleontology*, 22, 29-62.

Huber, B. T., 1990. Maestrichtian planktonic foraminifer biostratigraphy of the Maud Rise (Weddell Sea, Antarctica): ODP Leg 113 Holes 689B and 690C. In: *Proceedings of the Ocean Drilling Program, Scientific Results*, Volume 113 (P. F., Barker, J. P., Kennett and others, Eds). College Station: Ocean Drilling Program, 489-513.

Itzhaki, J., 1952. Séries de variabilité de *Pseudotextularia* (Rzehak) d'après la forme du test et ses tendances évolutives. *Comptes Rendus Sommaires de la Société Géologique de France*, 10, 187-189.

Jansen, H., Kroon, D., 1987. Maestrichtian foraminifers from Site 605, Deep Sea Drilling Project Leg 93, northwest Atlantic. In: *Initial Reports of the Deep Sea Drilling Project*, Volume 93 (van Hinte, J. E. and others, Eds). Washington, D.C.: United States Government Printing Office, 93, 555-575.

Kavary, E., Frizzell, D. L., 1963. Upper Cretaceous and Lower Cenozoic oraminifera from west central Iran. *University of Missouri School of Mines and Metallurgy Bulletin*, 102, 1-89.

Keller, G., 1988. Extinction, survivorship and evolution of planktic foraminifera across the Cretaceous/Tertiary boundary at El Kef, Tunisia. *Marine Micropaleontology*, 13, 239-263.

Keller, G., Adate, T., Stinnesbeck, W., Luciani, V., Karoui-Yaakoub, N., and ZaghbibTurki, D., 2002. Paleoecology of the Cretaceous-Tertiary mass extinction in planktonic foraminifera. *Palaeogeography, Palaeoclimatology, Palaeoecology*, 178, 257-297.

Keller, V. M., 1939. Foraminifers of the Upper Cretaceous deposits of the USSR. *Trudy Neftianogo Geologo-razvedochnogo Instituta*, A116, 7-28.

Kikoïne, J., 1948. Les Heterohelicidae du Crétacé supérieur pyrénéen. *Bulletin de la Société Géologique de France*, 18, 15-35.

Klasz, I., de, 1953. Einige neue oder wenig bekannte Foraminiferen aus der helvetischen Oberkreide der bayerischen Alpen südlich Traunstein (Oberbayern). *Geologica Bavarica*, 17, 223-240.

Korchagin, O. A., Upper Campanian-lower Maastrichtian planktonic foraminifers and biostratigraphy of the Moni Formation, Southern Cyprus. *Stratigraphy and Geological Correlation*, 19, 526-544.

Lehmann, R., 1966. Description des Globotruncanidés et Hétérohelicidés d'une faune maestrichtienne du Prérif (Maroc). *Eclogae Geologicae Helvetiae*, 59, 309-317.

Linares-Rodríguez, D., 1977. *Foraminiferos planctonicos del Cretacico superior de las Cordilleras Beticas (sector central)*. Universidad de Málaga, Departamento de Geología, 410 p.

Liu, K., 2007. Sequence stratigraphy and orbital cyclostratigraphy of the Mooreville Chalk (Santonian-Campanian), northeastern Gulf of Mexico area, US. *Cretaceous Research*, 28, 405-418.

Loeblich, A. R. Jr., Tappan, H., 1964. Sarcodina Chiefly "Thecamoebians" and Foraminifera. In: *Treatise on Invertebrate Paleontology. Part C* (R. C. Moore, Ed.). The Geological Society of America and The University of Kansas Press, 900 p.

Loeblich, A. R. Jr., Tappan, H., 1987. *Foraminiferal Genera and Their Classification*. New York: Van Nostrand Reinhold Company, 970 p.

Luciani, V., 1997. Planktonic foraminiferal turnover across the Cretaceous-Tertiary boundary in the Vajont valley (southern Alps, northern Italy). *Cretaceous Research*, 18, 799-821.

Malmgren, B. A., 1991. Biogeographic patterns in terminal Cretaceous planktonic foraminifera from Tethyan and warm transitional waters. *Marine Micropaleontology*, 18, 73-99.

Mancini, E. A., Puckett, T. M., Tew, B. H., 1996. Integrated biostratigraphic and sequence stratigraphic framework for Upper Cretaceous strata of the eastern Gulf Coastal Plain, US. *Cretaceous Research*, 17, 645-669.

Masters, B. A., 1976. Planktic foraminifera from the Upper Cretaceous Selma Group, Alabama. *Journal of Paleontology*, 50, 318-330.

Masters, B. A., 1977. Mesozoic planktonic foraminifera. A world-wide review and analysis. In: *Oceanic Micropaleontology* (A. T. S. Ramsay, Ed.). London-New York-San Francisco: Academic Press, 1, 301-731.

Montanaro Gallitelli, E., 1956. *Bronnimannella, Tappanina* and *Trachelinella*, three new foraminiferal genera from the Upper Cretaceous. *Contributions from the Cushman Foundation for Foraminiferal Research*, 7, 35-39.

Montanaro Gallitelli, E., 1957. A revision of the foraminiferal family Heterohelicidae. In: *Studies in foraminifera* (A. R. Jr. Loeblich, Ed.). Washington, D.C.: *United States National Museum History Bulletin*, 215, 133-154.

Mukhopadhyay, S. K., 2012. Morphogroups and small sized tests in *Pseudotextularia elegans* (Rzehak) from the Late Maastrichtian succession of Meghalaya, India as indicators of biotic response to paleoenvironmental stress. *Journal of Asian Earth Sciences*, 48, 111-124.

Nash, S., 1981. A neotype for the Cretaceous genus *Pseudotextularia* Rzehak, 1891. *Journal of Foraminiferal Research*, 11, 70-75.

Neagu, T., 1970. Micropaleontological and stratigraphical study of the Upper Cretaceous deposits between the upper valleys of the Buzău and Rîul Negru Rivers. *Memoriile Institului de Geologie și Geofizică*, 12, 1-109.

Nederbragt, A. J., 1989a. Chamber proliferation in the Cretaceous planktonic foraminifera Heterohelicidae. *Journal of Foraminiferal Research*, 19, 105-114.

Nedebragt, A. J., 1989b. Maastrichtian Heterohelicidae (planktic foraminifera) from the West North Atlantic. *Journal of Micropaleontology*, 8, 183-206.

Nedebragt, A. J., 1991. Late Cretaceous biostratigraphy and development of Heterohelicidae (planktic foraminifera). *Micropaleontology*, 37, 329-372.

Noth, R., 1951. Foraminiferen aus Unter- und Oberkreide des Österreichischen anteils an Flysch, Helvetikum und Vorlandvorkommen. *Jahrbuch der Geologischen Bundesanstaldt*, Sonderband, 3, 1-91.

Obaidalla, N. A., 2005. Complete Cretaceous/Paleogene (K/P) boundary section at Wadi Nukhul, southwestern Sinai, Egypt: inference from planktic foraminiferal biostratigraphy. *Revue de Paléobiologie*, 24, 201-224.

Olsson, R. K., 1960. Foraminifera of Latest Cretaceous and Earliest Tertiary age in the New Jersey coastal plain. *Journal of Paleontology*, 34, 1-58.

Omaña, L., 2006. Late Cretaceous (Maastrichtian) foraminiferal assemblage from the inoceramid beds, Ocozocoautle Formation, central Chiapas, SE Mexico. *Revista Mexicana de Ciencias Geológicas*, 23, 125-132.

Pandey, J., 1980. Cretaceous foraminifera of Um Sohryngkew River section, Meghalaya. *Journal of the Palaeontological Society of India*, 25, 53-74. [published in 1981].

Pérez-Rodriguez, I., Lees, J. A., Larrasoaña, J. C., Arz, J. A., Aremillas, I., 2012. Planktonic foraminiferal and calcareous nanofossil biostratigraphy and magnetostratigraphy of the uppermost Campanian and Maastrichtian at Zumaia, northern Spain. *Cretaceous Research*, 37, 100-126.

Perlmutter, N. M., Todd, R., 1965. Correlation and foraminifera of the Monmouth Group (Upper Cretaceous) Long Island, New York. *United States Geological Survey Professional Paper*, 483-I, 1-21.

Peryt, D., 1980. Planktic foraminifera zonation of the Upper Cretaceous in the middle Vistula River Valley, Poland. *Palaeontologia Polonica*, 41, 3-101.

Peryt, D., 1988. Maastrichtian extinctions of planktonic foraminifera in central and eastern Poland. *Revista Española de Paleontología*, 3, 105-115.

Pessagno, E. A. Jr., 1967. Upper Cretaceous planktonic foraminifera from the Western Gulf coastal plain. *Palaeontographica Americana*, 5(37), 243-445.

Pessagno, E. A. Jr., Brown, W. R., 1969. The microreticulation and sieve plates of *Racemiguembelina fructicosa* (Egger). *Micropaleontology*, 15, 116-117.

Premoli Silva, I., McNulty, C. L., 1984. Planktonic foraminifers and calpionellids from Gulf of Mexico sites, Deep Sea Drilling Project Leg 77. In: *Initial Reports of the Deep Sea Drilling Project,* Volume 77 (Buffler, R. T. and others, Eds.). Washington, D.C.: United States Government Printing Office, 77, 547-548.

Pringgoprawiro, H., 1965. Some significant Upper Cretaceous from Groisbach, Morzger Hügel, and Michelstetten, Austria. *Institute of Technology Bandung*, 60, 21-65.

Quilty, P. G., 1992. Upper Cretaceous planktonic foraminifers and biostratigraphy, Leg 120, southern Kerguelen Plateau. In: *Proceedings of the Ocean Drilling Program, Scientific Results*, Volume 120 (S. W. Jr., Wise, Ed.). College Station: Ocean Drilling Program, 371-392.

Rzehak, A., 1891. Die Foraminiferenfauna der alttertiären Ablagerungen von Bruderndorf in Nieder-Osterreich, mit Berüchsichtigung des angeblichen Kreidevorkommens von Leitzersdorf. *Annalen des K.K. Naturhistorischen Hofmuseums*, 10, 213-230.

Rzehak, A., 1895. Ueber einige merkwürdige Foraminiferen aus dem österreichischen Tertiär. *Annalen des K.K. Naturhistorischen Hofmuseums*, 6, 1-12.

Robaszynski, F., González Donoso, J.-M., Linares, D., Amédro, F., Caron, M., Dupuis, C., Dhont, A. V., Gartner, S., 1998. The Upper Cretaceous of the Kalaat Senan region, Central Tunisia. Integrated litho-biostratigraphy based on ammonites, planktonic foraminifera and nannofossil zones from upper Turonian to Maastrichtian. *Bulletin du Centre de Recherche, Exploration et Production, Elf-Aquitaine* 22, 359-489.

Saavedra, J. L. 1965. La evolución de los Globigerináceos. *Boletín de la Real Sociedad Española de Historia Natural*, 63, 317-349.

Sacal, V., Debourle, A., 1957. Foraminifères d'Aquitaine 2e partie. Peneroplidae a Victoriellidae. *Mémoires de la Société Géologique de France*, 78, 1-87.

Said, R., Kenawy, A., 1956. Upper Cretaceous and Lower Tertiary foraminifera from northern Sinai, Egypt. *Micropaleontology*, 2, 105-173.

Said, R., Kerdany, M. T., 1961. The geology and micropaleontology of the Farfara Oasis, Egypt. *Micropaleontology*, 7, 317-336.

Said, R., Sabry, H., 1964, Planktonic foraminifera from the type locality of the Esna Shale in Egypt. *Micropaleontology*, 10, 375-395.

Saito, T., Van Donk, J., 1974. Oxygen and carbon isotope measurements of Late Cretaceous and Early Tertiary foraminifera. *Micropaleontology*, 20, 152-177.

Salaj, J., 1980. *Microbiostratigraphie du Crétacé et du Paléogène de la Tunisie septentrionale et orientale (Hypostratotypes tunisiens)*. Bratislava: Geologický Ústav Dionýza Štúra, 238 p.

Salaj, J., 1983. Quelques problèmes taxinomiques concernant les foraminifères planctiques et la zonation du Sénonien supérieur d'El Kef. *Geologický Zborník*, 34, 187-212.

Salaj, J., Samuel, O., 1966. *Foraminifera der Westkarpaten-Kreide*. Bratislava: Geologický Ústav Dionýza Štúra, 291 p.

Seiglie, G. A., 1958. Notas sobre algunas especies de Heterohelicidae del Cretacico superior de Cuba. *Boletín de la Asociatión Mexicana de Geólogos Petroleros*, 11, 51-62.

Sellier de Civrieux, J. M., 1952. Estudio de la microfauna de la seccion–tipo del Miembro Socuy de la Formacion Colon Distrito Mara, Estado Zulia. *Ministerio de Minas e Hidrocarburos Direccion de Geologia*, 2, 231-310.

Shipboard Scientific Party, 1972. Site 111. In: *Initial Reports of the Deep Sea Drilling Project*, Volume 12 (Laughton, A. E. and others, Eds). Washington, D.C.: United States Government Printing Office, 12, 33-159.

Skinner, H. C., 1962. Arkadelphia foraminiferida. *Tulane Studies in Geology*, 1, 1-67.

Sliter, W. V., 1968. Upper Cretaceous foraminifera from Southern California and northwestern Baja California, Mexico. *The University of Kansas Paleontological Contributions*, 49, 1-141.

Smith, C. C., 1978. Taxonomic comments on some Upper Cretaceous planktonic foraminiferal genera. *Journal of Foraminiferal Research*, 8, 314-318.

Smith, C. C., Pessagno, E. A. Jr., 1973. Planktonic foraminifera and stratigraphy of the Corsicana Formation (Maestrichtian), north-central Texas. *Cushman Foundation for Foraminiferal Research, Special Publications*, 13, 5-68.

Thompson, L. B., 1991. Late Santonian to early Maastrichtian planktonic foraminiferal biostratigraphy and zonation of northeast Texas. *Micropaleontology Special Publication*, 5, 9-66.

Vassilenko, V. P., 1961. Upper Cretaceous foraminifera of the Mangyshlak Peninsula (descriptions, phylogenetical schemes for some groups and stratigraphic analysis). *Trudy VNIGRI*, 171, 1-487.

Villars, F., 1988. Progradation de la Formation de Wang dans les chaînes subalpines septentrionales (Alpes occidentales, France) au Maastrichtien supérieur: biostratigraphie et milieu de dépôt. *Eclogae Geologicae Helvetiae*, 81, 669-687.

Voorwijk, G. H., 1937. Foraminifera from the Upper Cretaceous of Habana, Cuba. *Proceedings of the Koninklijke Akademie van Wetenschappen te Amsterdam*, 40, 190-198.

Webb, P. N., 1973. Upper Cretaceous-Paleocene foraminifera from Site 208 (Lord Howe Rise, Tasman Sea), DSDP Leg 21. In: *Initial Reports of the Deep Sea Drilling Project* Volume 21 (R. E. Burns and others, Eds). Washington, D.C.: United States Government Printing Office, 541-573.

Weiss, W., 1983. Heterohelicidae (seriale planktonische Foraminiferen) der tethyalen Oberkreide (Santon bis Maastricht). *Geologisches Jahrbuch*, A72, 3-93.

Wessem, A., Van, 1943. *Geology and paleontology of Central Camaguey, Cuba*. Utrecht-Amsterdam: Drukkerij J. Van Boekhoven, 88 p.

White, M. P., 1929. Some index foraminifera of the Tampico Embayment area of Mexico. Part III. *Journal of Paleontology*, 3, 30-57.

Wille-Janoschek, U., 1966. Stratigraphie und Tektonik der Schichten der Oberkreide und des Alttertiärs im Raume von Gosau und Abtenau (Salzburg). *Jahrbuch der Geologisches Bundesanstalt*, 109, 91-172.

Witwika, E., 1958. Cretaceous stratigraphic micropaleontology from a borehole from Chełm. *Biuletyn Instytut Geologiczny*, 121, 177-232. [in Polish].

Wright, C. A., Apthorpe, M., 1976. Planktonic foraminiferids from the Maastrichtian of the northwest shelf, Western Australia. *Journal of Foraminiferal Research*, 6, 22-241.

Yassini, I., 1979. Maastrichtian-lower Eocene biostratigraphy and the planktic foraminiferal biozonation of Jordan. *Revista Española de Micropaleortología*, 11, 5-57.

Zapeda, M., 1998. Planktonic foraminiferal diversity, equitability and biostratigraphy of the uppermost Campanian-Maastrichtian, ODP Leg 122, Hole 762C, Exmouth Plateau, NW Australia, eastern Indian Ocean. *Cretaceous Research*, 19, 117-152.

PART 2: EVOLUTIONARY CLASSIFICATION FUNDAMENTS AND THE DEVELOPMENT OF A NEW ENGLISH-BASED NOMENCLATURE

In: Evolutionary Classification ...
Editors: M. Dan Georgescu and C. M. Henderson

ISBN: 978-1-63321-959-5
© 2015 Nova Science Publishers, Inc.

Chapter 7

EVOLUTIONARY CLASSIFICATION AND NOMENCLATURE OF THE CRETACEOUS PLANKTIC FORAMINIFERA WITH THE CHAMBERS ALTERNATELY ADDED WITH RESPECT TO THE TEST GROWTH AXIS

M. Dan Georgescu [*]

Department of Geosciences, University of Calgary,
Calgary, Alberta, Canada

ABSTRACT

Developments in the evolutionary classification of the Cretaceous planktic foraminifera with the chambers alternately added with respect to the test growth axis show that the lineage is the fundamental unit in this classification system. One lineage presents one or more stages or morphological relative stability, which are largely associated with the concept of composite paleontological species. The evolutionary nomenclature, which uses the English language, is developed for this classification system.

Keywords: Foraminifera, planktic, Cretaceous, evolutionary classification, evolutionary nomenclature

INTRODUCTION

A new level of understanding of the test morphology and stratigraphic distribution of the Cretaceous planktic foraminifera was achieved when the observation data were acquired with

[*] Corresponding author: M. Dan Georgescu. Department of Geosciences, University of Calgary, 2500 University Drive NW, Calgary, Alberta T2N 1N4, Canada. E-mail: dgeorge@ucalgary.ca.

the aid of the Scanning Electron Microscope (SEM) from throughout the stratigraphic range of various taxa. Therefore, it became possible to define the evolutionary continuum and group the species into lineages of various kinds (directional, branched, iterative and condensed), and the newly defined open systems at species and lineage levels led to a dynamic understanding of the taxonomic units used in the group classification. The process of definition of lineages is most advanced in the Cretaceous planktics in which at least one ontogenetic stage consists of chambers alternately added with respect to the test growth axis, in which 31 directional, branched and condensed lineages were revised between 2007 and 2014. This time interval represents a true "ultrastructure revolution" and it is characterized by the highest influx of new data in the group study history that began with the study by Ehrenberg (1838).

Throughout the "ultrastructure revolution" various concepts in classification were challenged and improved leading to the development of evolutionary classification system that provides a more accurate and comprehensive understanding of the fossil record when compared to the typological classification. The study of the lineages recognized within this group leads to the conclusion that the lineages rather than the species are fundamental units of organization in the evolutionary classification and what were considered species in the past largely are stages of morphological relative stability within a lineage evolution. This represents a shift at paradigm level in understanding the "tree of life" architecture and evolution and removes a contradiction existing till now in the evolutionary classification system. The contradiction is that lineages, which are dynamic units, were considered consisting of paleontological species, which are rather static and artificial units, and evolution cannot be accurately presented in classification and taxonomy as a succession of static units.

One of the effects of the "ultrastructure revolution" was the significant increase in number of lineages when compared to the genera in the typological classification and recognized stages of morphological relative stability when compared to the species respectively. Although in the initial stage of development of the evolutionary classification all the units of classification were named in Latin it is now possible to develop a new and more accurate nomenclatural system to accommodate such advances. The new nomenclature, which is herein named evolutionary nomenclature, is in the English language.

MATERIAL PROVENANCE

The fossil material on which this study is based comes from a variety of locations situated worldwide and seven collections that cover most of the original material used in the past to describe new taxonomical units in both evolutionary and typological classifications. This material was studied upon case with the aid of the scanning electron microscope (SEM), optical stereomicroscope and transmitted light microscope; the accent was put on the most accurate method of observation in all situations. Most of the assemblages were collected from the Deep Sea Drilling Project (DSDP) and Ocean Drilling Program (ODP) sites and holes but additional specimens from various other locations were included in the dataset used for this study.

Collection Material

- EHRENBERG COLLECTION is deposited at the Museum of Natural Sciences in Berlin; it was studied during two visits in 2010 and 2011. The collection includes the original material published by C.G. Ehrenberg between 1838 and 1854 and the specimens on which the work of 1856 was based could not be found and are most likely lost. This collection was studied in the past by Cushman (1927a, 1927b, 1938), Barr (1968), Masters (1980) and Georgescu (2012a, 2013a). The specimens collected from the Santonian-Maastrichtian sediments reported by Ehrenberg are preserved in Canada balsam slides; additional material came from 46 samples but only 19 of them yielded fossil specimens of Cretaceous planktics with chambers alternately added with respect to the test growth axis. The representatives of the group were found in the material from Lithuania (Puszkary), France (Meudon), Denmark (Moen Island), England (Gravesend), Germany (Rügen Island), Russia (Wolsk) and the US (South Dakota, Mississippi). The specimens in this collection are in general well preserved although in the case of those mounted in Canada balsam the possibilities of study are significantly limited.
- JACOB WHITMAN BAILEY COLLECTION is deposited at the Farlow Herbarium (Harvard University); the foraminiferal tests are mounted in Canada balsam on four glass slides. A detailed description of this material obtained through a temporary loan in 2011 was given by Georgescu (2013b). The fossil material is of Santonian age and proved of paramount importance in the assessment of the degradation of the original slides of the Ehrenberg Collection. The foraminifera in this collection were collected from South Dakota.
- CUSHMAN COLLECTION of the National Museum of Natural History, Smithsonian Institution in Washington, D.C. was examined during a 2005-2006 stage. The most part of the collection is represented by the original specimens of J.A. Cushman, which was published by him in the period 1926-1946. Additional specimens of relevance for this study were sent by F. Brotzen, N. Brown Jr., P. Brönnimann, I. de Klasz, B. Masters, A. Nederbragt and H.N. Tappan among others. The specimens are detached and in general in good state of preservation. The Cushman Collection includes the largest number of type specimens of Cretaceous planktic foraminifera. Some ESEM photographs of these specimens taken by the museum personnel and generously put at my disposal were used in a number of high resolution taxonomic studies (Georgescu 2007a, 2007b, 2008a, 2008b, 2009a, 2009b; Georgescu and Abramovich 2008a, 2008b; Georgescu and Huber 2009). Some of the designated type specimens of the new species described during the first three years of the "ultrastructure revolution" are deposited in this collection.
- LOEBLICH AND TAPPAN TOPOTYPE COLLECTION is deposited at the National Museum of Natural History, Smithsonian Institution in Washington, D.C. and consists of topotype material collected by A.R. Loeblich Jr. and H.N. Tappan or received by them from others specialists in the field (N. Brown Jr., P. Brönnimann, J. Kikoïne, I. de Klasz, E.A. Pessagno Jr., V. Pokorný, etc.). SEM illustrated specimens of this collection were used in several articles published during the "ultrastructure revolution" (Georgescu 2013; Georgescu and Almogi Labin 2008; Georgescu, Saupe and Huber 2008; Georgescu and Huber 2009). Most of the collection consists of

benthic foraminifera. This collection was examined during the 2005-2006 stage and a thorough study of it would be of most interest for foraminiferologists.

- VAN MORKHOVEN COLLECTION is deposited at the National Museum of Natural History, Smithsonian Institution in Washington, D.C. and consists of specimens from localities in the US and adjacent regions. Of particular interest are the samples from the Santonian-Campanian sediments of the Eureka 67-128 well (Gulf of Mexico). This collection was studied during the 2005-2006 stage at the National Museum of Natural History, Smithsonian Institution, Washington, D.C. Due to its general good preservation this material was used in several studies during the "ultrastructure revolution" (Georgescu 2010, 2014a; Georgescu and Abramovich 2008a, 2009a; Georgescu and Almogi Labin 2008).

- OCEAN MICROPALEONTOLOGY COLLECTION is hosted at the National Museum of Natural History, Smithsonian Institution in Washington, D.C. and consists of spot samples from the earlier sites and holes drilled under the auspices of the Deep Sea Drilling Program. These specimens were studied during the 2005-2006 stage at the National Museum of Natural History, Smithsonian Institution, Washington, D.C. The material from this collection was used practically in all the studies of the "ultrastructure revolution".

- EICHER AND WORSTELL COLLECTION is deposited within the Micropaleontology Collection of the University of Colorado at Boulder. From this collection there were studied the specimens reported by Eicher and Worstell (1970a, 1970b) from the US (Colorado, Kansas, Nebraska, South Dakota, Wyoming). This material was obtained through a temporary loan in 2011and was published by Georgescu (2013c).

- MCGUGAN COLLECTION is deposited at the University of Calgary and includes especially specimens from the British Chalk (England, Northern Ireland) of Campanian-Maastrichtian age and Demopolis Chalk, Ripley Formation and Prairie Bluff Chalk of the Southern US (Alabama). A part of the American material was used by Georgescu (2013a) as support specimens in the study of the foraminifera of the Ehrenberg Collection.

DSDP/ODP Sites and Holes

- DSDP LEG 10 SITE 95 is situated in the Gulf of Mexico (Yucatan outer shelf); geographic coordinates: 24° 09.00' N and 86° 23.85' W; the uppermost drilled sediments are of late Pleistocene age and a succession of more or less consolidated white chalks to which a Santonian-early Campanian age was assigned was recorded between Cores 13 and 17 (Shipboard Scientific Part 1973). The planktic foraminiferal taxa collected from these sediments were initially reported by McNeely (1973). Nederbragt (1991) revised the age of this succession and considered it of Coniacian-Santonian age. A new revision of the planktic foraminiferal zonation was made by Georgescu (2010) who assigned them a late Turonian-early Campanian age; this biostratigraphical framework is used herein.

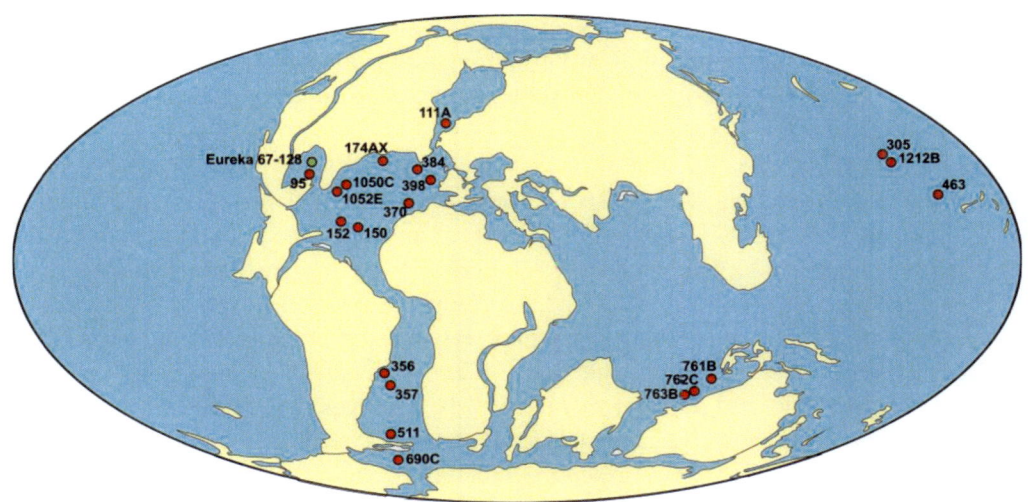

Figure 1. Location of the Deep Sea Drilling Project (DSDP) Sites and Ocean Drilling Program (ODP) Holes, which yielded the material for this study (red-filled circles); the Eureka 67-128 well (Gulf of Mexico) is given in green-filled circle. Base map is after Hay and others (1999).

- DSDP Leg 12 Hole 111A was drilled in the Northern Atlantic Ocean (Orphan Knoll, east of Newfoundland); geographic coordinates: 50° 25.57' N and 46° 22.05' W; samples started to be collected at a depth of 105 m in Pleistocene sediments and last cored segment (Core 11) yielded white and yellow nannofossil chalks, which were initially assigned to the Maastrichtian (Pessagno and Longoria in Shipboard Scientific Party 1972). Subsequent studies indicated that the lower part of the Maastrichtian is probably absent (Georgescu, Saupe and Huber 2008). Herein newly reported occurrences of rare tests of *Abathomphalus mayaroensis* in the lowermost section of Core 11 demonstrate that the whole sediment succession should be assigned a late Maastrichtian age. A complete set of samples from Core 11 was studied in the Ocean Micropaleontology Collection and this coverage was further improved by an additional sample request that reduced the sampling step to one half.

- DSDP Leg 15 Site 150 is situated in the Caribbean Region (Venezuelan Basin); geographic coordinates: 14° 30.69' N and 69° 21.35' W; the uppermost sediments at this site were cored starting at a depth of 49 meters and Cretaceous sediments were recovered in Cores 9 and 10 (Shipboard Scientific Party 1973). The age of the Cretaceous sediments given in the initial reports was Turonian-Santonian by Premoli Silva and Bolli (1973) with an addition by Pessagno and Longoria (1973). This biozonation was successively reviewed by Georgescu, Quinney and Anderson (2011) and Georgescu (2012b); the last review is used herein. The succession of planktic foraminiferal assemblages was studied in the spot samples from the Ocean Micropaleontology Collection and one additional complete sample request.

- DSDP Leg 15 Site 152 was drilled in the Caribbean Region (Nicaragua Rise); geographic coordinates: 15° 52.72' N and 74° 36.47' W; coring began at a depth of 153 meters with sediments of early Eocene age and the Cretaceous succession was encountered between Cores 10 and 24 (Shipboard Scientific Party 1973). The planktic foraminiferal biostratigraphy was given in the project Initial Reports by

Premoli Silva and Bolli (1973). Spot samples deposited in the Ocean Micropaleontology Collection from the middle Campanian-lower Maastrichtian stratigraphic interval were studied but not included in any of the studies of the "ultrastructure revolution".

- DSDP LEG 32 SITE 305 was drilled in the Central Pacific Ocean (Shatsky Rise); geographic coordinates: 32° 00.13' N and 157° 51.00' E; the first core collected Quaternary sediments at a depth of 1.2 metres and a sediment succession consisting mostly of white and yellow nannofossil chalks of late Albian-Maastrichtian age was encountered between Cores 15 and 48 (Shipboard Scientific Party 1975). Caron (1975) reported the planktic foraminiferal biostratigraphy for the Coniacian-Maastrichtian stratigraphic interval. The Campanian-Maastrichtian planktic foraminiferal biostratigraphy was reviewed by Georgescu (2012b) and is followed herein; this review was based on a sample request between Core 15 and Core 28.

- DSDP LEG 39 SITE 356 was drilled in the South Atlantic Ocean (São Paulo Plateau); geographic coordinates: 28° 17.22' S and 41° 05.28' W; coring began at a depth of 9.5 meters in Pliocene sediments and the Cretaceous ones were recorded in the interval between Cores 29 and 44 (Shipboard Scientific Part 1977a). The initial planktic foraminiferal biozonation was given by Premoli Silva and Boersma (1977) and then reviewed by Nederbragt (1991); the reviewed version is used herein. Spot samples from the Santonian-Maastrichtian stratigraphic interval were studied in the Ocean Micropaleontology Collection and the well-preserved foraminiferal tests yielded by these samples were used by Georgescu and Abramovich (2008a) and Georgescu (2010) for high resolution taxonomic studies.

- DSDP LEG 39 SITE 357 was drilled in the South Atlantic Ocean (Rio Grande Plateau); geographic coordinates: 30° 00.25' N and 35° 33.59' W; the uppermost sediments are of early Pleistocene age and the Cretaceous sediments were drilled between Cores 31 and 51 (Shipboard Scientific Part 1977b). The first planktic foraminiferal biozonation was given by Premoli Silva and Boersma (1977), and it was reviewed by Nederbragt (1991). Georgescu (2010) used the tests from the spot samples of this site in the Ocean Micropaleontology Collection in the taxonomic review of the Santonian planktics with chamber proliferation in the adult stage.

- DSDP LEG 41 SITE 370 is situated in the North Atlantic Ocean (Deep Basin off Morocco); geographic coordinates: 32° 50.20' N and 10° 46.60' W; the uppermost sediments at a depth of 0.9 meters were assigned a Pleistocene age and the succession between Core 20 and Core 50 was assigned to the Cretaceous and of it the upper part (Cores 20-26) was considered of late Albian-early Cenomanian age (Shipboard Scientific Party 1978). Pflaumann and Krasheninnikov (1978) considered that only the interval between Core 20 and Core 24 is of late Albian in age. The planktic foraminiferal biostratigraphy was reviewed by Georgescu (2012b) who conferred a late Albian-early Cenomanian age to the stratigraphic interval between Core 20 and Core 24, which from a lithological point of view consists of dark green and grey claystones with thin levels of argillaceous limestones. The representatives of the study group are rare in this stratigraphic interval. Notably, the earliest representatives of the *Praeplanctonia* directional lineage were recognized in this section.

- DSDP LEG 43 SITE 384 is situated in the North Atlantic Ocean (*J*-Anomaly Ridge); geographic coordinates: 40° 21.65' N and 51° 39.80' W; the uppermost sediments have a middle Eocene age and started to be drilled at a depth of 50.8 meters; Cretaceous sediments occur in the interval between Cores 13 and 22 of which the interval between Cores 13 and 15 yielded sediments of Maastrichtian age (Shipboard Scientific Party 1979a). The foraminiferal content was presented in the Initial Reports by McNulty (1979). Spot samples of the Maastrichtian sediments deposited in the Van Morkhoven Collection yielded well-preserved specimens that helped in the clarification of the taxonomic status of planktic foraminifera with early planispiral coil and adult stage with globular chambers alternately added with respect to the test growth axis (Georgescu 2013a).

- DSDP LEG 47B SITE 398 is located in the North Atlantic Ocean (southern part of the Vigo Seamount); geographic coordinates: 40° 57.60' N and 10° 43.10' W; the uppermost sediments started to be cored at a depth of 15.5 m and are Pleistocene in age and the Cretaceous (Hauterivian-Maastrichtian) sediment succession was recorded in the interval between Cores 41 and 138 (Shipboard Scientific Party 1979b). The planktic foraminiferal initial report was made by Sigal (1979). A complete sample request that covered the interval between Cores 57 and 69 yielded foraminiferal assemblages of middle Albian-early Cenomanian age (Georgescu 2012b); the representatives of the study group occur sporadically in the terminal Albian and lowermost part of the lower Cenomanian sediments.

- DSDP LEG 62 SITE 463 was drilled in the Central Pacific Ocean (western Mid-Pacific Mountains); geographic coordinates: 21° 21.01' N and 174° 40.07' E; the uppermost sediments drilled are of Pleistocene age and the Cretaceous (Barremian-Maastrichtian) succession was recorded between Cores 7 and 92 (Shipboard Scientific Party 1981). The first planktic foraminiferal zonation for the Cretaceous was given by Boersma (1981); Georgescu (2012b) reviewed this zonation based on a sample request from Core 7 to 53, which covers the upper Albian-lower Maastrichtian stratigraphic interval. The well-preserved material of this site was used in high resolution taxonomic studies by Georgescu (2007a, 2007b, 2009a, 2011a, 2013a, 2013c), Georgescu, Saupe and Huber (2008), Georgescu and Huber (2009), Georgescu, Quinney and Anderson (2011) and Georgescu and others (2013).

- DSDP LEG 71 SITE 511 was drilled in the South Atlantic Ocean (Falkland Plateau); geographic coordinates: 51° 00.28' S and 46° 58.30' W; the uppermost drilled sediments are of Quaternary age and the Cretaceous succession (Barremian-Maastrichtian) is recorded between the Cores 23 and 60 (Shipboard Scientific Party 1983). The first report of Cretaceous planktic foraminifera together with the biozonation based on the representatives of this group was given by Krasheninnikov and Basov (1983); it was reviewed by Huber (1992) and Huber, Hodell and Hamilton (1995) and the reviewed framework is used herein. The material from the upper Turonian-Campanian (interval between Cores 22 and 47) was used in high resolution taxonomic studies during the "ultrastructure revolution" by Georgescu (2009a, 2013a, 2013c), Georgescu and Abramovich (2008a) and Georgescu and Huber (2009).

- ODP Leg 113 Hole 690C is situated in the South Atlantic Ocean (southwestern flank of the Maud Rise); geographic coordinates: 65° 09.62' S and 1° 12.28' E; the uppermost sediments started to be drilled at a depth of 2914.3 meters and the Cretaceous sediment succession was recorded between Cores 15 and 22 (Shipboard Scientific Party 1988). The first report of the Cretaceous foraminiferal assemblages was given by Huber (1990); this sample set is herein used in the high resolution taxonomic study of the planktic foraminifera with chambers alternately added with respect to the test growth axis.
- ODP Leg 122 Hole 761B was drilled in the East Indian Ocean (Wombat Plateau); geographic coordinates: 16° 44.23' S and 115° 32.10' E; the uppermost sediments drilled at this site are of Quaternary age and the Cretaceous (Berriasian-Maastrichtian) sediment succession was recorded in the interval between Cores 21 and 30 (Shipboard Scientific Party 1990). The planktic foraminifera and first biozonation based on the representatives of this group were presented by Wonders (1992). A sample request from the white and yellowish nannofossil chalks of late Santonian-Maastrichtian age of the stratigraphic interval between Cores 21 and 27 yielded in general well-preserved representatives of the study group, which were used in high resolution taxonomic studies by Georgescu, Saupe and Huber (2008) and Georgescu (2013a).
- ODP Leg 122 Hole 762C is situated in the East Indian Ocean (Exmouth Plateau); geographic coordinates: 19° 53.23' S and 112° 15.24' E; the uppermost sediments (late Eocene-early Oligocene in age) started to be drilled at a depth of 170 metres and the Cretaceous (Berriasian-Maastrichtian) sediments were encountered between the Cores 43 and 91 (Shipboard Scientific Party 1990). The first report of planktic foraminifera and planktic foraminiferal biostratigraphy was made by Wonders (1992). A sample request covering the upper Albian-Maastrichtian sediments consisting of clayey chalks and claystones in the upper Albian-Cenomanian interval and white and yellowish chalks in the Turonian-Maastrichtian stratigraphic interval yielded the representatives of the study group; these tests were used in taxonomic studies of high resolution by Georgescu (2013a) and Georgescu and others (2013).
- ODP Leg 122 Hole 763B was drilled in the East Indian Ocean (Exmouth Plateau); geographic coordinates: 20° 35.21' S and 112° 12.51' E; uppermost drilled sediments are of late Eocene age and started to be drilled at a depth of 190 meters and the Cretaceous (Berriasian-Campanian) sediment succession was recorded between the Cores 8 and 46 (Shipboard Scientific Party 1990). The sediments at this location present resemblances with the coeval ones from the nearby ODP Hole 762C. The first report of planktic foraminifera from this location was made by Wonders (1992). A sample request from the interval between Cores 8 and 30 yielded well-preserved foraminiferal assemblages of late Albian-Campanian age; Georgescu (2012b) revised the planktic foraminiferal biozonation and used the material from these samples in one high resolution taxonomic study.
- ODP Leg 171B Hole 1050C is situated in the North Atlantic Ocean (Blake Nose); geographic coordinates: 30° 05.99' N and 76° 14.10' W; the uppermost sediments that started to be drilled at a depth of 327.1 metres are of late Paleocene age and the succession of Cretaceous (upper Albian-Maastrichtian) sediments was reported

between Cores 10 and 29 (Shipboard Scientific Party 1998). The planktic foraminiferal biozonation was given for the first time by Bellier and others (2000). The representatives of the planktic foraminifera with chambers alternately added with respect to the test growth axis are in general well preserved and were used in the high resolution taxonomic studies by Georgescu (2009b), Georgescu and Huber (2009) and Georgescu, Burke and Heikkinen (2012).

- ODP Leg 171B Hole 1052E was drilled in the North Atlantic Ocean (Blake Nose); geographic coordinates: 29° 57.07' N and 76° 37.60' W; the uppermost sediments are of middle Eocene age and started to be drilled at a depth of 140 meters and the Cretaceous (upper Albian-Maastrichtian) sediment succession was recorded in the interval between Cores 18 and 58 (Shipboard Scientific Party 1998). The late Albian and early Cenomanian planktics with chambers alternately added with respect to the test growth axis were used in high resolution taxonomic studies by Georgescu (2009b) and Georgescu and Huber (2009).

- ODP Leg 174AX Bass River Site is situated in the New Jersey coastal plain; geographic coordinates: 39° 36' 42" N and 74° 26' 12" W; the Cretaceous (upper Cenomanian-Maastrichtian) sediment succession was recorded in the stratigraphic interval 384.1-596.3 meters (Miller and others 1998). The first planktic foraminiferal biozonation was given by Olsson in Miller and others (1998). The well-preserved tests of planktic foraminifera with chambers alternately added with respect to the test growth axis were used in high resolution taxonomic studies by Georgescu (2009a, 2010, 2013a, 2013c), Georgescu and Abramovich (2008a) and Georgescu and others (2013).

Additional Material

Five additional sources of fossil material are considered.

- Several hundred samples from more than sixty prospection, exploration and development wells drilled in the Romanian western Black Sea offshore; the specimens are of middle Cenomanian-middle Maastrichtian age and were included in the studies by Georgescu (1995, 1997).

- Twenty samples from the Maastrichtian Kemp Clay from the wells Mullinax-1 and Mullinax-3 of Texas; the samples were put to my disposal by Dr Sigal Abramovich (Ben Gurion University of the Negev).

- One sample from the Kemp Clay taken from the surroundings of the Limestone City (Texas) and deposited in the Huber Collection of the National Museum of Natural History, Smithsonian Institution, Washington, D.C.

- One sample from the Santonian Gingin Chalk of Western Australia from the type locality of the species *Costellagerina pilula* and *C. bulbosa* described by Belford (1960) and deposited in the Huber Collection of the National Museum of Natural History, Smithsonian Institution, Washington, D.C.

- Picked specimens from the Maastrichtian white chalks from the ODP Leg 198 Hole 1212B drilled in the Central Pacific Ocean (Shatsky Rise); the specimens were put at my disposal by Dr Sigal Abramovich (Ben Gurion University of the Negev) and were used in the high resolution taxonomic study by Georgescu and Abramovich (2008b) in the description of *Lipsonia lipsonae*.

TEST MORPHOLOGICAL FEATURES

The terminology associated with the typological classification in the case of the Cretaceous representatives of the planktic foraminifera with chambers alternately with respect to the growth test axis was based on few characters and remained relatively stable over a long period of time. Terms and expressions such as chamber, suture, biserial arrangement, early planispiral stage, adult multiserial stage, chamber backward extension, main aperture, supplementary apertures, costate ornamentation, porosity and pore mounds were defined and used when the observations on the test morphology were made exclusively in transmitted light in mica slides or thin section or with the classical optical stereomicroscope in reflected light. The introduction of the scanning electron microscope (SEM) in the late 1960s did not bring a significant increase in understanding the group morphology and evolution; although tests were illustrated with the aid of the SEM the new and more accurate data on test morphology were rarely introduced in the description of the typological units at any level of classification (Schreiber 1979; Weiss 1983).

The "ultrastructure revolution" led to the description of more morphological features that can be used in taxonomy and classification and also led to the re-evaluation of the pre-existing ones; the increased accuracy led to the development of an evolutionary framework for the representatives of the studied grouped and in parallel it was possible a re-evaluation of the previously described and used morphological features. Twenty five morphological features are used herein to describe and classify the lineages of Cretaceous planktic foraminifera with chambers alternately added with respect to the test growth axis: test symmetry/asymmetry, peripheral structures symmetry/asymmetry, chamber arrangement and growth stages, occurrence and degree of development of the early planispiral coil, occurrence and features of the adult stage with multichamber growth, flange/chamber ratio in the adult stage with multichamber growth, chamber shape in lateral view, chamber transversal elongation, sutures and sutural ridges, chamber extensions, periphery shape, peripheral structures, aperture shape, periapertural structures morphology, occurrence of transversal walls, occurrence of the perforate central plate, simple ornamentation patterns, transversally zoned ornamentation, longitudinally zoned ornamentation, costae ornamentation, costae characteristics, occurrence of the pustulose periapertural area, wall characteristics, pore characteristics and proloculus size.

This set of morphological features is herein presented as an open system to which new morphological features or possible new databases with relevance in the group taxonomy and classification can be added in the future. The twenty five morphological features are presented following the order used in general in the description of the lineages and stages of morphological relative stability within one lineage.

TEST SYMMETRY AND ASYMMETRY (Figure 2).

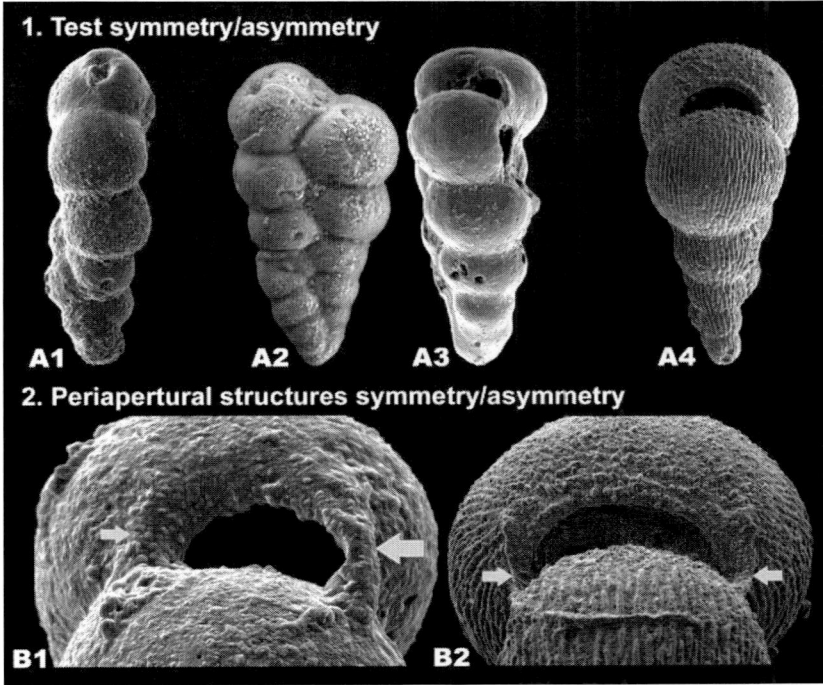

Figure 2. Test symmetry and asymmetry (1) and periapertural structures symmetry/asymmetry (2). A1 Asymmetrical test in the benthic ancestor *Praeplanctonia globifera*. A2 and A3 Asymmetrical tests as seen in lateral view (A2) and edge view (A3). A4 Symmetrical tests. B1 Asymmetrical periapertural structures. B2 Symmetrical periapertural structures. No scale is implied.

The tests of the Cretaceous planktics with chambers alternately added with respect to the test growth axis were traditionally considered symmetrical in edge view. The earliest studies on the representatives of the group have no reference on the test symmetry but the detached specimens figured in lateral and apertural views indicate that they are bilaterally symmetrical (Reuss 1845, 1854; Brown 1853). Tappan (1940) described the earliest species of the group from the late Albian of US mentioning the small size of its tests, which present a length around two times smaller when compared to those of the average Cretaceous tests; seemingly the test size impeded in observing that the earliest tests of this group are asymmetrical. Test asymmetry can be observed especially in lateral view and occasionally in edge view and Georgescu (2009b) who first reported and figured asymmetrical tests of late Albian age demonstrated that the test asymmetry is a vestige of the group's smaller benthic ancestry and occurs only in its earliest representatives.

PERIAPERTURAL STRUCTURES SYMMETRY AND ASYMMETRY (Figure 2). Periapertural structures were mentioned for the first time by Cushman (1938) and ever since were generally considered symmetrically developed on both sides of the test; Georgescu (2007a) made the first direct mention of the periapertural structures symmetry. Asymmetrical periapertural structures were described from the representatives of the group that evolved in the late Albian by Georgescu (2009b) and together with the test asymmetry represent vestiges of the group's benthic ancestry. All the taxa that evolved above the Albian/Cenomanian boundary present symmetrical periapertural structures.

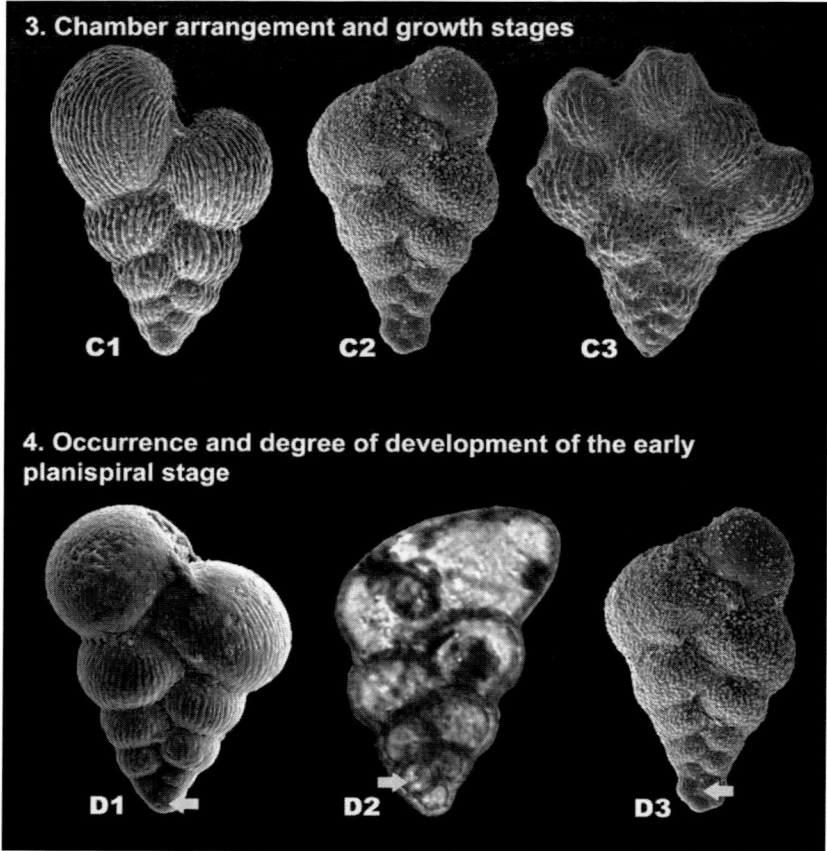

Figure 3. Chamber arrangement and growth stages (3) and occurrence and degree of development of the early planispiral stage (4). C1 Test with the chambers alternately added throughout the ontogenetic development. C2 Test with early planispiral stage. C3 Test with the adult stage with chamber proliferation. D1 Ahelicid test. D2 Hysterohelicid test. D3 Holohelicid test.

CHAMBER ARRANGEMENT AND GROWTH STAGES (Figure 3). Ehrenberg (1838) illustrated the earliest Cretaceous planktic foraminifera with the test consisting of chambers alternately added with respect to the test growth axis; subsequently such chamber arrangement was referred to as 'biserial'. In an attempt to connect the terminology to the process rather than to form Georgescu (2010) renounced at the term 'biserial' in favour of the more accurate expression 'chambers alternately added with respect to the test growth axis. One early planispiral coil was reported for the first time by Ehrenberg (1841) and the genus *Spiroplecta* was described by Ehrenberg (1844) to accommodate the tests that present such a feature; this is the earliest mention on the occurrence of multiple growth stages in the representatives of the group. Georgescu (2013a) demonstrated that *Heterohelix* was a provisional name and *Spiroplecta* is the correct generic name for the tests with one early planispiral coil described by Ehrenberg (1841); subsequent studies demonstrated that such an early coiled growth stage evolved four times in distinct lineages in the history of the group. The third test architecture is characterized by chamber proliferation in the adult stage and was first mentioned by Rzehak (1895); Georgescu (2010) proposed the replacement of the inaccurate expression "multiserial stage" with "multichamber growth stage", which indicates more accurately the kind of growth in the adult stage in such foraminiferal tests, and this terminology is followed herein.

OCCURRENCE AND DEGREE OF DEVELOPMENT OF THE EARLY PLANISPIRAL STAGE (Figure 3). The occurrence of an early planispiral coil in the representatives of the group was demonstrated by Ehrenberg (1841, 1844). Brown (1969) in the revision of the group mentioned that the early planispiral coil is a late development in the heterohelicid group history; four distinct lineages that led to the evolution of such a feature were recognized by Georgescu and Abramovich (2008a) and Georgescu (2013a, and this study). Georgescu (2013a) recognized three types of tests function of the occurrence and degree of development of the early planispiral coil: ahelicid (without early planispiral coil and proloculus in apical position), hysterohelicid (early planispiral coil in incipient state of development and proloculus eccentric, adjacent to the test margin) and holohelicid (fully developed early planispiral coil and proloculus completely enclosed by one whorl).

OCCURRENCE, DEVELOPMENT AND FEATURES OF THE ADULT STAGE WITH MULTICHAMBER GROWTH (Figure 4). The variability of the adult stage with multichamber growth was convincingly illustrated by Egger (1899) soon after the earliest report of such a feature by Rzehak (1895). The most frequent chamber proliferation is in the plane in which chambers were added in the juvenile stage. There are two morphological versions of this kind of chamber proliferation: incipient and well-developed, which were recognized for the first time by Plummer (1931) and Egger (1899) respectively. A second kind of chamber proliferation is by adding the chambers of the successive sets in planes perpendicular to the test growth axis; such kind of chamber proliferation was often referred to as tridimensional and the new term 'multiplane' is herein proposed for it. The incipient and well-developed versions of this kind of chamber proliferation were firstly recognized by de Klasz (1953) and Egger (1899) respectively. The differences between the two kinds of chamber proliferation are not apparent only in the pattern of chamber addition; there is another difference in the way the chamber proliferation is initiated. Van Hinte (1965) defined the progressive chamber as the first chamber with a second aperture. The adult stage with multichamber growth in the case of chamber proliferation in the plane in which chambers were added in the juvenile stage initiates with the progressive chamber, whereas such a chamber does not occur in the case the tests with multi-plane chamber proliferation.

FLANGE/CHAMBER RATIO IN THE ADULT STAGE WITH MULTICHAMBER GROWTH (Figure 5). A new morphological feature was described by de Klasz (1953) who noted in the study of the species of the genus *Gublerina* that the two rows of divergent chambers are separated by a "calcareous lamella". Nederbragt (1989a) showed that this structure is formed by the expanded periapertural structures of the successive chambers and additional data about these morphological structures were brought in the revision of the genus by Georgescu, Saupe and Huber (2008) that led to its transformation into one evolutionary classification unit. As continuation of these developments it is now possible by examining the tests in X-ray micrograph to recognize three kinds of tests function of the degree of development of the periapertural structures and the flange/chamber ratio in the adult stage with chamber proliferation.

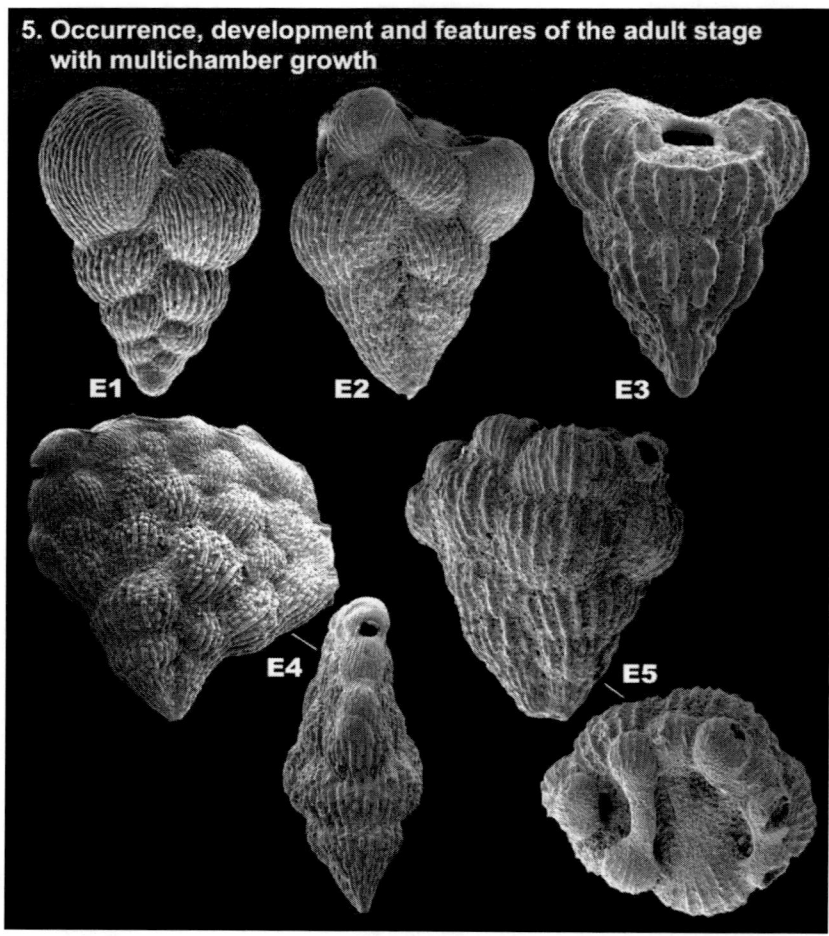

Figure 4. Occurrence, development and features of the adult stage with multichamber growth (5). E1 Test without adult stage with chamber proliferation. E2 Test with incipient chamber proliferation in the plane in which the chambers of the juvenile stage are alternately added with respect to the test growth axis. E3 Test with incipient multi-plane chamber proliferation. E4 Test with well-developed chamber proliferation in the plane in which the chambers of the juvenile stage are alternately added with respect to the test growth axis. E5 Test with well-developed multi-plane chamber proliferation.

The first kind is that in which in the adult stage the area occupied by flanges dominates over that occupied by chambers and corresponds largely to the classical gublerinid test architecture. The second kind is that in which the chambers dominate over the flange area; this kind occurs in most of the Late Cretaceous lineages that evolved chamber proliferation in the adult stage. The third kind is that in which the flanges are very small or appear absent and is the only known in the tests with multi-plane chamber proliferation in the adult stage.

CHAMBER SHAPE IN LATERAL VIEW (Figure 6). This feature was recognized since the pioneering study by Ehrenberg (1838) when the tests were observed in transmitted light; chamber shape was described for the first time as globular but the subsequent studies indicated that this kind also includes subglobular chambers. One additional kind was described by Cushman (1938) as "...chamber much lower than broad..." for which nowadays the term 'reniform' is used. Also Cushman (1938) described a new kind as chambers "...with the sides nearly parallel..." and the modern term of 'subrectangular' is used today for it. The fourth kind was reported by Eicher and Worstell (1970a) in the diagnosis of their genus *Lunatriella* as "...vertically elongated chambers, which form an irregularly uniserial pattern..."; Georgescu (2013c) refined the terminology as "...[chambers] with the elongation almost parallel to the test growth axis..."

CHAMBER TRANSVERSAL ELONGATION (Figure 6). Chamber transversal elongation is one of the two major morphological features discovered at the beginning of the last decade of the nineteenth century by Rzehak (1891); the two species in which Rzehak (1891, 1895) recognized this feature present well-developed chamber transversal elongation. Sandidge (1932) recognized the incipient stage of this feature and in the same year Voorwijk recognized the extreme case of chamber transversal elongation in his species *Gümbelina nuttalli*. Further observations on the feature variability were made by Cushman (1938) who noted the compressed chambers in *Gümbelina planata* and spherical chambers in *G. globulosa*.

SUTURES AND SUTURAL RIDGES (Figure 7). Sutures between two adjacent chambers are depressed in most of the Cretaceous representatives of the planktic foraminifera with chambers alternately added with respect to the test growth axis. Sutures lined with ridges were reported for the first time by Sigal (1952) and this feature was of paramount importance when Nederbragt (1991) recognized two Santonian lineages that evolved adult stage with multichamber growth. Georgescu (2010) recognized two kinds of sutural ridges and proposed a distinct terminology for this feature: calyptoridges are the sutural ridges in incipient state of development, whereas the well-developed sutural ridges were named phaneroridges.

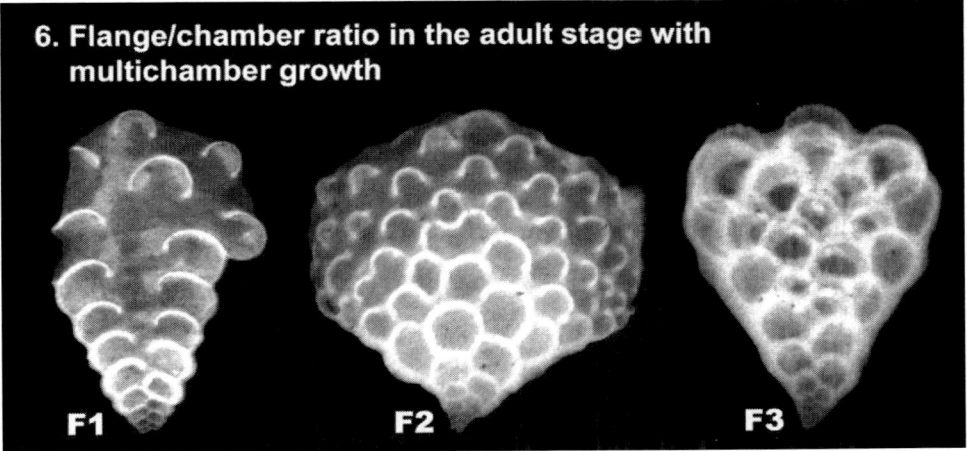

Figure 5. Flange/chamber ratio in the adult stage with multichamber growth (6). F1 Test in which the flanges dominate the adult stage with chamber proliferation. F2 Test in which the chambers dominate the adult stage with chamber proliferation. F3 Test with strongly reduced or absent flanges in the adult stage.

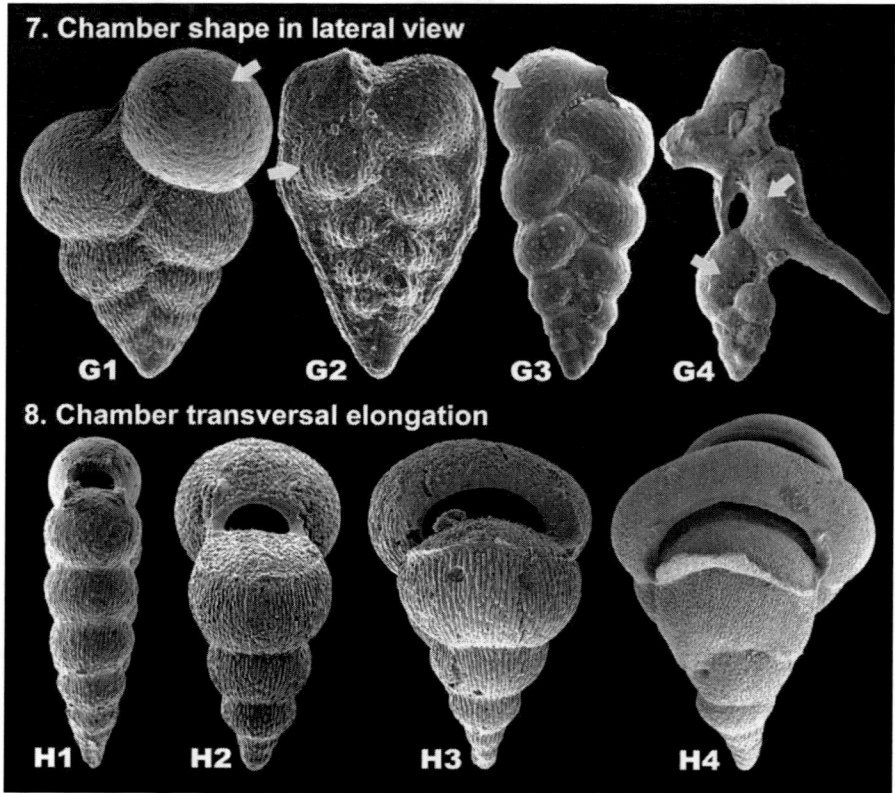

Figure 6. Chamber shape in lateral view (7) and chamber transversal elongation (8). G1 Globular or subglobular. G2 Subrectangular. G3 Reniform. G4 Chamber with the elongation axis almost parallel to the test growth axis. H1. Laterally compressed. H2 Spherical. H3 With incipient transversal elongation. H4 Extreme case of chamber transversal elongation.

CHAMBER EXTENSIONS (Figure 8). Due to the nature of this morphological feature its description and terminology occasionally overlaps or appears complementary to the chamber shape in edge view. Originally the chambers in the tests of the representatives of this group were considered globular (Ehrenberg 1838) but subaculeate chambers in the distal portion of the tests were described in as *Textilaria americana* by Ehrenberg (1841) and figured by Bailey (1844) and Ehrenberg (1854). Frerichs and Gaskill (1978) in the first report of this species after circa 130 years misinterpreted this feature and considered that spines occur at the peripheral margin of the last-formed chambers in this species and a similar conclusion was drawn by Masters (1980) after the on-site re-examination of the original specimens illustrated by C.G. Ehrenberg. Georgescu (2013a) did not confirm the occurrence of spines in the original specimens and new ones collected from the Ehrenberg's original bulk samples; instead the term used was 'laterally-pinched', which is closer to the original term, used by Ehrenberg (1841) of 'subaculeate' and this terminology is used herein. Two species with double backward extension, one on each side of the test, were described by Cushman (1926, 1938) and an accurate description of this feature was given by Brönnimann and Brown (1953) who mentioned the occurrence of accessory apertures at the proximal end of the double chamber backward elongation; Brönnimann and Brown (1953) used this feature to describe the genus Pseudoguembelina. Georgescu (2007a) and Georgescu, Quinney and Anderson

(2011) demonstrated that supplementary apertures at the posterior end of the double backward extensions are late achievements in the group history and the earlier tests of the Turonian-lower Campanian stratigraphic interval lack such openings. The incipient stages of development of the double backward extension were for the first time reported by Georgescu (2007a) for one Turonian directional lineage. The earliest chamber extension in the group's evolution was in the lateral position and oriented backward; it was reported in incipient stage by Masella (1959) and well-developed stage by Eicher and Worstell (1970a, 1970b).

Figure 7. Sutures and sutural ridges (9). I1 Test without sutural ridges. I2 Calyptoridges. I3 Phaneroridges.

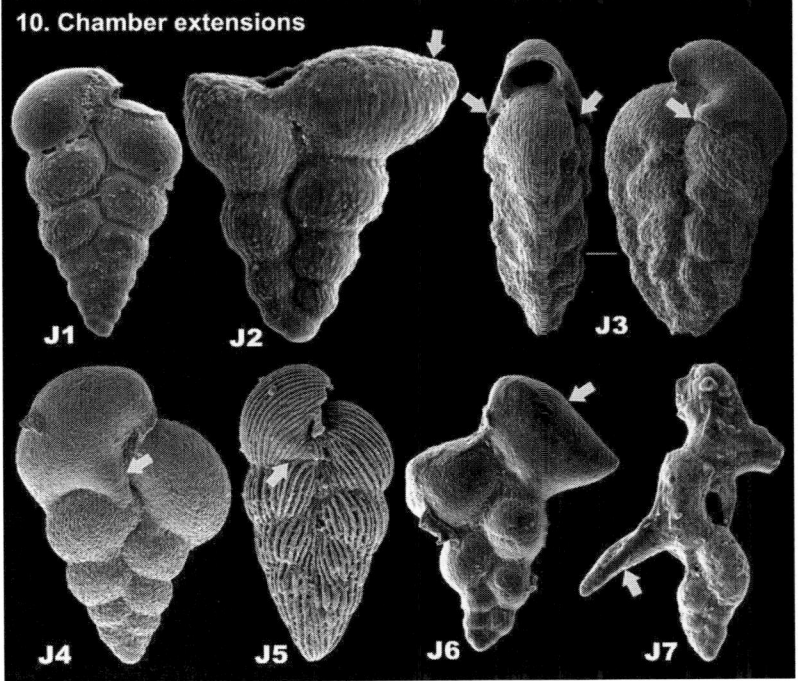

Figure 8. Chamber extensions (10). J1 Chambers without peripheral extensions. J2 Laterally-pinched chamber. J3 Chambers with well-developed double backward extensions with supplementary apertures at the posterior end. J4 Chambers with well-developed double backward extensions without supplementary apertures at the posterior end. J5 Chambers with incipient double backward extensions. J6 Incipient peripheral lateral and backward extension. J7 Well-developed lateral and peripheral backward extension.

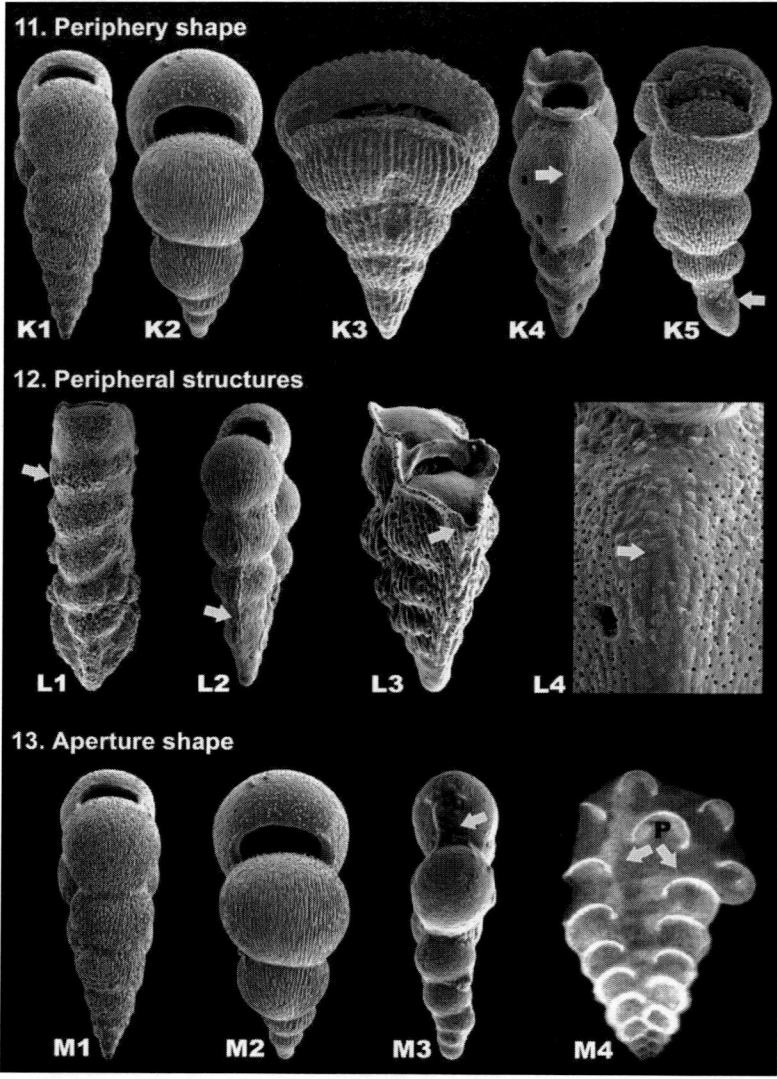

Figure 9. Periphery shape (11), peripheral structures (12) and aperture shape (13). K1 Rounded. K2 Broadly rounded. K3 Extreme case of broadly rounded periphery. K4 Subacute. K5 Mixed, subacute in the early stage and broadly rounded in the adult stage. L1 Transverse keels across the periphery. L2 Incipient peripheral wall flexure. L3 Well-developed peripheral wall flexure. L4 Longitudinal imperforate peripheral band or one band with low pore density. M1 Low. M2 Medium-high. M3 High. M4 Biaperturate progressive chamber (P).

PERIPHERY SHAPE (Figure 9). Cushman (1938) started to describe the test periphery using terms such as 'whole', 'indented' and 'keeled'; it is evident that the periphery characteristics were those observed in lateral view and Cushman referred to what is today considered as test outline. The modern use of this feature involves the examination of the test in edge view. Petters (1983) used the term 'rounded' to characterize the test periphery is edge view and Nederbragt (1991) referred the periphery of a small number of taxa as 'subacute' probably to separate them from the most part of the representatives of this group in which this feature has a rounded appearance. A new and consistent terminology was developed with the extensive observations associated with the study of the peripheral structures. 'Rounded' periphery occurs is the tests with lateral compression, whereas 'broadly rounded' is applied to the tests with globular or spherical chambers or chambers that increase rapidly in thickness. Compressed tests may present 'subacute' periphery and the term 'subrounded' should be considered its synonym. Mixed periphery, subacute in the earlier portion of the test and rounded to broadly rounded in the adult, was reported by Georgescu (2013a); this should not be considered one synonym to that reported by Cushman (1938) in which the observations were made in lateral view.

PERIPHERAL STRUCTURES (Figure 9). The first mentions of peripheral structures in the representatives of the Cretaceous planktics with chambers alternately added with respect to the test growth axis were those of Cushman (1938) who described the periphery of certain species as keeled, either in the earlier portion of the test or throughout the periphery. Observations with the aid of the SEM showed that longitudinal keels along the periphery do not occur in the representatives of this group. Georgescu and Abramovich (2008a) showed that the species *Gümbelina carinata* described by Cushman (1938) as keeled presents a well-developed peripheral wall flexure; the early stage of development of this structure was reported by Georgescu and Abramovich (2008a) in the species *Hendersonia hendersoni*. Keeled periphery evolved in through the fusing across the periphery of the sutural ridges on the two sides of the test; this feature was described for the first time by Georgescu and Abramovich (2008b). A new peripheral structure is herein reported for the first time in occasional specimens of *Gümbelina reussi* from the Woodbury Clays of New Jersey (US) and *Gümbelina glabrans* from the Kemp Clay of Texas (US); this structure is represented by one longitudinal imperforate band or one band with low pore density. The occurrences of this new morphological feature cannot be associated with a peculiar stage of morphological stability or instability of one lineage and therefore it does not appear to have taxonomic significance.

APERTURE SHAPE AND NUMBER (Figure 9). Test apertures in the representatives of this group were described in a more or less accurate by many authors; one of the most accurate descriptions is that given by Sandidge (1932) for his species *Ventilabrella plummerae*. The experience accumulated in the last years when the first consistent evolutionary framework of the group was developed shows that now is possible to provide a classification of this feature that can be used throughout the group lineages. In all the specimens without chamber proliferation in the adult stage the aperture is situated at the base of the last-formed chamber; three qualitative terms forming a continuous transition, namely 'low', 'medium' and 'high arch' are proposed and illustrated. The definition of these three categories of aperture shape function of the height/width ratio creates the possibility of using this feature consistently throughout group. Occurrence of multiple apertures in the adult stage was first mentioned by Plummer (1931) in the description of her new species *Ventilabrella carseyae*. One further development in the study of the apertures of the adult stage with chamber proliferation includes that by van Hinte (1965) who defined as progressive chamber the first biaperturate chamber of the test. Nederbragt (1989a) recognized three types of chambers function of the number of apertures within the adult stage with chamber proliferation: relapsed, bi-derivative and tri-apertured. This classification appears of no use in the description of such a foraminiferal test and for this reason it is not taken in consideration herein.

PERIAPERTURAL STRUCTURES MORPHOLOGY AND ORIENTATION (Figure 10). Periapertural structures played a minor role in the description of the tests of Late Cretaceous planktics with chambers alternately added with respect to the test growth axis. General terms such as 'lamella', 'rim', 'lip' or 'flange' were used by various authors for this morphological feature in the period before the extensive use of the SEM in recognizing the test features, but they were not presented in a consistent morphological and terminological framework. Understanding the variability and evolutionary significance of the periapertural structures began when Georgescu (2009b) demonstrated that the earliest representatives of the group are asymmetrical: one rim on one side and one ridge on the other side of the aperture. Delicate lamellae commonly referred to as flanges evolved only in the latest Albian according to Georgescu and Huber (2009).

Georgescu (2010) provided the first terminology of the periapertural structures based on the high resolution SEM observation; this terminology is followed and further improved herein. The asymmetrical peripheral structures were named archaeoflanges. Orthoflanges differ from the archaeoflanges by being symmetrically developed, one on each side of the test. Metaflanges are also symmetrically developed on each side of the test but differ from orthoflanges by having a rim at the distal end. The last kind of periapertural structures recognized by Georgescu (2010) are the leptoflanges, which are symmetrical structures attached to the previous chamber creating false supplementary apertures along the test central suture. Leptoflanges were originally recognized as structures with rimmed margin both in anterior and posterior sides; leptoflanges without a rim were described by Eicher and Worstell (1970a). All these periapertural structures are oriented from the last-formed chamber towards the penultimate one. Further evolution of the periapertural structures led to the occurrence of flanges oriented backwards, towards the last-formed chamber; such periapertural structures are herein named retroflanges. Retroflanges are mostly ornamented structures and can be rimmed or not. A distinct periapertural structure was described by Georgescu (2007b) as imperforate band that borders the aperture, which is developed between the two flange-type structures developed on each side of the aperture. Smith and Pessagno (1973) described a kind of periapertural structures called ponticuli (singular ponticulus), which occurs in taxa with multi-plane chamber proliferation. Such structures were named bridges and sieve-plates by Nederbragt (1989a, 1989b, 1991) who considered them microperforate.

TRANSVERSAL WALLS (Figure 11). This structure was reported by Georgescu, Saupe and Huber (2008) occasionally in specimens of the species *Praegublerina robusta* of the late Campanian and Maastrichtian; it was interpreted as one structure developed for the reinforcement of the thin test wall in the adult portion of the test.

PERFORATE CENTRAL PLATE (Figure 11). Georgescu (2014a) reported this feature, which is represented by a calcareous plate formed over the central suture of the adult portion of the test; this structure is developed on test both sides. The only species where the perforate central plate occurrence is demonstrated is *Gümbelina semicostata*.

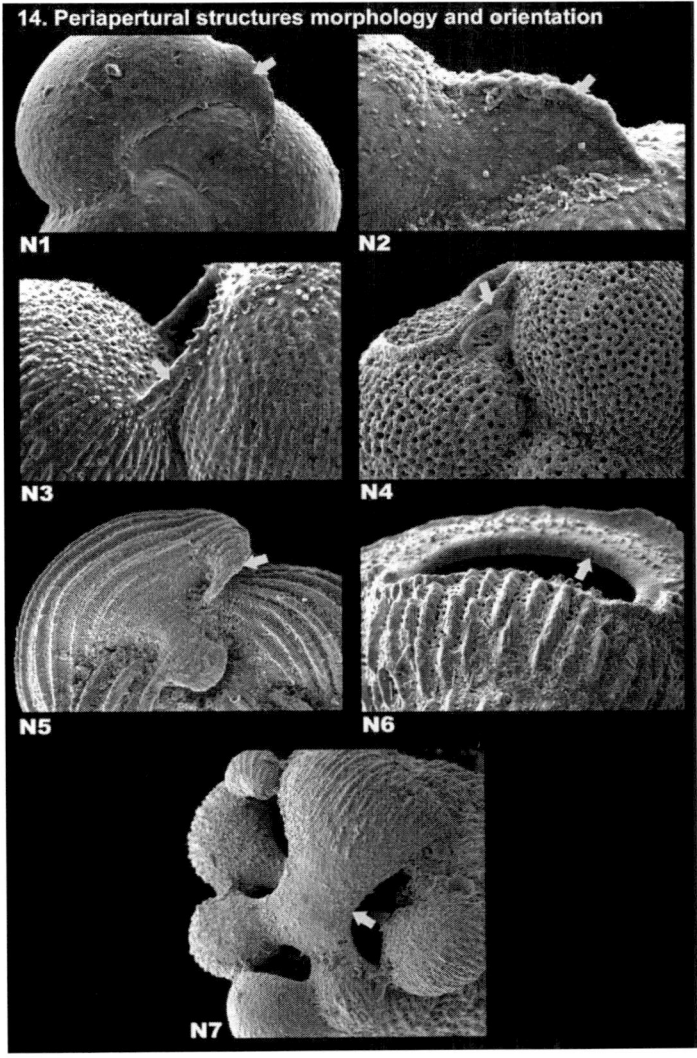

Figure 10. Periapertural structure morphology and orientation (14). N1 Orthoflanges. N2 Wide metaflanges. N3 Narrow metaflanges. N4 Rimmed leptoflanges. N5 Retroflanges. N6 Imperforate band bordering the aperture. N7 Bridge.

SIMPLE ORNAMENTATION (Figure 12). Ehrenberg (1838) used chamber ornamentation among other test features to recognize the first species of the group. Two patterns were recognized by this author, namely rough and striate, whereas in one other the test surface was mentioned as smooth; such ornamentation features were used to define and name the species *Textularia aspera*, *T. striata* and *T. globulosa* respectively among others. Georgescu (2013a) demonstrated based on the restudy of the material from the Ehrenberg Collection that the rough (hispid) and smooth ornamentation are the result of preservation in Canada balsam and all the tests reported by Ehrenberg from the Cretaceous chalky rocks are ornamented with longitudinal costae. Pessagno (1967) reported one complex ornamentation pattern in the species *Gümbelina punctulata* where each pore is surrounded by fused ridges; Georgescu (2007b) applied for the first time the term reticulate to this ornamentation pattern. Vermicular ornamentation was reported and illustrated by Martin (1972) from late Campanian-Maastrichtian tests with multichamber growth in the adult stage. Masters (1977) described a new ornamentation pattern in which the pores are situated in the central portion of a pore mound. Fused costae often result in the formation of zones with irregular ornamentation, which were mentioned for the first time by Nederbragt (1991) as "irregular ridges".

TRANSVERSALLY ZONED ORNAMENTATION (Figure 13). Tests in which the ornamentation is coarser over the earlier portion of the test were described for the first time by Cushman (1938), which contrast to those in which the ornamentation is equally developed over the entire test surface as described by Ehrenberg (1838) and generally accepted in the subsequent studies.

LONGITUDINALLY ZONED ORNAMENTATION (figure 13). This feature was described by Georgescu, Saupe and Huber 2008 in the case of the species *Gublerina acuta*. Apparently the lineage to which this species belongs is the only one that evolved such a feature.

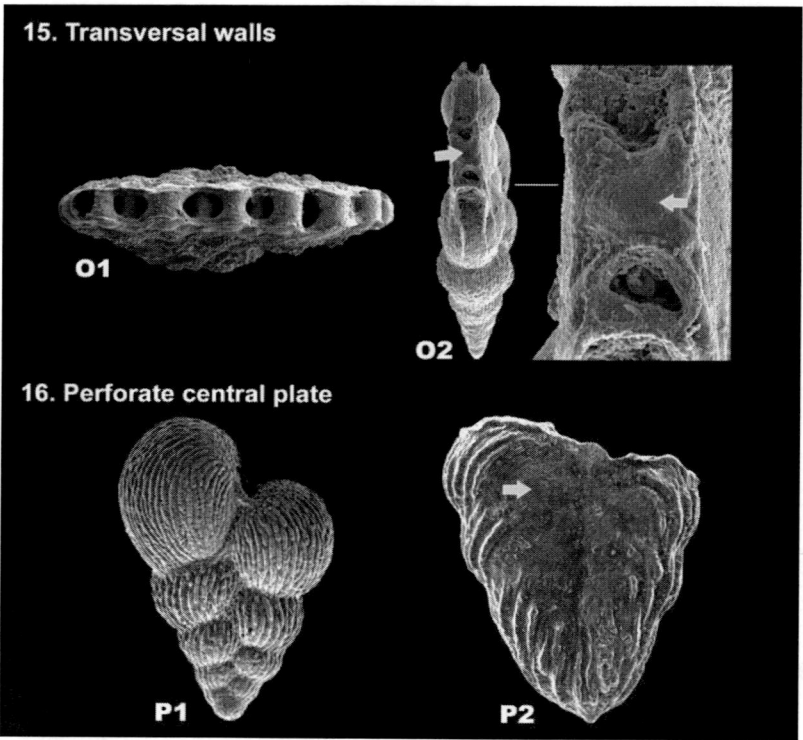

Figure 11. Transversal walls (15) and perforate central plate (16). O1 Test without transversal walls. O2 Test with transversal walls. P1 Test without perforate central plate. P2 Test with perforate central plate.

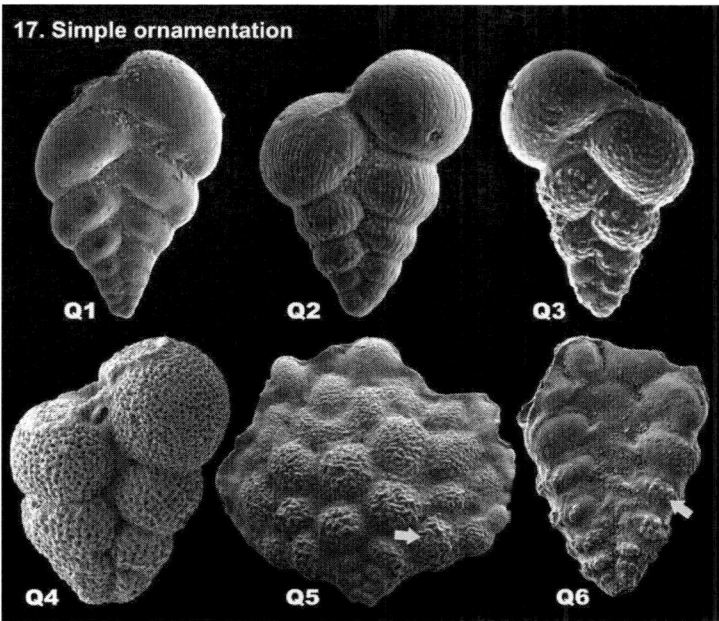

Figure 12. Simple ornamentation (17). Q1 Smooth test surface. Q2 Costate ornamentation. Q3 Pore-mounded ornamentation. Q4 Reticulate ornamentation. Q5 Vermicular ornamentation. Q6 Zones with irregular ornamentation.

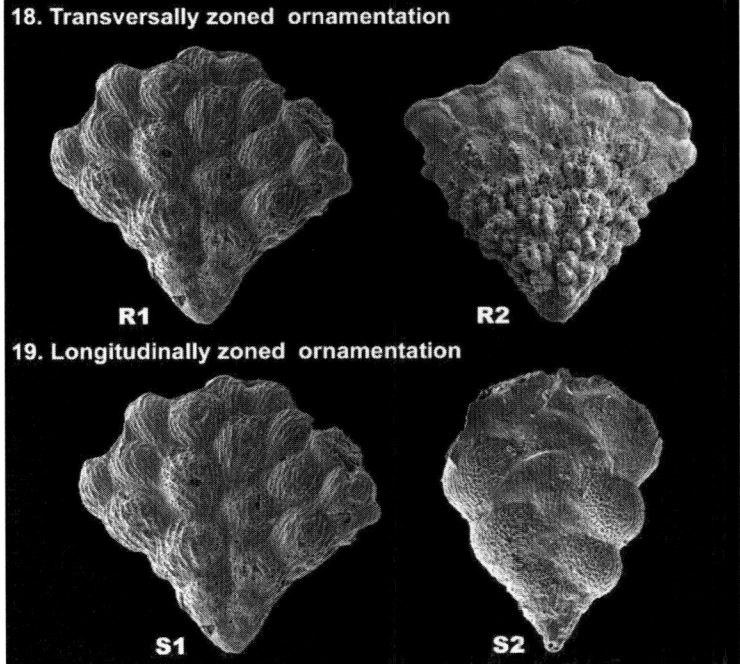

Figure 13. Transversally zoned ornamentation (18) and longitudinally zoned ornamentation (19). R1 Test without transversally zoned ornamentation. R2 Test with transversally zoned ornamentation. S1 Test without longitudinally zoned ornamentation. S2 Test with longitudinally zoned ornamentation.

COSTAE ORIENTATION (Figure 14). It was generally accepted that most representatives of this group are ornamented with costae of various thickness, which are longitudinally arranged, more or less parallel to the periphery and test growth axis. High resolution observations on tests from throughout the stratigraphic range of the group indicate that this is not the only pattern of costae ornamentation. Such a costae distribution pattern in which they are parallel to the periphery in the test marginal regions and parallel to the test growth axis in the test central portion is termed unimodal ornamentation; this kind of ornamentation orientation is that illustrated by Ehrenberg (1838). Bimodal costae orientation is a pattern that occurs in three lineages of the Santonian-Maastrichtian stratigraphic interval; in this case the costae are parallel to the periphery in the marginal portions of the test and oblique to the test growth axis in the central region (Georgescu 2014a, 2014b).

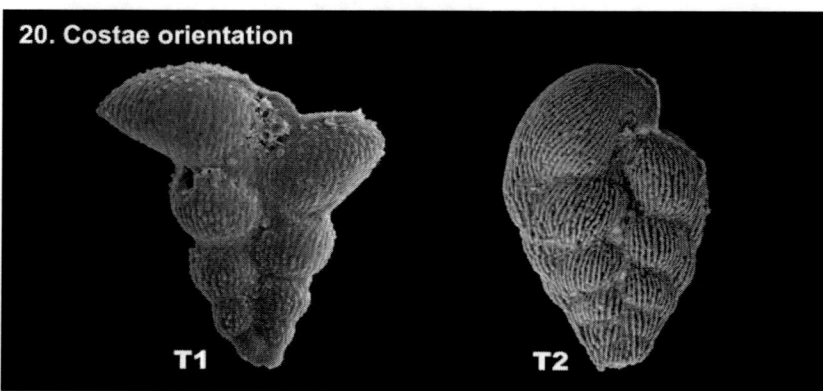

Figure 14. Costae orientation. T1 Unimodal orientation. T2 Bimodal ornamentation.

COSTAE APPEARANCE AND THICKNESS (Figure 15). Little or no attention was given to the costae appearance and thickness before the extensive use of the SEM in the study of this group. Occasional mentions of thick, thin or fine costae exist in the literature but in general such terms were used in descriptive, inconsistent and subjective way. Georgescu and Huber (2009) described granular costae consisting of aligned pustules, which differ in appearance from the costae with solid appearance. Georgescu (2010) provided a classification of the costae with solid appearance function of their thickness in leptocostae, which are thinner, and pycnocostae, which are thicker. Pycnocostae evolved in the late Santonian and the transition between the leptocostate and pycnocostate ornamentation is gradual.

PUSTULOSE PERIAPERTURAL AREA (Figure 15). Georgescu (2009a) described for the first time a zone ornamented with small-sized dome-like scattered pustules in the anterior portion of the chamber; this feature is the pustulose periapertural area. Subsequent studies showed that this ornamentation feature occurs consistently in certain lineages and there are a small number of lineages in which this occurs occasionally. The thorough study of the ornamentation of the lineages of the Cretaceous planktic foraminifers with chambers alternately added with respect to the test growth axis shows that this feature is absent in the early evolution of the group.

WALL CHARACTERISTICS (Figure 15). Test wall in the representatives of this group was not studied in detail until recently and most of the references to this feature almost invariably were related to the occurrence of pores. Georgescu (2009c) demonstrated in the case of some trochospirally coiled planktics of the Albian-Turonian stratigraphic interval that ornamentation and wall ultrastructure are two features with independent evolution; such observations led to the first classification of the test wall in Cretaceous planktic foraminifera by Georgescu (2011a). Examination of the test wall characteristic throughout the group of Cretaceous planktics with chambers alternately added with respect to the test growth axis shows that most of the tests present simple wall. Georgescu (2011a) reported simple-ridged test wall in the species *Hendersonites pacificus* and this is the only one known with a wall structure other than simple.

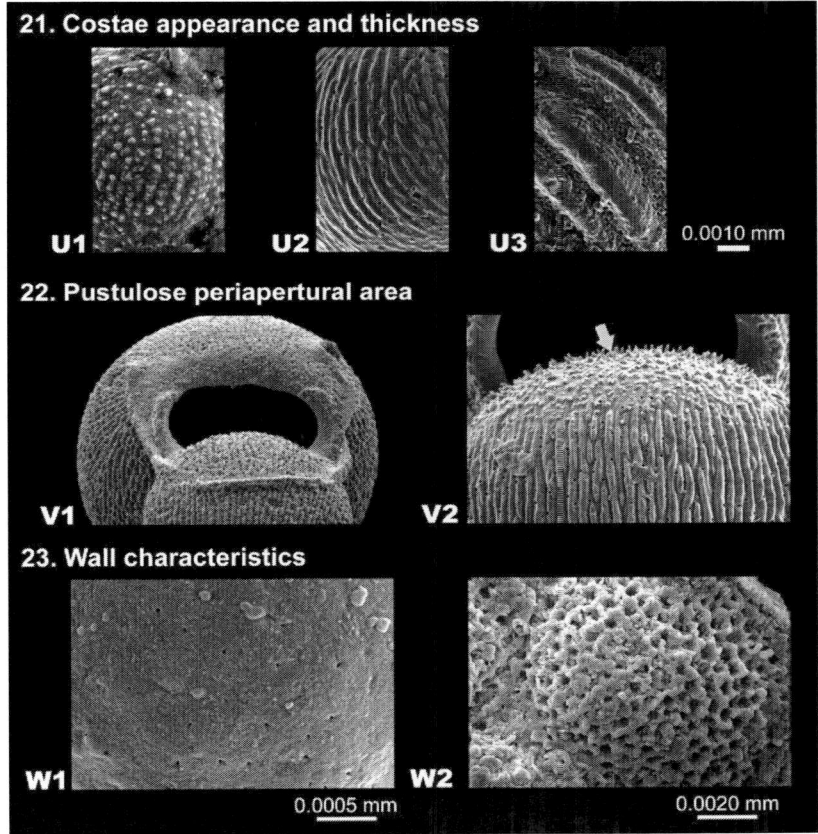

Figure 15. Costae appearance and thickness (21), pustulose periapertural area (22) and wall characteristics (23). U1 Pustulose (granular) costae. U2 Leptocostae. U3 Pycnocostae. V1 Chamber without pustulose periapertural area. V2 Chamber with pustulose periapertural area. W1 Simple wall. W2 Simple-ridged wall.

PORE CHARACTERISTICS (Figure 16). Pores were considered major features of the test wall by most taxonomists but no high detail observations on the pore size were consistently made until the introduction of the SEM as main tool to acquire data on the test morphology. The separation of the pores into three size categories (microperforate, finely perforate and macroperforate) was used between 2007 and 2009; nannopores were recognized by

Georgescu (2009b) in the primitive representatives of the group and pore small sizes were considered indicative for the small benthic ancestry of the group. Such separation of the pores into categories based on their size was rejected subsequently by Georgescu (2010) when its artificiality became evident and was demonstrated that pore ranges for one specific taxon frequently cross the boundaries between these categories. The numerical values were used ever since to describe and quantify the pore dimensions. A new type of pores in the representatives of the group was reported by Georgescu and Abramovich (2008a); such pores have crater-like appearance and were renamed scalaropores by Georgescu (2010). Vuggy pores present irregular shape and apparently are the result of the successive addition of layers of calcite throughout the ontogenetic development and were recognized by Georgescu and Abramovich (2008b). Such advances in the knowledge of the pore morphology resulted in the first morphological classification of the pores into simple, scalaropores and vuggy (Georgescu 2010).

PROLOCULUS SIZE (Figure 17). The proloculus size was used as taxonomic criterion by Georgescu and Huber (2009) in the description of the monospecific genus *Globoheterohelix*. Further studies showed that this taxon is the only one with such a feature in the entire Albian-Maastrichtian evolutionary history of the group.

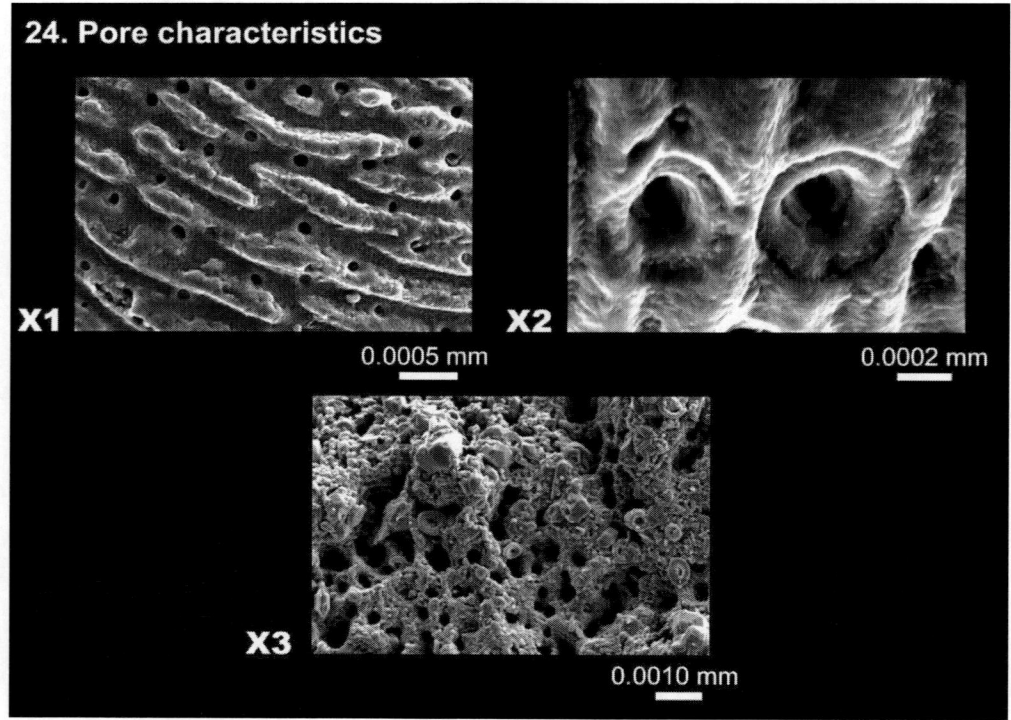

Figure 16. Pore characteristics (24). X1 Simple pores. X2 Scalaropores. X3 Vuggy pores.

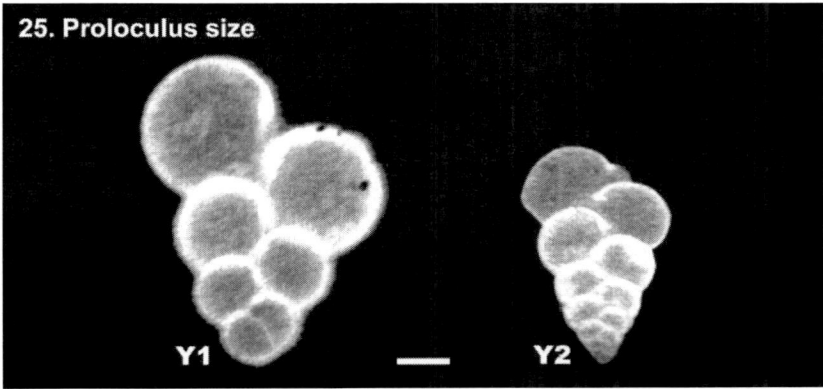

Figure 17. Proloculus size. Y1 Test with large-sized proloculus. Y2 Test with small proloculus.

Four periods of variable duration in which the test morphological features were recognized in the representatives of the study group are herein recognized and in addition their magnitude quantified (Figure 18). The pioneering studies by C.G. Ehrenberg (1838-1844) led to the basic features including chamber shape and arrangement, ornamentation and occurrence of multiple growth stage; the observations were made using the transmitted light microscope. It was following by a 47 years period in which no other test morphological features were discovered but in which there is encountered a dominance of the observations made with the aid of the reflected light microscope. The higher accuracy observations on foraminiferal test led to the advance in a short period close to the end of the nineteenth century (1891-1892) by the studies of A. Rzehak who discovered and reported new morphological features to which he conferred taxonomic significance, such as the chamber transversal elongation and adult stage with chamber proliferation respectively. This second period of accumulation was followed by a 39 years period in which no other test morphological features were discovered. The third period of accumulation lasted 22 years and can be largely associated with the foraminiferal studies required in the exploration for hydrocarbons; new morphological features were discovered by H.J. Plummer, J.A. Cushman, I. de Klasz and J. Sigal; by the end of this period the number of newly discovered test morphological features was equal to that by C.G. Ehrenberg from the pioneering period. A 56 years period in which no new morphological features of the group representatives were discovered followed and within its timeframe we record the earliest observations based on the scanning electron microscope (SEM); the fact that the SEM was used as an illustrating tool rather than one to acquire new information is demonstrated by the fact that no new morphological features were discovered until the beginnings of the "ultrastructure revolution". The extensive use of the SEM in the "ultrastructure revolution" led to the discovery of a large number of new test morphological features, which is almost double when compared to those described by C.G. Ehrenberg in the pioneering period and on the overall is comparable to all the discoveries in the three previous periods of accumulation together. The high level of magnitude of new discoveries can be explained by the high accuracy observations allowed by the use of such tool. The four periods of data accumulation can also be recognized when the number of various kinds of test morphological features is analyzed. The pioneering period (1838-1844) brought the earliest kinds of morphological features. The second period of data accumulation lasted between 1891 and 1899 and followed a 47 years period of stagnation; it presents a longer extent due reports of taxa with chamber proliferation by Egger (1899). The third period of accumulation began after 39 years of stagnation and lasted between 1931 and 1991, namely between the studies of H.J. Plummer and A. Nederbragt. New kinds of morphological features were added to those already recognized intermittently, with a break period generated probably by World War II; this period lasted longer when compared to the corresponding time period defined according to the test morphological features and this is considered indicative for an increase refinement on the observations based especially on the use of the optical stereomicroscope and occasionally SEM. "Ultrastructure revolution" (2007 to present) began 16 years after the end of the previous period of accumulation and led to the description of a large number of new kinds of test morphological features, which surpasses those described in the 60 years of the previous period of accumulation. All these data show the significant increase in the amount of new data generated during the "ultrastructure Revolution" through the extensive use of the SEM. The next step in understanding further the significance of this scientific event is to study the nature and number of the newly defined taxonomical units, which were described since the beginning of the "ultrastructure revolution", and then compare these to those described before it.

EVOLUTIONARY CLASSIFICATION

The new level of knowledge that is apparent in test morphology, stratigraphic distribution and evolutionary relationships between the different components of the study group allows the development of an evolutionary framework and classification. Probably the most important practical achievement of the new system is that with few exceptions the specimens of the fossil record can be included within a lineage. This contrasts with the typological classification in which a significant number of specimens cannot be assigned to one species or another and many of the specimens of the fossil record cannot be accurately classified.

Fundamentals and Brief History

Evolutionary classification was initiated with the release of the Theory of Evolution (Darwin, 1859, p. 413-414): "Such expressions as that famous one of Linnaeus, and which we often meet with in a more or less concealed form, that the characters do not make the genus, but the genus gives the characters, seem to imply that something more is included in our classification, than mere resemblance. I believe that something more is included; and the propinquity of descent, -the only known cause of the similarity of organic beings,- is the bond, hidden as it is by various degrees of modification, which is partially revealed to us by our classification."

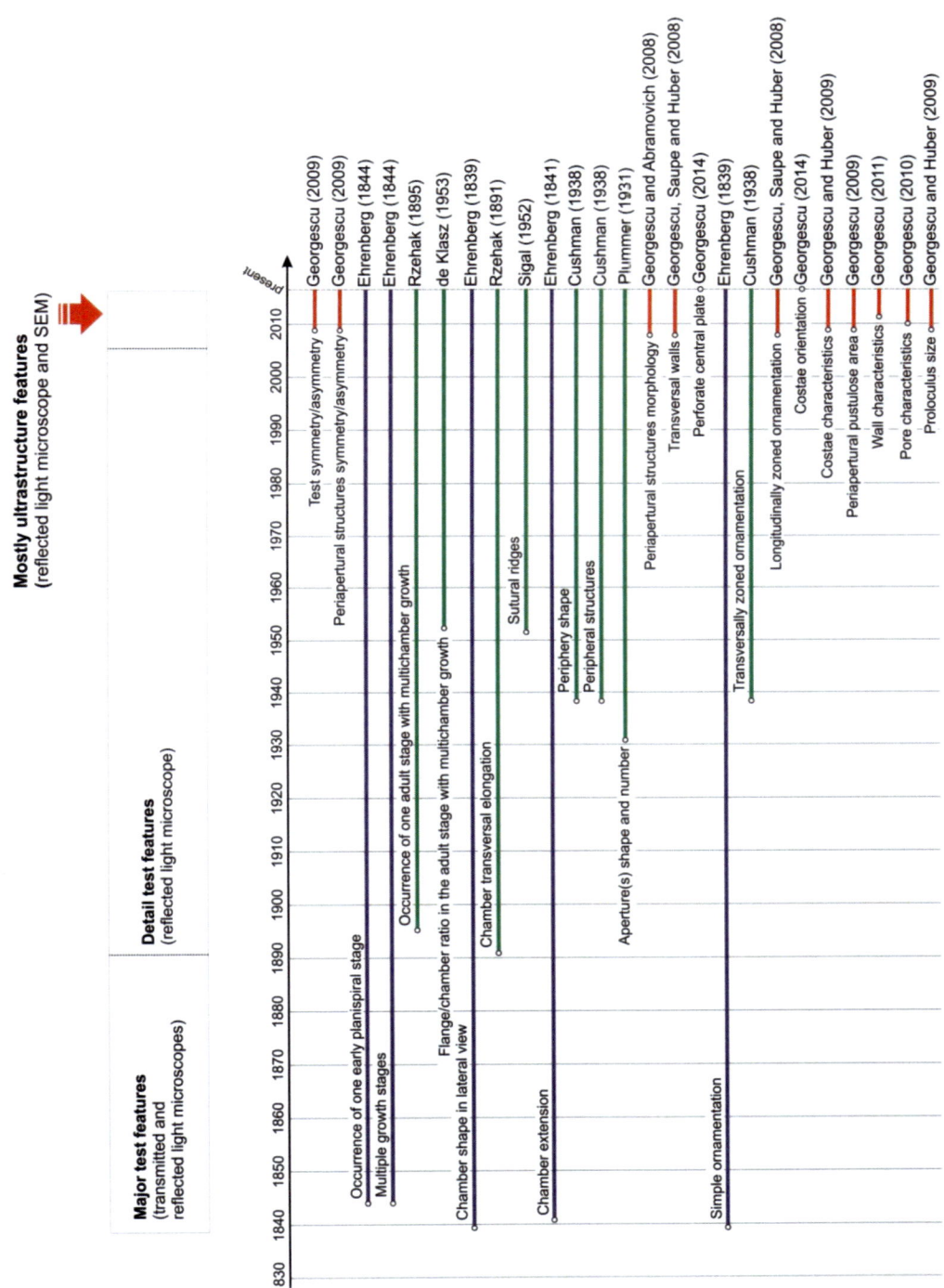

Figure 18. Synthetic presentation of the test main morphological features and when they were first mentioned in the Cretaceous planktic foraminifera with chambers alternately added with respect to the test growth axis. The features are presented in the order used in general in the test description. The features recognized in the earliest period are given in blue, those from the second period in green and those from the "ultrastructure revolution" in red.

Darwin did not elaborate further on this idea and the development of a new classification framework based not only on the resemblance was left to the studies of subsequent authors. Although this idea was mentioned frequently afterwards by taxonomists no significant advance was made for more than one hundred years.

Simpson (1961) considered an evolutionary classification in a practical way, showing that the evolutionary classification should be "... *based on* phylogeny or is a *partial* expression of it." (*ibid.*, p. 113). Based on this observation the author drew the conclusion that an evolutionary classification does not come in contradiction with the views of a taxonomist on a particular group phylogeny. During this period evolutionary classification was not considered a distinct method in classification, but rather an advanced stage of development of the typological classification.

Mayr (1968, p. 546-548) in a brief presentation of the five classification methods (i.e., Essentialism, Nominalism, Empiricism, Cladism, and Evolutionary Classification) mentions with clarity the fundamental principle and strength of the evolutionary classification (*ibid.*, p. 548): "The theory of evolutionary classification, first proposed by Darwin, delimits taxa on the basis of two considerations-common ancestry and subsequent divergence. Its method is to infer relationship on the basis of an *a posteriori* weighting of similarity. It would be going too far to discuss such methods of weighting (...), but they are essentially those which the great masters of taxonomists have practised for more than one hundred years. To me it seems that the evolutionary approach combines the best features of the phenetic and of the cladistic approaches. By not being committed to any one-sided dogma, such as that all characters have equal weight or that there is only one process in evolution (the splitting of branches), it is able to evaluate all available evidence and arrive at balanced conclusions." Theoretical and practical considerations on the evolutionary classification were subsequently analyzed by various authors, such as Bock (1973), Mayr (1974, 1981), Nelson (1974), Padian (1999) etc. Probably the most comprehensive account on the evolutionary classification was given by Mayr and Ashlock (1991).

Iterative Evolution: A Challenge for Typological Classification

It is accepted in general that the beginnings of questioning the typological classification in planktic foraminifera happened in the sixties and early seventies with the advances in the knowledge of the Cenozoic representatives of the group (Parker 1962, 1967; Lipps 1966; Berggren 1968; McGowran 1968, 1971; Steineck 1971; Fleisher 1974). Such studies based on high-resolution observations on the gross test architecture, ornamentation and wall structure demonstrated the recurrent evolution of various test architectures in the Cenozoic planktic foraminiferal history. This evolution process is known as iterative evolution and was subsequently demonstrated to occur throughout the Cretaceous-Cenozoic evolution of the planktic foraminifera (Frerichs 1971).

Such advances determined Steineck and Fleisher (1978) to advocate the reclassification of the Cenozoic globigerinaceans based on the classical evolutionary method in an article that became classical. By contrast, there were only a few studies on the test ultrastructure in the Mesozoic representatives of the group during this period and the dominant trend in the Mesozoic foraminiferal classification was towards producing a framework applicable to both detached specimens and specimens studied thin sections.

Georgescu (2013a) reviewed the advances in the study of the Cretaceous planktic foraminifera and showed that the earliest report of an iterative evolution process was given by Cushman (1938); the iterative evolution was demonstrated in the case of the planktic genera with chambers alternately added with respect to the test growth axis in the early stage and adult proliferating stage *Planoglobulina* Cushman 1927 of Campanian-Maastrichtian age and *Ventilabrella* Cushman 1928 of Santonian age.

A new iterative evolution pattern was mentioned by van Hinte (1963) who showed that the evolution from trochospiral to planispiral test architecture happened several times in the planktic foraminiferal history during the Cretaceous times. In addition the iterative evolution of meridional ornamentation was demonstrated by Petters and others (1983) and further developed by Georgescu and Huber (2006).

A major change in the classification of the Cretaceous planktics happened when high-resolution data from throughout the stratigraphic ranges of various taxa at species and genus level started to be acquired with the aid of the SEM. Besides the recognition of new high-detail morphology elements of the foraminiferal tests it became also possible to identify new iterative processes (Georgescu 2007a, 2009a, 2010, 2011a, 2011b, 2012, 2013a, 2013c, 2014a, 2014b, 2014c, 2014d, 2014e; Georgescu and Carrigy 2012, etc.). The taxonomic solution adopted in these cases was to assign one distinct name for each grouping of species defined according to the ancestor-descendant relationships. Such advances in understanding iterative evolution resulted in a significant increase of the number of taxa, but also revealed with more accuracy the evolutionary continuum among the representatives of the Cretaceous planktics, and especially those presenting at least one stage with chambers alternately added with respect to the test growth axis.

Advances in Evolutionary Classification Unit Definition

Newly acquired data based on the high-resolution SEM observations resulted in a significant advance in the evolutionary classification. Such advance happened when Georgescu (2009c) renounced at grouping species in ancestor-descendant relationship into genera but into lineages, conferring the lineage a formal status in evolutionary classification. According to Georgescu (2009c) a lineage is a unit with significance in evolutionary classification in which the species are included based on a mixture of morphological resemblances derived from the common ancestry and differences resulted from the divergence that occurs during the evolution process; the idea was not new and it occurred in the works of Darwin (1859), Mayr (1968) and Mayr and Ashlock (1991). The total novelty of this advance is represented by the replacement of the static concept of genus (typological classification) with the dynamic concept of lineage (evolutionary classification). The studies by Georgescu (2009b) and Georgescu and others (2011) showed that similar data can be used to recognize lineages in the benthic foraminifera.

In a subsequent study on the planktic lineages with chambers alternately added with respect to the test growth axis in the early stage and adult proliferating stage of the Santonian age by Georgescu (2010) the lineages were differentiated based on their branching pattern into directional and branched lineages. Both directional lineages (DL) and branched lineages (BL) received formal status in evolutionary status, whereas the 'lineage' level remained to be used rather in informal sense. This advance in evolutionary classification in which there are different kinds of units at one taxonomic level further separated this classification method from the other four. New kinds of lineages were further described: iterative lineage (IL) by Georgescu (2013) and condensed lineage (CL) by Georgescu (2014a). Therefore, an open system of evolutionary classification units was defined at one taxonomic level and it contrasts to the closed system of the typological classification in which the units at one taxonomic level are identical throughout the living world (Georgescu 2011c). A presentation of the four kinds of known lineages is given below together with the original definitions (Figure 19).

- DIRECTIONAL LINEAGE (Georgescu 2010). "A directional lineage is characterized by the continuous evolution of two or more features in one direction and is a monophyletic-linear succession of species."
- BRANCHED LINEAGE (Georgescu 2010). "... at least one feature shows divergent evolution in a branched lineage resulting in a monophyletic-branched succession of species."
- ITERATIVE LINEAGE (Georgescu 2013a). "... is characterized by the successive and independent evolution of the same feature in the descendant species."
- CONDENSED LINEAGE (Georgescu 2014a). "...consists of only the initiating species and is characterized by the rapid evolution of a new morphological feature."

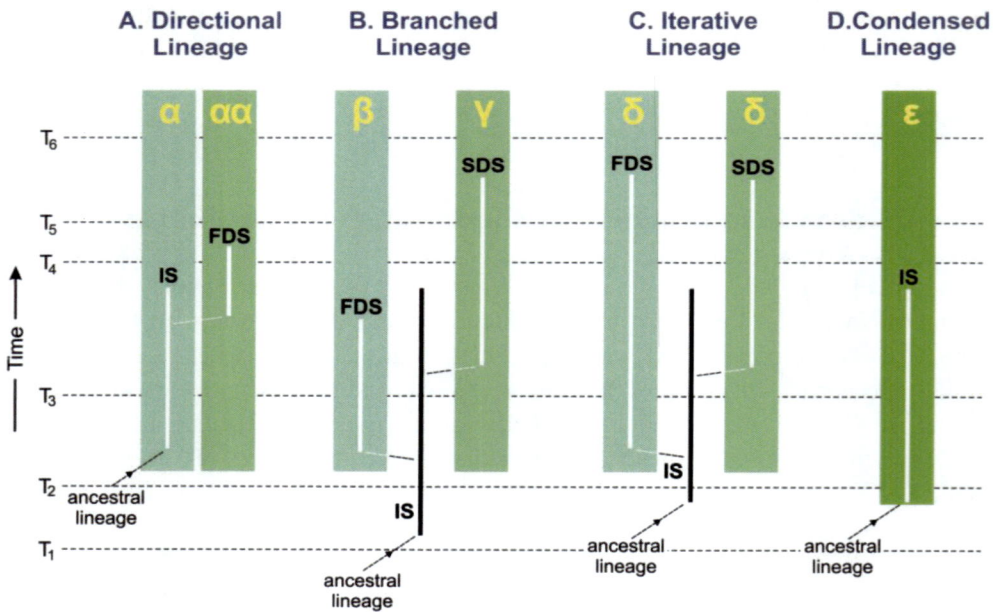

Figure 19. Kinds of lineages recognized in evolutionary classification. Species abbreviations: IS-initiating species, FDS-first descendant species, SDS-second descendant species. Morphological characters: α, β, γ, δ and ε. After Georgescu 2014a, this volume.

The evolutionary history of the Cretaceous planktics with at least one stage in which the chambers are alternately added with respect to the test growth axis shows that only directional, branched and condensed lineages occur, whereas the iterative lineages are known only from the planktics with the chambers added along a coiled axis. Lineage branching in the study group, which is herein assembled for the first time since the beginnings of the "ultrastructure revolution", shows that there is not a specific pattern or preferential position of a certain kind of lineage in the group's general architecture (Figure 20).

A second open system in the evolutionary classification was defined when Georgescu (2011b) labeled the species within one function of their timing of occurrence. According to this system the earliest species of the lineage is the initiating species (IS), its first descendant the first descendant species (FDS), second descendant in the lineage second descendant species (SDS), etc.

The openness of this system of species labelling is given by the fact that the number of species in one lineage is not *a priori* fixed and can increase as new species are discovered. This system was originally defined to help in comparing the evolution of praeglobotruncanid test appearance in three distinct directional lineages of late Albian-Santonian age (Georgescu 2011b) but was thoroughly used afterwards. Herein are provided the successive occurrences of the component species of all the lineages recognized in the study group.

The Fundamental Unit in Evolutionary Classification

The species represents the fundamental unit of the typological and phylogenetic classifications. *Typological concept of species* or *morphospecies*, which is based entirely on test morphological features, was the only used before the significant influx of new data during the "ultrastructure revolution". It is the oldest concept of species that is fundamentally based on the reference to a type and has its roots in the Idealist philosophy of the Ancient Greece. Although time-honoured it cannot accommodate the knowledge level of modern science, and especially the dynamic perspective of the species in the light of the Theory of Evolution; therefore, additional concepts of species were developed to emphasize the evolutionary character and morphological heterogeneity of such entities. Georgescu and Huber (2007, 2009) developed such a concept named *composite paleontological species* in order to having a working concept in the classification practice. According to this concept: "A composite paleontological species is the basic unit with taxonomic significance in the fossil record, and has the following characteristics: (1) it is monophyletic; (2) it has a distinct range of morphological variability, showing relative stability over a definable period of time and presenting relatively discrete evolutionary changes; (3) it is a morphologically heterogeneous and discontinuous entity, consisting of one or (mostly) more morphological and/or paleoecological varieties; (4) it has its own and continuous developmental history traceable in space and time, which can be directly derived from the fossil record; and (5) its existence and integrity can be tested not only by comparative morphological distinctiveness, but also by its response to paleoenvironmental and geological factors (e.g., paleoclimatic changes, sea-level fluctuations), as inferred from paleontology and related geological disciplines." (Georgescu and Huber 2009, p. 360). The concept of composite paleontological species allowed assigning more specimens to one species or another within a lineage, and therefore, was extensively used in the developments in evolutionary classification. Probably the most important of such developments was the use of the composite paleontological species in the definition of the various kinds of lineages function of their evolution pattern.

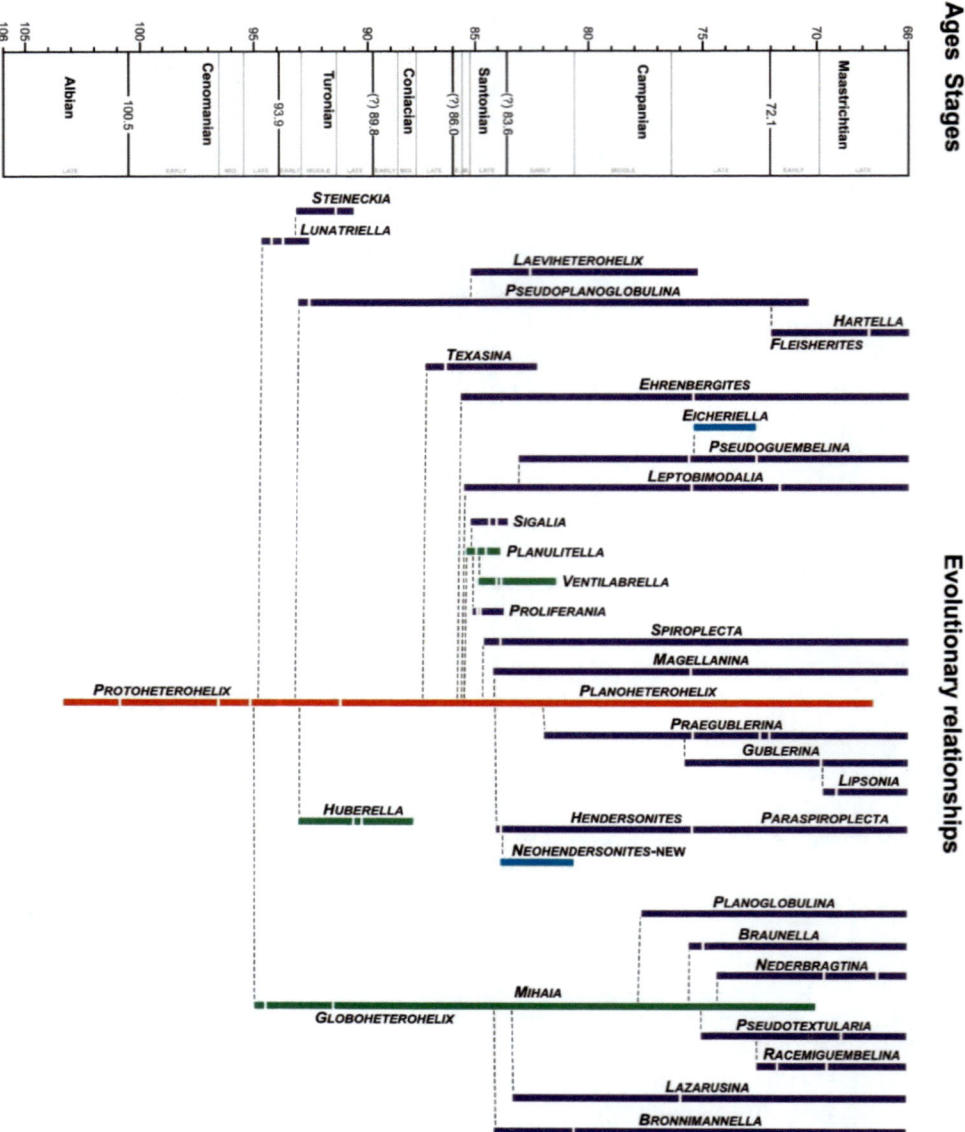

Figure 20. Lineages of planktic foraminifera with at least one growth stage in which the chambers are alternately added with respect to the test growth axis and evolutionary relationships between them. The earliest occurrences of the stages of morphological relative stability within each lineage are marked with a horizontal white line. Ages are after Gradstein and others (2012). The stem lineage is given in red, directional lineages in dark blue, branched lineages in green and condensed lineages in light blue.

The new advances in evolutionary classification were the effect of a significant influx of data and this impacted the way species are considered. The multitude of specimens with intermediate morphological features occurring throughout the stratigraphic range of a species and not only in the proximity of the speciation event can only be interpreted as one proof that species do not exist as distinct morphological entities. Therefore, the *lineage* is herein recognized as *fundamental unit of organization in evolutionary classification.*

One lineage has the following characteristics:

- Presents a variable degree of morphological distinctiveness when compared to other similar units;
- Is a succession of individuals or groupings of individuals (e.g., populations, etc) in ancestor-descendant relationships;
- There are morphological similarities between the individuals and groupings of individuals as a result of the common ancestry;
- Differences in morphology between the individuals and groupings of individuals of a lineage are of evolutionary nature or not;
- There are one or more stages of relative morphological stability during the lifespan of a lineage;
- Evolutionary history in space and time of a lineage can be reconstructed from the data of the fossil and rock records; neontological data can be added when a lineage has living representatives.

The concept of lineage is herein applied for thirty one such entities of Cretaceous planktic foraminifera with at least one stage with chambers alternately added with respect to the test growth axis. Each of the studied lineages consists of one to five composite paleontological species, which are considered stages of morphological relative stability (SMRS). The SMRS of one lineage are labelled according to the system given by Georgescu (2011b): ISMRS-initiating stage of morphological relative stability, FSMRS-first stage of morphological relative stability, SSMRS-second stage of morphological relative stability, etc.

EVOLUTIONARY NOMENCLATURE

The change in the nature of the fundamental unit in evolutionary classification from species to lineage provides new opportunities of development for this method. One of them is of nomenclatural nature and leads to the development of a new system of lineage naming. The new nomenclatural system developed herein is practical and based on the extensive occurrence of iterative evolution in the Cretaceous planktic foraminifera that have at least one ontogenetic stage in which the chambers are alternately added with respect to the test growth axis.

Lineage Branching and Naming

The reconstruction of the evolutionary history of the group shows that the oldest lineage is also the longest and presents the highest number of stages of morphological relative stability (Figure 21). This lineage is herein named Stalk in the evolutionary classification. The Stalk is the most prolific lineage in the group history; it directly developed into twelve other lineages, and every lineage of the group is its direct or indirect descendant. The succession of SMRSs of the Stalk was for the first time presented by Georgescu and Huber (2009) as a succession of two genera (*Protoheterohelix* and *Planoheterohelix*).

All the other lineages are named in the new system function of their morphological achievements, which are mainly apparent in the lineage terminal stages. Lineage names are given by a combination of one number and one name. The name represents a one-word description of the main morphological achievement, and four such names occur in the study group: alternate (chambers alternately added with respect to the test growth throughout the ontogeny), backextended (chambers present one or two backward extension/extensions), multichamber (adult stage presents chamber proliferation) and planispiral (early stage presents the chambers with one planispiral coil). The number, which is situated in front of the name, indicates the order of lineage evolution. For example the early planispiral coil evolved in four distinct lineages, which are named function of their initiation 1planispiral, 2planispiral, 3planispiral and 4planispiral. The evolution of the group indicates that alternate architecture occurs in seven lineages, the backextended one in four, the multichamber one in thirteen and the planispiral one in four. A different name is used for the lineages that evolved two or more features used in the nomenclature of the other lineages; this name is mixed. There are two mixed lineages, which are labeled 1mixed and 2mixed (Figure 21).

Iterative evolution in the study group is not restricted to the gross test architecture features used in the new nomenclatural system developed herein. For example, sutural ridges evolved independently in three lineages: 5multichamber, 6multichamber and 2mixed, ornamentation consisting of pore mounds in 2alternate and 4multichamber, leptoflanges in 1backextended, 6alternate and 12multichamber, etc. Therefore, the nomenclatural system presented herein is not unique, and the multiple iterative achievements may be used in further refining the group's evolutionary nomenclature.

Stages of Morphological Relative Stability Nomenclature

The stages of morphological relative stability (SMRS) are herein named by one letter in front of the lineage name. The initiating stage (ISMRS) of the lineage 3alternate for example is named ISMRS-3alternate or I-3alternate in shortened form. Similarly the first stage of morphological relative stability (FSMRS) is named FSMRS-3alternate or F-3alternate. Not all the specimens of one lineage can be assigned with precision to one stage of morphological relative stability or another.

The stage of morphological relative stability of one lineage from which another lineage evolved is herein named precursor stage of morphological relative stability or simply the precursor. Each lineage has one precursor stage of morphological relative stability and one such stage can be the potential precursor of more than one lineage (e.g., S-1alternate, which is the precursor of six lineages).

Nomenclature Compression Rate

Comparing the typological and evolutionary classification nomenclatures shows another advantage of using the evolutionary nomenclature. Only six English names are used in evolutionary classification to define the group's systematic, and this contrasts to the 110 Latin names used in the typological classification system. The rate of compression is higher than 18:1 (=one English word for more than 18 Latin names). The Appendix presents the taxonomic units that cannot be related to the study group through a direct ancestor-descendant relationship, and are probably the result of different originations.

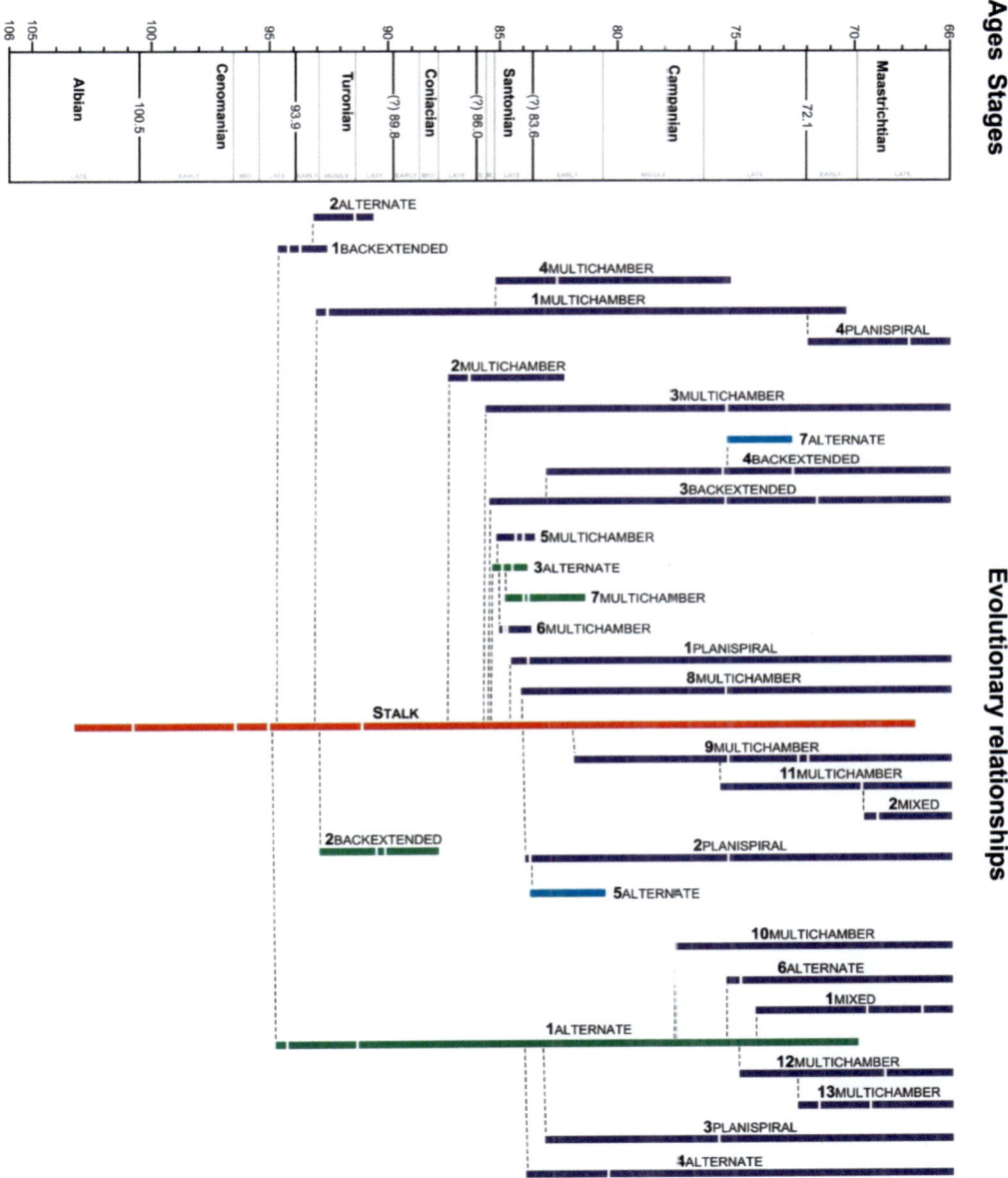

Figure 21. Evolutionary nomenclature of the Cretaceous planktic foraminiferal lineages with at least one growth stage in which the chambers are alternately added with respect to the test growth axis. Ages are after Gradstein and others (2012). The stem lineage is given in red, directional lineages in dark blue, branched lineages in green and condensed lineages in light blue.

LINEAGE DESCRIPTIONS

A description of the lineages of Cretaceous planktic foraminifera with chambers alternately added with respect to the test growth axis is given herein with emphasis on the lineage descent and main morphological changes along each lineage. Linnaean nomenclature is used in the case of the benthic ancestral lineage and the newly developed evolutionary classification and nomenclature for the representatives of the study group.

The Benthic Ancestors
(Figure 22)

The origin of the heterohelicid planktic foraminifera from the small-sized calcareous benthics was demonstrated by Georgescu (2009b) together with the report and description of the late Albian ancestor genus *Praeplanctonia* Georgescu 2009b consisting of two SMRS: *P. globifera* Georgescu 2009b and *P. quasiplanctonica* Georgescu 2009b. The general test architecture of the SMRS of *Praeplanctonia* shows an early triserial stage followed of a biserial adult. The test in the proximity of the subterminal aperture in the SMRS of *Praeplanctonia* also presents one supplementary suture that connects the aperture margin to the suture between the last-formed chambers, which is a characteristic feature of the SMRS included within the genus *Praeplanctonia*.

The supplementary suture was interpreted by Georgescu (2009b) as the result of the test wall wrapping around the aperture, and this process indicates that the ancestor of *P. globifera*, which was the oldest SMRS of *Praeplanctonia* known at that time, is among the pleurostomellid foraminifera such as *Pleurostomella obtusa* Berthelin 1880, a SMRS with large aperture at the base of the last-formed chamber.

Notably, the two SMRS of *Praeplanctonia* have smooth test wall and porosity consisting of a mixture of circular and elongate pores with a diameter or maximum dimension of 0.0002-0.0005 mm. The precise morphological significance of the elongate pores is not sufficiently understood, but they were interpreted by Georgescu (2009b) as formed in the course of the evolution process from pleurostomellid to praeplanctonid foraminifera. Evolution of *Praeplanctonia* was initially reported from the upper Albian sediments of the western North Atlantic Ocean, namely from the ODP Hole 1050C (offshore Florida), but subsequently SMRS of *Praeplanctonia* were encountered in the upper Albian sediments of the eastern North Atlantic Ocean (DSDP Sites 370 and 398) and eastern Indian Ocean (ODP Holes 762C and 763B).

Genus *Praeplanctonia* was transformed into a directional lineage by Georgescu and Burke in Georgescu, Burke and Heikkinen (2014) who described a new SMRS from the lower part of the upper Albian sediments of the eastern North Atlantic Ocean (DSDP Site 370, offshore Morocco): *P. atlantis*. The combination of large proloculus and deeply incised sutures was considered diagnostic for this SMRS. The occurrence of the supplementary suture between the subterminal ovate aperture and the suture between the last-formed two chambers indicates that this feature evolved early in the history of *Praeplanctonia*. Test porosity consisting of a mixture of circular and elongate pores with a diameter or maximum dimension of 0.0002-0.0007 mm further documents the position of this SMRS within the *Praeplanctonia* directional lineage as the elongate pores do not occur in the pleurostomellid taxa. Morphological features and stratigraphic position of *P. atlantis* further support the *Pleurostomella-Praeplanctonia* direct ancestor-descendant relationship.

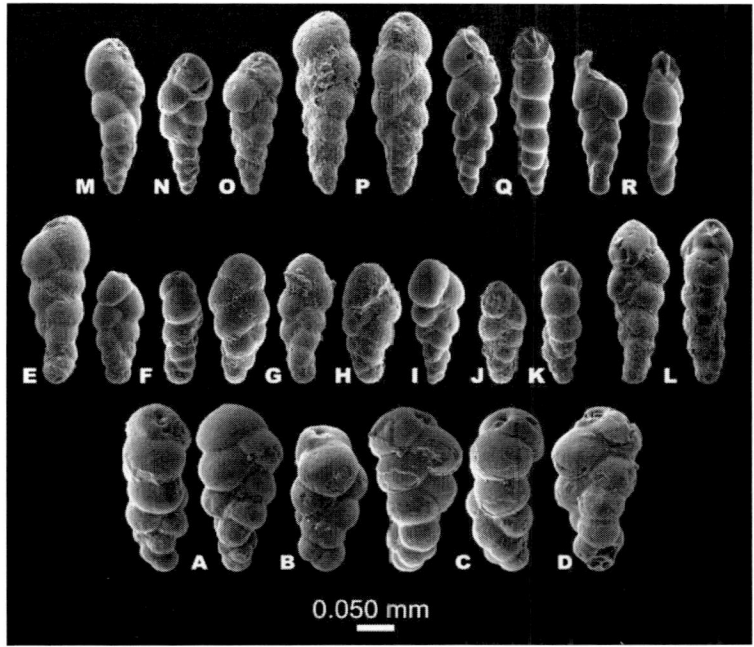

0.050 mm

Figure 22. Specimens of the benthic foraminiferal directional lineage *Praeplanctonia* Georgescu 2009; Linnaean nomenclature is thoroughly followed. **A-D** ISMRS: *Praeplanctonia atlantis* Burke and Georgescu in Georgescu, Burke and Heikkinen 2014; all specimens are of early late Albian age, were collected from Sample DSDP 41-370-26-4, 101-102 cm and previously illustrated by Georgescu, Burke and Heikkinen (2014) as follows: A-Figure 3:1; B-Figure3:2; C-Figure 3:3 and D-Figure 3:4. **E-L** FSMRS: *Praeplanctonia globifera* Georgescu 2009b. E-late Albian, Sample 171B-1050C-28-4, 66-69 cm, figured by Georgescu (2009b, pl. 2, Figure 7); F-late Albian, Sample 171B-1050C-28-4, 66-69 cm, figured by Georgescu (2009b, pl. 3, figures 1-2), G- late Albian, Sample 171B-1050C-28-4, 66-69 cm, figured by Georgescu (2009b, pl. 3, figures 5-6); H-late Albian, Sample 171B-1050C-28-4, 66-69 cm, figured by Georgescu (2009b, pl. 3, Figure 8); I-late Albian, Sample 171B-1050C-28-4, 66-69 cm, figured by Georgescu (2009b, pl. 3, Figure 10); J-late Albian, Sample 171B-1050C-28-4, 66-69 cm, pl. 3, Figure 12); K-late Albian, Sample 171B-1050C-27-3, 139-141 cm, figured by Georgescu (2009b, pl.4, Figure 1); L-late Albian, Sample 171B-1050C-27-4, 80-83 cm, figured by Georgescu (2009b, pl. 4, figures 5-6). **M-R** SSMRS: *Praeplanctonia quasiplanctonica* Georgescu 2009b. M-late Albian, Sample 171B-1050C-28-4, 66-69 cm, figured by Georgescu (2009b, pl. 4, Figure 9) N-late Albian, Sample 171B-1050C-28-4, 66-69 cm, figured by Georgescu (2009b, pl. 4, Figure 12); O-late Albian, Sample 171B-1050C-28-4, 66-69 cm, figured by Georgescu (2009b, pl. 4, Figure 13); P-late Albian, Sample 171B-1050C-28-4, 66-69 cm, figured by Georgescu (2009b, pl. 4, figures 14-15); Q-late Albian Sample 171B-1052E-40-3, 71-75 cm, figured by Georgescu (2009b, pl. 4, figures 16-17); R-late Albian Sample 171B-1052E-40-3, 71-75 cm, figured by Georgescu (2009b, pl. 4, figures 19-20).

As the result of such advances in the knowledge of the *Praeplanctonia* directional lineage it becomes now possible to define the three SMRS position: ISMRS =*Praeplanctonia atlantis*, FSMRS =*P. globifera* and SSMRS =*P. quasiplanctonica*. The analysis of the morphological features along the *Praeplanctonia* directional lineage shows that the twisted test axis, occurrence of the supplementary suture and porosity consisting of a mixture of circular and elongate pores are conservative features that occur throughout this directional lineage. Moreover, they also occur throughout the benthic directional lineage *Haigella* Georgescu 2009b, which evolved from *P. globifera*, and gradually developed chamber lateral projections. A significant morphological change in the *Praeplanctonia* lineage is observed in the proloculus size, which gradually decreases in diameter from 0.0396-0.0483 mm in ISMRS to 0.0280-0.0440 mm in FSMRS and to 0.0110-0.0140 mm in SSMRS; notably, the proloculus diameter in *Pleurostomella obtusa*, which is the presumed ancestor of the directional lineage *Praeplanctonia*, is of 0.0420-0.0550 mm. Another evolutionary trend observed throughout the stratigraphic range of the *Praeplanctonia* directional lineage is the reduction in the suture indentation.

According to Georgescu (2009b) the transition from a benthic to a planktic habitat involved morphological changes in chamber arrangement, aperture position and size, periapertural structures and pore size and shape. Two distinct lineages that led to the evolution of planktic foraminiferal habitat emerged from the benthic triserial-biserial *P. globifera* in the late Albian. The evolution of the two lineages was recorded in ODP Hole 1050C (Georgescu 2009b). The earliest lineage consists of tests with triserial throughout chamber arrangement and was formalized as genus *Archaeoguembelitria* Georgescu 2009b; it evolved through the loss of the adult biserial stage of *P. globifera*. The second lineage resulted in the evolution of heterohelicid planktics, which have the test consisting entirely of chambers alternately added with respect to the test growth axis resulting in biserial appearance; the earliest heterohelicid planktics were assigned to the genus *Protoheterohelix* Georgescu and Huber 2009 and they evolved through the loss of the early triserial stage of *P. globifera*. Georgescu, Burke and Heikkinen (2014) suggested that the evolution of the late Albian planktics from benthic ancestors is the result of the initiation of the Oceanic Anoxic Event OAE 1d in the terminal Albian and the evolutionary occurrences of the two lineages correspond to the earliest accumulation of black shales related to the OAE 1d in the Western North Atlantic Ocean.

The Stalk Lineage (DL: Stalk)
(Figure 23)

The initiation of the group of planktic foraminifera with chambers alternately added with respect to the test growth axis happened in the late Albian; the precursor SMRS is *Praeplanctonia globifera* as demonstrated by Georgescu (2009b) and Georgescu, Burke and Heikkinen (2014). Only one lineage existed at the beginning of the group evolution till the earliest group diversification that happened in the late Cenomanian. The representatives of this lineage occur throughout the evolutionary history of the group, namely from the late Albian to the late Maastrichtian. The general architecture of the lineage branching of this group shows that this lineage, which is herein named stalk lineage (S), is the precursor for most of the other of the group's lineages. The pattern of the stalk lineage is directional and it consists of five SMRSs: ISMRS=*Gümbelina washitensis* Tappan 1940 (S1), FSMRS=*Protoheterohelix obscura* Georgescu and Huber 2009 (S2), SSMRS=*Gümbelina moremani* Cushman 1938 (S3), TSMRS=*Planoheterohelix postmoremani* Georgescu and Huber 2009 (S4) and FoSMRS= *Gümbelina planata* Cushman 1938 (S5). The stalk lineage of this group is herein recognized for the first time. The succession of the five SMRS was given by Georgescu and Huber (2009), which separated them as genera *Protoheterohelix* Georgescu and Huber 2009 and *Planoheterohelix* Georgescu and Huber 2009 to include the asymmetrical and symmetrical tests respectively; such separation can be made in typological classification in order to distinguish between two types of test with such different appearance but is subjective in evolutionary classification and therefore cannot be accepted.

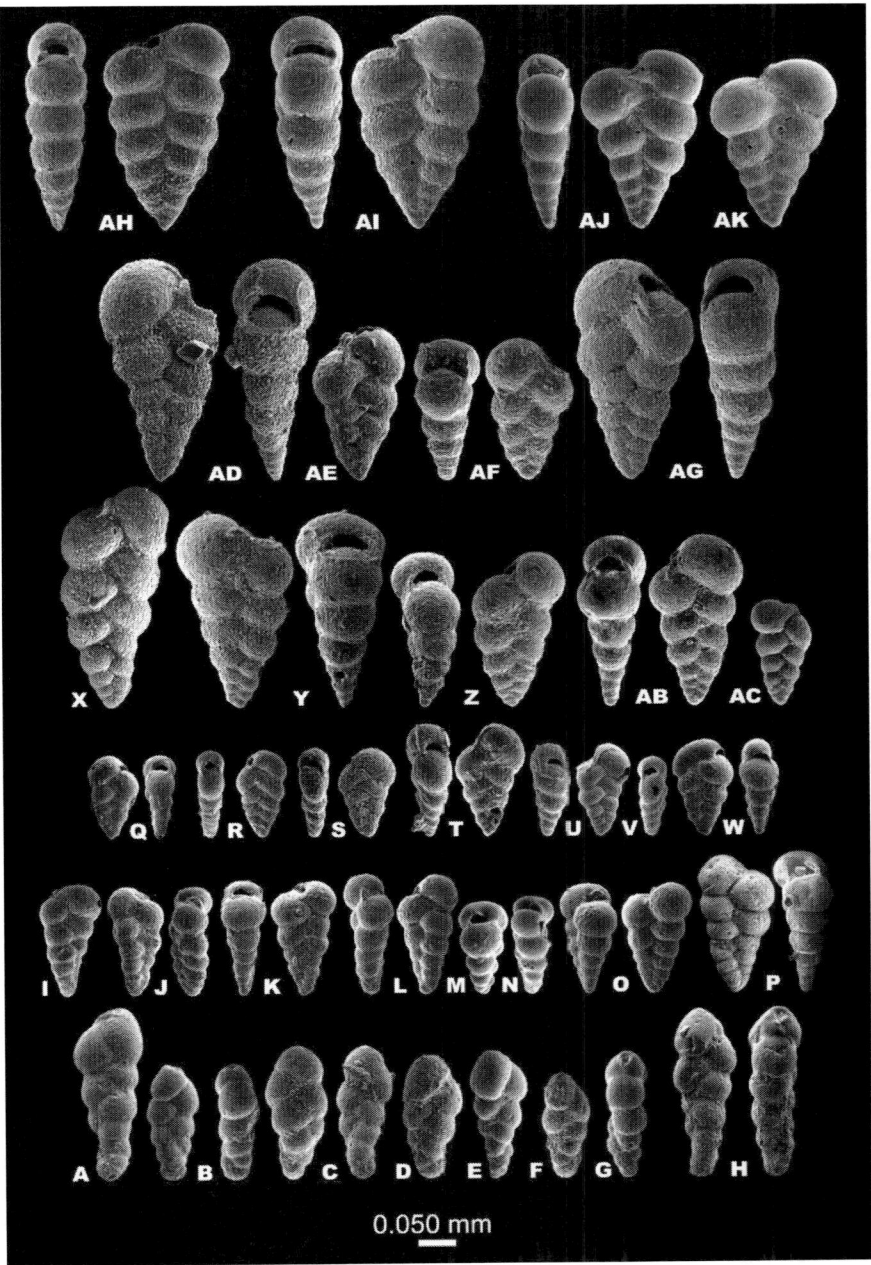

Figure 23. Continued.

Figure 23. Specimens of the stalk directional lineage of the Cretaceous planktic with chambers alternately added with respect to the test growth axis; Linnaean nomenclature is used only for the precursor benthic SMRS. **A-H** Precursor SMRS: *Praeplanctonia globifera* Georgescu 2009. A-late Albian, Sample 171B-1050C-28-4, 66-69 cm, figured by Georgescu (2009b, pl. 2, Figure 7); B-late Albian, Sample 171B-1050C-28-4, 66-69 cm, figured by Georgescu (2009b, pl. 3, figures 1-2); C-late Albian, Sample 171B-1050C-28-4, 66-69 cm, figured by Georgescu (2009b, pl. 3, figures 5-6); D-late Albian, Sample 171B-1050C-28-4, 66-69 cm, figured by Georgescu (2009b, pl. 3, Figure 8); E-late Albian, Sample 171B-1050C-28-4, 66-69 cm, figured by Georgescu (2009b, pl. 3, Figure 10); F-late Albian, Sample 171B-1050C-28-4, 66-69 cm, figured by Georgescu (2009b, pl. 3, Figure 12); G-late Albian, Sample 171B-1050C-27-3, 139-141 cm, figured by Georgescu (2009b, pl.4, Figure 1); H-late Albian, Sample 171B-1050C-27-4, 80-83 cm, figured by Georgescu (2009b, pl. 4, figures 5-6). **I-P** ISMRS: S1. I-late Albian, Sample 171B-1050C-28-1, 70-73 cm, figured by Georgescu and Huber (2009, pl. 1, Figure 14); J-late Albian, Sample 171B-1050C-27-3, 30-32 cm, figured by Georgescu and Huber (2009, pl. 1, figures 12-13); K-early Cenomanian, Sample 171B-1050C-26-1, 7-9 cm, figured by Georgescu and Huber (2009, pl. 1, figures 8-9); L-early Cenomanian, Sample 171B-1050C-27-3, 30-32 cm, figured by Georgescu and Huber (2009b, pl. 1, figures 6-7); M-late Albian, Sample 171B-1052E-40-2, 118-121 cm, figured by Georgescu (2009b, pl. 1, Figure 7) and Georgescu and Huber (2009, pl. 1, Figure 5); N-late Albian, Sample 171B-1052E-40-2, 118-121 cm, figured by Georgescu (2009, pl. 1, Figure 8) and Georgescu and Huber (2009, pl. 1, Figure 10); O-late Albian, Del Rio Formation (Austin, Travis County, Texas), figured by Georgescu and Huber (2009, pl. 1, figures 3-4); P-late Albian, Grayson Bluff, Texas, originally illustrated by Tappan (1940) and refigured by Georgescu (2009b, pl. 1, figures 3-4) and Georgescu and Huber (Pl. 1, figures 1-2). **Q-W** FSMRS: S2. Q-early Cenomanian, Sample 171B-1050C-25-2, 70-72 cm, figured by Georgescu and Huber (2009, pl. 3, figures 10-11); R-early Cenomanian, Sample 171B-1050C-25-2, 70-72 cm, figured by Georgescu and Huber (2009, pl. 3, figures 7-8); S-early Cenomanian, Sample 171B-1050C-25-2, 70-72 cm, figured by Georgescu and Huber (2009, pl. 3, figures 4-5); T-early Cenomanian, Sample 171B-1050C-25-2, 83-87 cm, figured by Georgescu and Huber (2009, pl. 3, figures 1-2); U-early Cenomanian, Sample 171B-1050C-25-2, 70-72 cm, figured by Georgescu and Huber (2009, pl. 2, figures 9-10); V-early Cenomanian, Sample 171B-1050C-25-2, 70-72 cm, figured by Georgescu and Huber (2009, pl. 2, Figure 11); W-early Cenomanian, Sample 171B-1050C-25-2, 70-72 cm, figured by Georgescu and Huber (2009, pl. 2, figures 5-6). **X-AC** SSMRS: S3. X-late Cenomanian, lower part of the Eagle Ford Shale (Itasca, Hill County, Texas), figured by Georgescu and Huber (2009, pl. 4, Figure 3); Y-late Cenomanian, Sample 171B-1050C-21-1, 105-106 cm, figured by Georgescu and Huber (2009, pl. 4, figures 4-5); Z-late Cenomanian, Sample 171B-1050C-21-1, 105-106 cm, figured by Georgescu and Huber (2009, pl. 4, figures 10-11); AB-late Cenomanian, Sample 171B-1050C-21-1, 105-106 cm, figured by Georgescu and Huber (2009, pl. 4, figures 6-7); AC-early Turonian, Sample 171B-1050C-20-4, 110-111 cm, figured by Georgescu and Huber (2009, pl. 4, Figure 9). **AD-AG** TSMRS: S4. AD-early Turonian, Sample 62-463-35-1, figures 17-19 cm, figured by Georgescu and Huber (2009, pl. 5, figures 4-5); AE-early Turonian, Sample 62-463-35-1, figures 17-19 cm, figured by Georgescu and Huber (2009, pl. 5, Figure 9); AF-early Turonian, Sample 171B-1050C-20-4, 110-111 cm, figured by Georgescu and Huber (2009, pl. 5, figures 7-8); AG-early Turonian, Sample 62-463-35-1, figures 17-19 cm, figured by Georgescu and Huber (2009, pl. 5, figures 1-2). **AH-AK** FoSMRS: S5. AH-early Maastrichtian, Sample 113-690C-19-1, 119-123 cm. AI-early Maastrichtian, Sample 113-690C-20-5, 75-76 cm. AJ-late Santonian, Sample 174AX, 505.35-505.38 m, figured by Georgescu, Saupe and Huber (2008, pl. 3, Figure 2). AK-late Santonian, Sample 174AX, 505.35-505.38 m.

The ISMRS is the only known completely asymmetrical SMRS in the entire history of the planktic foraminifera with chambers alternately added with respect to the test growth axis. The test growth axis is gently twisted and this feature is one vestige that demonstrates the group's praeplanctonid ancestry. Periapertural structures are asymmetrical: one rim on one side of the aperture and one subrounded ridge on the other; such periapertural structures are morphologically closer to the praeplanctonid precursor than to any other of the SMRS that evolved subsequently in the group history. The asymmetrical morphological features (i.e., test twisted axis and periapertural structures) of S1 were observed only after extensive study under the SEM (Georgescu 2009; Georgescu and Huber 2009). Chamber surface is smooth, without any ornamentation features.

Pores are circular, rarer than in the praeplanctonid ancestor and with a diameter of 0.0004-0.0007 mm; the pores with elongate shape that are frequent occurrences in the precursor SMRS are no longer encountered in S1 and Georgescu (2009b) inferred that the loss of elongate pores a phenomenon associated with the transition from the benthic to planktic habitat. The next evolutionary step in the stalk lineage evolution is the achievement of symmetrical tests, which occurs for the first time in S2. Periapertural structures are still asymmetrical and Georgescu (2010) defined them as archaeoflanges; S2 is the only known SMRS of Cretaceous planktic foraminifera with chambers alternately added with respect to the test growth axis that present this kind of periapertural structures. Chamber surface in S2 is smooth, without ornamentation elements. Pores are circular and simple and present an increase in diameter when compared to S1, from 0.0004-0.0007 mm to 0.0004-0.0009 mm.

The first occurrence of completely symmetrical tests and periapertural structures is known in the middle Cenomanian in the SMRS S3. In this SMRS the periapertural structures consist of orthoflanges, which are symmetrically developed, one on each side of the test. Chamber surface in the tests of this SMRS is smooth and with circular pores with a diameter of 0.0005-0.0007 mm.

Georgescu and Huber (2009) mentioned the occurrence of aligned pustules and incipient costae that occasionally occur over the earlier portion of the test; the re-evaluation of these structures indicate that they are rather preservation artifacts and the chamber surface in S3 should be considered smooth. The next evolutionary step in the stalk directional lineage evolution is represented by the development of pustulose or granular leptocostae, which evolved in S4; in this stage the primitive character of ornamentation, is given by the aligned and non-fused pustules. Pores are circular and situated in the space between the pustulose leptocostae, rarely interrupting them; pores diameter is of 0.0005-0.0009 mm. In the tests of S4 the periapertural structures consist of symmetrically developed orthoflanges, one on each test side. The terminal evolutionary stage (S5) of the stalk directional lineage marks the evolution of leptocostate ornamentation in which the longitudinal leptocostae have a solid, discontinuous appearance and a thickness of 0.0018-0.0025 mm; this morphological achievement is first recorded in the stalk lineage in the late Turonian. Periapertural structures in S5 are symmetrical, one on each side of the test and consist of a mixture of orthoflanges and metaflanges. Pores are simple and circular in shape and have a diameter of 0.0006-0.0009 mm.

Earliest Evolution of Globular Chambers (BL: 1alternate)
(Figure 24)

Globular-chambered heterohelicids were reported in the past from throughout the Upper Cretaceous sediments. The earliest known occurrence is of late Cenomanian age (Georgescu and Huber 2009). A taxonomic solution for the earliest heterohelicids that evolved globular chambers was given by Georgescu in Georgescu and others (2013) who reviewed all the previous reports and based on them and newly collected material recognized the directional lineage *Mihaia* consisting of two SMRS: ISMRS=*M. mihaii* Georgescu in Georgescu and others 2013 and FSMRS =*M. reussi* (Cushman 1938). This lineage is herein revised and transformed into a branched lineage that includes the monospecific genus *Globoheterohelix* Georgescu and Huber 2009 together with the directional lineage *Mihaia*, as follows: ISMRS=*Mihaia mihaii* Georgescu in Georgescu and others 2013 (I-1alternate), FSMRS =*Globoheterohelix paraglobulosa* Georgescu and Huber 2009 (F-1alternate) and SSMRS =*Gümbelina reussi* Cushman 1938 (S-1alternate).

Branched lineage 1alternate evolved during the late Cenomanian from the earliest costate SMRS of the stalk, namely S4 (Georgescu and Huber 2009). The ornamentation in S4 is formed of longitudinal leptocostae and the ultrastructural study of these ornamentation elements shows that they consist of aligned pustules. Georgescu and Huber (2009) used the term 'granular costae' to describe their appearance; the ornamentation in the descendant

SMRS I-1alternate is almost identical to that in S4. Pores are circular in shape are situated in the space between the pustulose leptocostae and have a diameter of 0.0005-0.0009 mm in S4 and 0.0006-0.0008 mm in I-1alternate.

Figure 24. Specimens of the branched lineage 1alternate. **A-D** Precursor SMRS: S4. A-early Turonian, Sample 62-463-35-1, figures 17-19 cm, figured by Georgescu and Huber (2009, pl. 5, figures 4-5); B-early Turonian, Sample 62-463-35-1, figures 17-19 cm, figured by Georgescu and Huber (2009, pl. 5, Figure 9); C-early Turonian, Sample 171B-1050C-20-4, 110-111 cm, figured by Georgescu and Huber (2009, pl. 5, figures 7-8); D-early Turonian, Sample 62-463-35-1, figures 17-19 cm, figured by Georgescu and Huber (2009, pl. 5, figures 1-2). **E-J** ISMRS: I-1alternate. E-late Cenomanian, Pueblo Section, Colorado, figured by Georgescu in Georgescu and others (2013, pl. 6, Figure 11). F-late Cenomanian, Pueblo Section, Colorado, figured by Georgescu in Georgescu and others (2013, pl. 6, Figure 10). G-early Turonian, Sample 62-463-34-1, 79-81 cm, figured by Georgescu in Georgescu and others (2013, pl. 6, Figure 9). H-early Turonian, Sample 62-463-34-1, 79-81 cm, figured by Georgescu in Georgescu and others (2013, pl. 6, figures 2-3). I-early Turonian, Sample 62-463-34-2, 53-55 cm, figured by Georgescu in Georgescu and others (2013, pl. 6, figures 7-8). J-early Turonian, Sample 62-463-34-2, 53-55 cm, figured by Georgescu in Georgescu and others (2013, pl. 6, figures 4-5). **K-M** FSMRS: F-1alternate. K-early Turonian, Sample 62-463-35-1, 17-19 cm, figured by Georgescu and Huber (2009, pl. 7, Figure 1 only). L-early Turonian, Sample 62-463-35-1, 17-19 cm, figured by Georgescu and Huber (2009, pl. 7, figures 4-5). M-early Turonian, Sample 62-463-35-1, 17-19 cm, figured by Georgescu and Huber (2009, pl. 6, figures 9-10). **N-Q** SSMRS: S-1alternate. N-Coniacian to early Santonian, 122-763B-18-1, 72-73 cm, figured by Georgescu in Georgescu and others (2013, pl. 7, Figure 1 only). O-early Campanian, Sample 71-511-38-6, 13-17 cm, figured by Georgescu in Georgescu and others (2013, pl. 7, Figure 5 only). P-late Santonian, Sample 1595b, Ehrenberg Collection, Missouri River Basin. Q-Santonian, Toolonga Calcilutite, western Australia, Sample Huber, figured by Georgescu in Georgescu and others (2013, pl. 7, Figure 9 only).

In addition the periapertural structures of S4 and I-1alternate consist of symmetrically developed orthoflanges. Morphological differences between S4 and I-1alternate are mostly apparent in the test lateral compression and less prominent sutures in the ancestor and subglobular chambers separated by deeper incised sutures in the descendant; it was also observed an increase in proloculus size from 0.0082-0.0143 mm in the ancestor to 0.0156-0.0174 mm in the descendant.

Georgescu and Huber (2009) reported from the upper Cenomanian sediments of the ODP Hole 1050C a lower occurrence of F-1alternate when compared to those of S4 and I-1alternate. This bioevent succession appears incomplete at this site and lower occurrences of I-1alternate and F-1alternate from below that of S4 are herein reported from DSDP Site 463.

Evolution from I-1alternate towards the two descendants happened in the late Cenomanian for F-1alternate and late Turonian for S-1alternate; both processes are marked by an increase in test size. A pustulose periapertural area in the anterior portion of the chamber occurs in most of the specimens of the S-1alternate SMRS The 1alternate lineage presents a somewhat dual feature: it is iterative in the development of longitudinal solid leptocostae, metaflanges and globular chambers and divergent in the proloculus size (0.0156-0.0174 mm in I-1alternate, 0.0310-0.0560 mm in F-1alternate and 0.0137-0.0220 mm in S-1alternate). Morphological features changes along the 1alternate lineage indicate that F-1alternate is rather a collateral unit that became extinct without leaving descendants, whereas S-1alternate gave birth to a large number of other lineages.

Earliest Evolution of Peripherally Elongate Chambers (DL: 1backextended)
(Figure 25)

The earliest lineage that evolved backward peripherally elongate chambers in the group evolutionary history was *Lunatriella*, which was initially defined as monotypic genus in typological classification by Eicher and Worstell (1970a). It was redefined as directional lineage in evolutionary classification by Georgescu (2013a) and in this form includes three SMRS: ISMRS=*Heterohelix fayose* Petters 1983 (I-1backextended), FSMRS =*H. digitata* Masella 1959 (F-1backextended) and SSMRS =*Lunatriella spinifera* Eicher and Worstell 1970a (S-1backextended). Directional lineage 1backextended evolved from the smooth S3 which is its precursor SMRS and appears the only lineage that evolved from the last unornamented SMRS of the stalk. The representatives of 1backextended are of late Cenomanian-early Turonian age and the stratigraphical interval in which tests with lateral backward chamber extension occur stretches the Cenomanian/Turonian boundary. Directional lineage 1backextended occurrences are restricted to the southern Northern Atlantic Ocean, United States Western Interior and western Tethyan Realm (Georgescu 2013a).

The direct evolutionary relationship from the ancestor S3 to the descendant I-1backextended is demonstrated by the similarities in the periapertural structures consisting of orthoflanges, smooth chamber surface and pore size (0.0005-0.0007 mm in S3 and 0.0009-0.0010 mm in I-1backextended). The morphological differences are represented by the development of more compressed tests and a change in appearance of the chamber arrangement in the adult stage of the descendant where presents loose biserial in appearance.

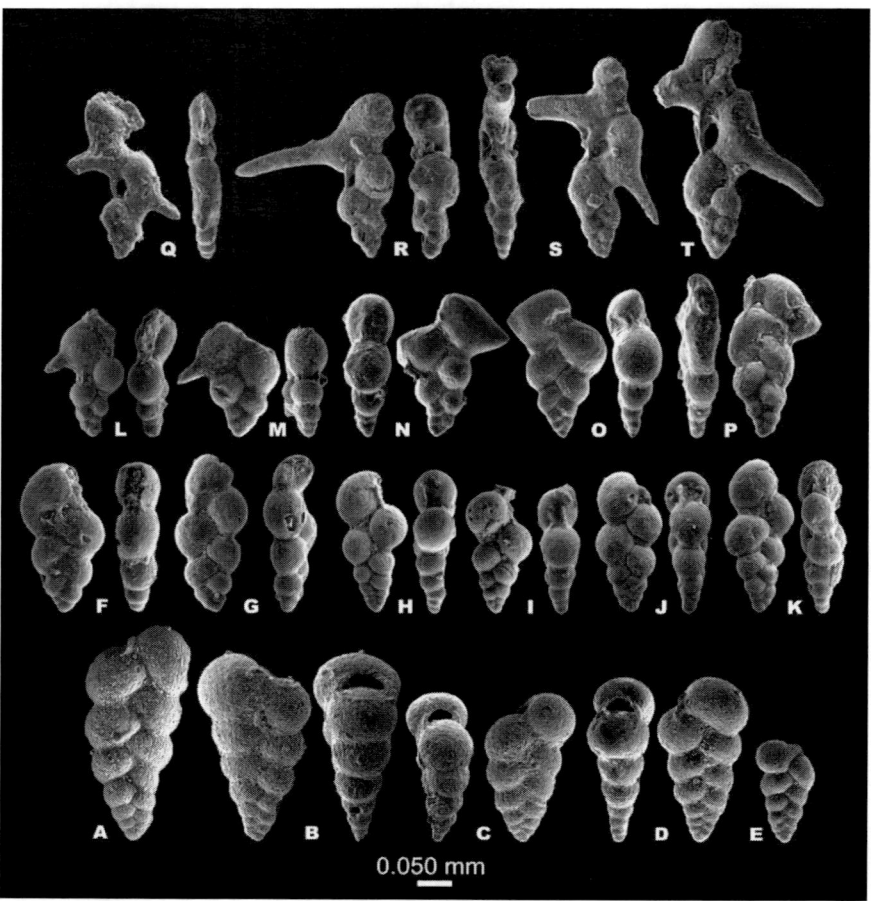

Figure 25. Specimens of the directional lineage 1backextended. **A-E** Precursor SMRS: S3. A-late Cenomanian, lower part of the Eagle Ford Shale (Itasca, Hill County, Texas), figured by Georgescu and Huber (2009, pl. 4, Figure 3); B-late Cenomanian, Sample 171B-1050C-21-1, 105-106 cm, figured by Georgescu and Huber (2009, pl. 4, figures 4-5); C-late Cenomanian, Sample 171B-1050C-21-1, 105-106 cm, figured by Georgescu and Huber (2009, pl. 4, figures 10-11); D-late Cenomanian, Sample 171B-1050C-21-1, 105-106 cm, figured by Georgescu and Huber (2009, pl. 4, figures 6-7); E-early Turonian, Sample 171B-1050C-20-4, 110-111 cm, figured by Georgescu and Huber (2009, pl. 4, Figure 9). **F-K** ISMRS: I-1backextended. F-early Turonian, Sample 62, locality 5, Hamilton County, Kansas of Eicher and Worstell (1970b), figured by Eicher and Worstell (1970a, pl. 1, Figure 1), Eicher and Worstell (1970b, pl. 8, Figure 9) and Georgescu (2013c, Figure 2: 1-2). G-early Turonian, Sample 62, locality 5, Hamilton County, Kansas of Eicher and Worstell (1970b). H-early Turonian, Sample 62, locality 5, Hamilton County, Kansas of Eicher and Worstell (1970b), figured by Georgescu (2013c, Figure 2: 14 only). I-early Turonian, Sample 62, locality 5, Hamilton County, Kansas of Eicher and Worstell (1970b). J-early Turonian, Sample 62, locality 5, Hamilton County, Kansas of Eicher and Worstell (1970b), figured by Georgescu (2013c, Figure 2: 8-9). K-early Turonian, Sample 62, locality 5, Hamilton County, Kansas of Eicher and Worstell (1970), figured by Georgescu (2013c, Figure 2:6-7). **L-P** FSMRS: F-1backextended. L-early Turonian, Sample 62, locality 5, Hamilton County, Kansas of Eicher and Worstell (1970b). M-early Turonian, Sample 62, locality 5, Hamilton County, Kansas of Eicher and Worstell (1970b). N-latest Cenomanian to earliest Turonian, Sample 1651-1660 m, well R2d, Senegal, figured by Georgescu and Huber (2009, pl. 6, figures 1-2) and Georgescu (2013c, Figure 3: 6-7). O-latest Cenomanian to earliest Turonian, Sample 1651-1660 m, well R2d, Senegal, figured by Georgescu and Huber (2009, pl. 6, figures 4-5) and Georgescu (2013c, Figure 3: 10-11). P-early Turonian, Sample 62, locality 5, Hamilton County, Kansas of Eicher and Worstell (1970b), figured by Eicher and Worstell (1970a, pl. 1, Figure 4) and Georgescu (2013c, Figure 3: 2-3). **Q-T** SSMRS: S-1backextended. Q-early Turonian, Sample 62, locality 5, Hamilton County, Kansas of Eicher and Worstell (1970b). R-early Turonian, Sample 62, locality 5, Hamilton County, Kansas of Eicher and Worstell (1970b), figured by Eicher and Worstell (1970a, pl. 1, Figure 12). S-early Turonian, Sample 62, locality 5, Hamilton County, Kansas of Eicher and Worstell (197b0), figured by Eicher and Worstell (1970a, pl. 1, Figure 7; 1970b, pl. 8, Figure 12) and Georgescu (2013c, Figure 4: 11-12). T-early Turonian, Sample 62, locality 5, Hamilton County, Kansas of Eicher and Worstell (1970b), figured by Eicher and Worstell (1970a, pl. 1, Figure 6a) and Georgescu (2013c, Figure 4: 2).

Along the 1backextended lineage there is a gradual development of the lateral backward chamber extension, which is in incipient stage in F-1backextended and best developed in S-1backextended. Notably, well-developed leptoflanges evolved in the S-1backextended.

There is a decrease in test size in the evolution from the precursor S3 to I-1backextended as the 1backextended directional lineage initiated. There are no apparent changes in size in the course of the 1backextended lineage evolution, namely between I-1backextended and F-1backextended. Some larger tests are recorded in the S-1backextended and seemingly these are produced by the well-developed loose biserial chamber arrangement in the adult test portion.

Earliest Evolution of Pore-Mounded Ornamentation (DL: 2alternate)
(Figure 26)

This lineage evolved chamber ornamentation consisting of pore mounds for the first time in the history of the planktic foraminifera with alternately added chambers and was formally described by Georgescu (2009a) as a monospecific genus: *Steineckia*, which included the SMRS *S. steinecki* Georgescu 2009a. It was recognized from *Laeviheterohelix* Nederbragt 1991 primarily by the larger pore mounds and stratigraphical ranges that do not overlap. The origin of *Steineckia* was considered from a taxon designated as "*Heterohelix*" sp. (Georgescu 2009a, Figure 17), which was later assigned to a new species and genus, *Protoheterohelix obscura* by Georgescu and Huber (2009); the new name of *P. obscura* in evolutionary classification nomenclature is herein given as S2. New material especially from the United States Western Interior allows the reassessment of this lineage. Therefore, it becomes possible to redefine *Steineckia* in evolutionary classification as the directional linage 2alternate, which consists of two SMRS: ISMRS=*Steineckia* sp. Georgescu 2013c (I-2alternate) and FSMRS =*Steineckia steinecki* Georgescu 2009a (F-2alternate).

Directional lineage 2alternate evolved in the latest Cenomanian and specimens assignable to I-2alternate were illustrated by Eicher and Worstell (1970a, 1970b) as *Heterohelix pulchra* (Brotzen); re-examination of the illustrated specimens allowed Georgescu (2013c) to assign them to *Steineckia* sp. The tests present a loose biserial chamber arrangement in the adult stage and the last-formed chambers are oblique to the test growth axis, which is a chamber arrangement similar to that known in I-1backextended and helps in understanding the origin of 2alternate; the morphological features that separate I-2alternate of its ancestor I-1backextended are the larger number of chambers and ornamentation consisting of pore mounds rather than having smooth chamber surface.

Therefore, the SMRS I-1backextended represents the precursor of 2alternate directional lineage. A decrease in pore diameter is recorded with the evolution of the 2alternate lineage, from 0.0009-0.0010 mm in I-1backextended to 0.0005-0.0008 mm in I-2alternate. Two specimens of I-2alternate with multichamber growth in the adult stage were found during the re-study of the material from the United States Western Interior; the specimens occur in the Blue Hill Shale Member of the Carlile Shale in Kansas (early Turonian in age).

At the present state of knowledge it appears premature to assign them to a distinct SMRS and consequently transform the directional lineage 2alternate into a branched one; more material is necessary in order to recognize these tests as distinct SMRS. Specimens with multichamber growth in the adult stage reported from Gabon by de Klasz and others (1969) and Senegal by de Klasz and others (1995) present different general test architecture and chamber ornamentation and cannot be included within the directional lineage 2alternate; additional studies are necessary in order to clarify their taxonomic status.

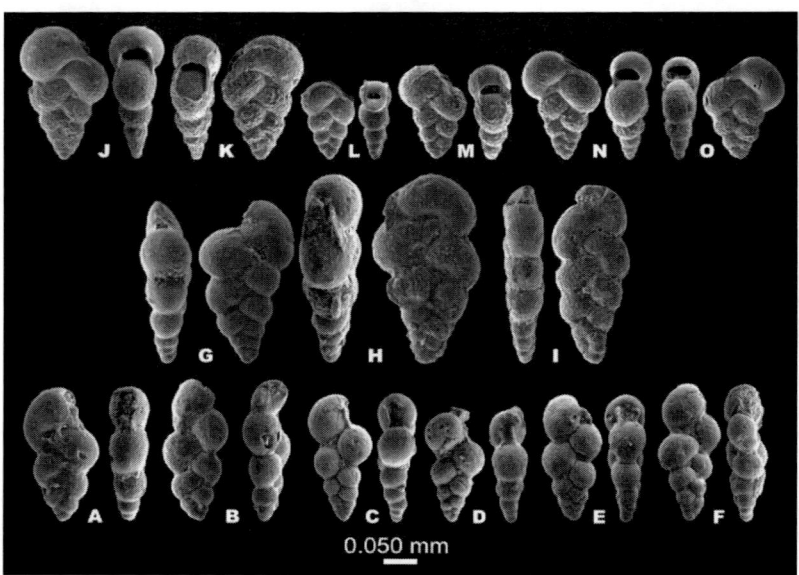

Figure 26. Specimens of the directional lineage 2alternate. **A-F** Precursor SMRS: I-1backextended. A-early Turonian, Sample 62, locality 5, Hamilton County, Kansas of Eicher and Worstell (1970b), figured by Eicher and Worstell (1970a, pl. 1, Figure 1), Eicher and Worstell (1970b, pl. 8, Figure 9) and Georgescu (2013c, Figure 2: 1-2). B-early Turonian, Sample 62, locality 5, Hamilton County, Kansas of Eicher and Worstell (1970b). C-early Turonian, Sample 62, locality 5, Hamilton County, Kansas of Eicher and Worstell (1970b), figured by Georgescu (2013c, Figure 2: 14 only). D-early Turonian, Sample 62, locality 5, Hamilton County, Kansas of Eicher and Worstell (1970b). E-early Turonian, Sample 62, locality 5, Hamilton County, Kansas of Eicher and Worstell (1970b), figured by Georgescu (2013c, Figure 2: 8-9). F-early Turonian, Sample 62, locality 5, Hamilton County, Kansas of Eicher and Worstell (1970b), figured by Georgescu (2013c, Figure 2:6-7). **G-I** ISMRS: I-2alternate. G-early Turonian, Sample 62, locality 5, Hamilton County, Kansas of Eicher and Worstell (1970b), figured by Georgescu (2013c, Figure 5:11-12). H-early Turonian, Sample 62, locality 5, Hamilton County, Kansas of Eicher and Worstell (1970b), figured by Eicher and Worstell (1970a, pl. 1, Figure 3), Eicher and Worstell (1970b, pl. 8, Figure 10) and Georgescu (2013c, Figure 5: 7-8). I-early Turonian, Sample 62, locality 5, Hamilton County, Kansas of Eicher and Worstell (1970b), figured by Georgescu (2013c, Figure 5: 2-3). **J-O** FSMRS: F-2alternate. J-late Turonian, Sample 71-511-47-6, 24.5-25.5 cm. K-late Turonian, Sample 71-511-47-6, 24.5-25.5 cm. L-late Turonian, Sample 71-511-47-5, 13.5-14.5 cm. M-late Turonian, Sample 71-511-47-5, 13.5-14.5 cm, figured by Georgescu (2009c, Figure 9: 2). N-late Turonian, Sample 71-511-47-5, 13.5-14.5 cm, figured by Georgescu (2009c, Figure 9: 2). O-late Turonian, Sample 71-511-47-6, 24.5-25.5 cm, figured by Georgescu (2009c, Figure 9: 1).

The only known occurrence of F-2alternate is in the late Turonian of Southern Atlantic Ocean (DSDP Site 511). Therefore, there is a gap between the last known occurrence of I-2alternate, which is in the lower Turonian sediments and the evolutionary occurrence of F-2alternate from the sediments situated in a higher stratigraphical position; this shows that new data are necessary in order to understand better the evolution within the directional lineage 2alternate and the transition between its two SMRS. Test general architecture in the two SMRS of this directional lineage shows that both present alternate chamber addition with respect to the test growth axis throughout the ontogeny; the loose biserial chamber arrangement in I-2alternate is not so well-developed in F-2alternate.

Chamber surface is ornamented with scattered pore mounds with a diameter of 0.0031-0.0050 mm in I-2alternate and 0.0040-0.0064 mm in F-2alternate; rare pore mounds with smaller diameters occasionally occur especially in I-2alternate. Notably, none of the specimens assigned to the lineage 2alternate evolved a periapertural pustulose area.

Earliest Evolution of Multichamber Growth (DL: 1multichamber)
(Figure 27)

Initiation of the earliest lineage that will lead to the evolution of multichamber growth in the adult stage happened in the early Turonian; this lineage is herein named 1multichamber and includes two SMRS: ISMRS=*Gümbelina turonica pseudotesseriformis* Agalarova in Djafarov and others 1951 (I-1multichamber) and FSMRS=*Pseudoplanoglobulina nakhitschevanica* Aliyulla 1977 (F-1multichamber). The genus *Pseudoplanoglobulina* Aliyulla 1977 was defined to accommodate the taxa with incipient multichamber growth in the adult stage and was validated by Loeblich and Tappan (1987).

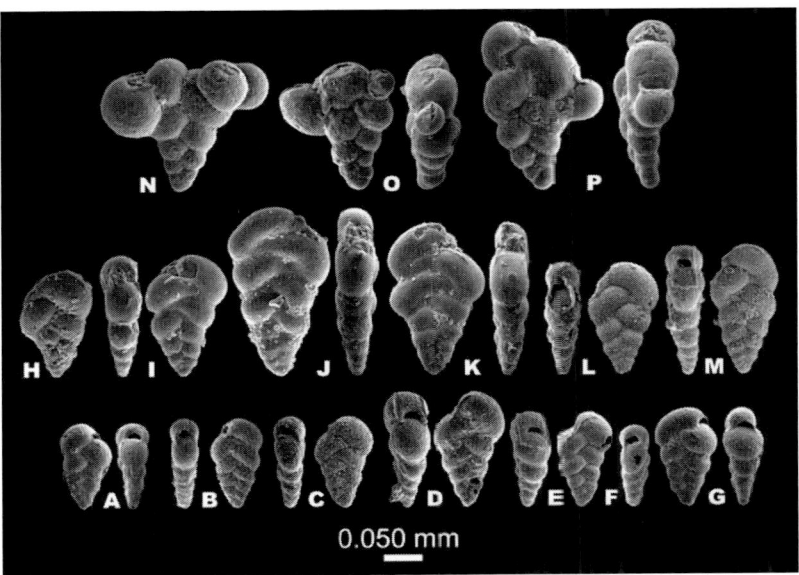

0.050 mm

Figure 27. Specimens of the directional lineage 1multichamber. **A-G** Precursor SMRS: S2. A-early Cenomanian, Sample 171B-1050C-25-2, 70-72 cm, figured by Georgescu and Huber (2009, pl. 3, figures 10-11); B-early Cenomanian, Sample 171B-1050C-25-2, 70-72 cm, figured by Georgescu and Huber (2009, pl. 3, figures 7-8); C-early Cenomanian, Sample 171B-1050C-25-2, 70-72 cm, figured by Georgescu and Huber (2009, pl. 3, figures 4-5); D-early Cenomanian, Sample 171B-1050C-25-2, 83-87 cm, figured by Georgescu and Huber (2009, pl. 3, figures 1-2); E-early Cenomanian, Sample 171B-1050C-25-2, 70-72 cm, figured by Georgescu and Huber (2009, pl. 2, figures 9-10); F-early Cenomanian, Sample 171B-1050C-25-2, 70-72 cm, figured by Georgescu and Huber (2009, pl. 2, figures 11); G-early Cenomanian, Sample 171B-1050C-25-2, 70-72 cm, figured by Georgescu and Huber (2009, pl. 2, figures 5-6). **H-M** ISMRS: I-1multichamber. H-early Maastrichtian, Sample 2404b, Ehrenberg Collection, Rügen Island. I-late Santonian, Sample 1595b, Ehrenberg Collection, Missouri River Basin, figured by Georgescu (2013c, Figure 6: 14-15). J-late Santonian, Sample 1595b, Ehrenberg Collection, Missouri River Basin, figured by Georgescu (2013c, Figure 6: 12-13). K-late Santonian, Sample 1595b, Ehrenberg Collection, Missouri River Basin, figured by Georgescu (2013c, Figure 6: 10-11). L-Coniacian to early Santonian, 62-463-26-5, 53-58 cm, figured by Georgescu (2013c, Figure 6: 3-4). M-Coniacian to early Santonian, 62-463-26-5, 53-58 cm, figured by Georgescu (2013c, Figure 6: 5-6). **N-P** FSMRS: F-1multichamber. N-early Turonian, Blue Hill Member of the Carlile Shale, Locality 47 of Hattin (1962). O-early Turonian, Blue Hill Member of the Carlile Shale, Locality 48 of Hattin (1962). P-early Turonian, Blue Hill Shale Member of the Carlile Shale, Locality 48 of Hattin (1962).

Transforming this genus into a unit with significance in evolutionary classification raised significant difficulties especially due to the lack of knowledge on the test wall ultrastructure and ornamentation in the original material reported by Aliyulla (1965, 1977). This gap was filled by Georgescu (2013) who defined the directional lineage *Pseudoplanoglobulina* based on material collected worldwide and this emendation is followed herein.

There are two major differences between S2, which is the precursor of the directional lineage 1multichamber, and I-1multichamber. Probably the most significant morphological change is the evolution of completely symmetrical tests in I-1multichamber from S2, which has the aperture bordered by archaeoflanges, whereas in I-1multichamber the aperture is bordered by orthoflanges. This demonstrates a second independent evolution of completely symmetrical tests through divergent evolution from S2.

Initiation of the directional lineage 1multichamber is associated with a significant increase in size from S2 to I-1multichamber; the latter SMRS frequently presents tests with a length at least double when compared to those of S2. There are rather trivial changes in pore size with the initiation of the 1multichamber; pore diameter in S2 is of 0.0004-0.0009 mm and of 0.0004-0.0010 mm in I-1multichamber. The extinction of I-1multichamber was considered by Georgescu (2013a) in the late Santonian; new material from the Ehrenberg Collection yielded by the samples from the Rügen Island (northern Germany), which complements that reported by Georgescu (2012a), and indicates that I-1multichamber occurs at a stratigraphical level as high as lower Maastrichtian.

Evolution of the multichamber growth happened in the early Turonian with the evolutionary occurrence of F-1multichamber. The young stage of this SMRS consists of proloculus followed by chambers alternately added with respect to the test growth axis, whereas the adult stage begins with one biaperturate progressive chamber followed by one pair of chambers simultaneously added on each of its side. F-1multichamber is the earliest known occurrence of one SMRS with multichamber addition of chambers. The tests of F-1multichamber are completely symmetrical in edge view and the aperture is bordered by symmetrically developed orthoflanges. Chamber surface is smooth and the observed pore diameter is of 0.0006-0.0009 mm.

Earliest Evolution of the Double Backward Chamber Extensions (BL: 2backextended)
(Figure 28)

Earliest evolution of the double backward chamber extension happened above the Cenomanian/Turonian boundary and the tests that evolved this feature present thoroughly an alternate chamber addition with respect to the test growth axis throughout the ontogeny. They are included in the newly named branched lineage 2backextended, which became extinct before and in the proximity of the Coniacian/Santonian boundary. This lineage was originally described as genus *Huberella*, which accommodated a directional lineage by Georgescu (2007), and was emended as branched lineage in evolutionary classification by Georgescu, Quinney and Anderson (2011).

Three SMRS are included in the 2backextended branched lineage: ISMRS=*Huberella praehuberi* Georgescu 2007a (I-2backextended), FSMRS=*Huberella huberi* Georgescu 2007a (F-2backextended) and SSMRS=*Huberella yucatanensis* Georgescu, Quinney and Anderson 2011 (S-2backextended).

2backextended is the second lineage in the evolutionary history of the planktic foraminifera with chambers alternately added with respect to the test growth axis that presents a branched pattern. The tests of the 2backextended branched lineage are completely symmetrical when seen in edge view.

Branched lineage 2backextended initiated in the early Turonian and its earliest SMRS has incipient double backward chamber extensions, one on each test side. Ornamentation in I-2backextended consisting of granular leptocostae consisting of aligned pustules is similar to that of S4, which is considered the precursor SMRS of the 2backextended branched lineage; leptocostae thickness in I-2backextended is of 0.0022-0.0030 mm. An increase in test length is encountered in the evolution from S4 to I-2backextended; the latter SMRS often presents tests twice as long as those of its ancestor. The double backward chamber extensions are attached to the previous chambers but they never cover the test central suture; a shift towards the central portion of the test is recorded in rare specimens. Aperture in I-2backextended is bordered by symmetrically developed orthoflanges. Wall is simple and perforate; pore diameter is of 0.0007-0.0009 mm in I-2backextended, which is almost identical to the pore diameter range observed in the ancestor S4: 0.0005-0.0009 mm. Evolution within the branched lineage 2backextended happened through divergence from I-2backextended.

F-2backextended evolved in the late Turonian and became extinct at the Turonian/ Coniacian boundary. Evolution from I-2backextended to F-2backextended is apparent in changes both in gross test architecture and ultrastructure features. There is an increase in the rate of chamber increase in width in F-2backextended, which is accompanied by a general increase of the test width. In addition the two symmetrical backward chamber extensions, one on each test side are narrower and more elongate longitudinally when compared to the corresponding structures of the ancestor I-2backextended; they are distinctly shifted towards the central suture with to they are frequently adjacent. The most apparent evolutionary development is the evolution of solid longitudinal leptocostae in F-2backextended from granular ones consisting of aligned pustules of the I-2backextended; this process is associated with a slight increase in leptocostae thickness from 0.0022-0.0030 mm in I-2backextended to 0.0024-0.0032 mm in F-2backextended. Pores are circular in shape and with a diameter of 0.0007-0.0010 mm in F-2backextended and are situated in the space between the leptocostae, rarely interrupting them; their diameters are almost identical to those known in the ancestral I-2backextended (0.0007-0.0009 mm). SMRS S-2backextended evolved in the latest Turonian and became extinct just before the Coniacian/Santonian boundary. The general test appearance with low values of the apical angle and relatively slow rate of chamber width increase throughout ontogeny indicates that it evolved from I-2backextended.

The two symmetrical backward chamber extensions are distinctly shifted towards the test central suture but they appear rather blunt and lack the short cylindrical aspect these structures have in F-2backextended. There is an increase in the leptocostae thickness from 0.0022-0.0030 mm in I-2backextended to 0.0024-0.0032 mm in S-2backextended. Pore diameter varies between 0.0007-0.0010 mm in S-2backextended, which is close to the sizes observed in I-2backextended: 0.0007-0.0009 mm. The morphological features of the tests in the 2backextended indicate that the two symmetrical backward chamber extensions are attached to the previous chambers and no additional apertures are developed. Evolution of the two symmetrical backward chamber extensions is apparently associated with a gradual increase in test size in the I-2backextended to F-2backextended, but a decrease in size is recorded with the evolution from I-2backextended to S-2backextended.

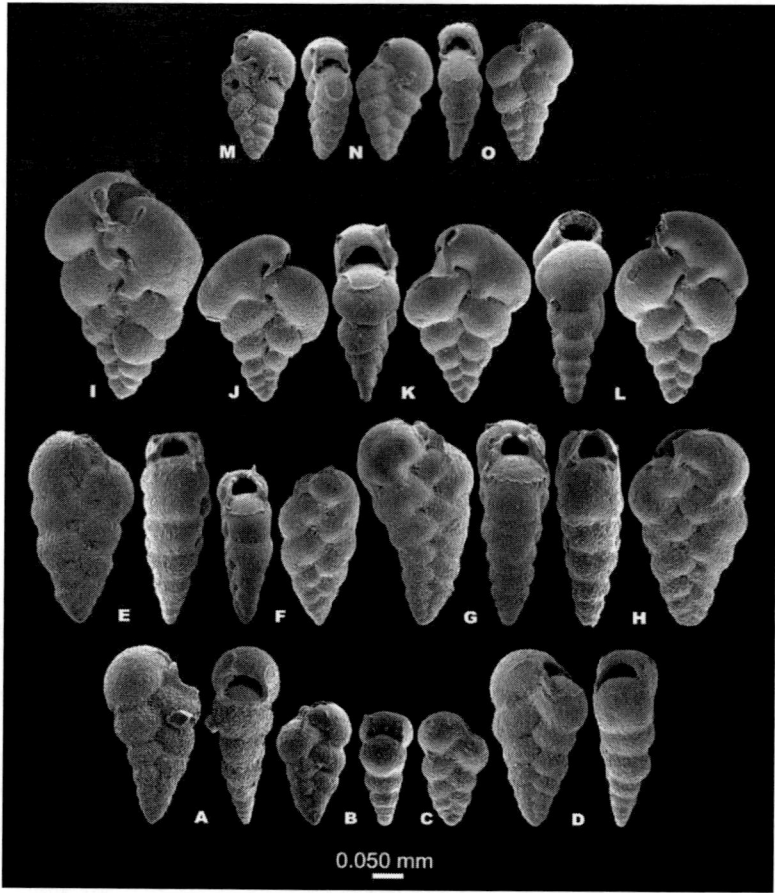

Figure 28. Specimens of the branched lineage 2backextended. **A-D** Precursor SMRS: S4. A-early Turonian, Sample 62-463-35-1, figures 17-19 cm, figured by Georgescu and Huber (2009, pl. 5, figures 4-5); B-early Turonian, Sample 62-463-35-1, figures 17-19 cm, figured by Georgescu and Huber (2009, pl. 5, Figure 9); C-early Turonian, Sample 171B-1050C-20-4, 110-111 cm, figured by Georgescu and Huber (2009, pl. 5, figures 7-8); D-early Turonian, Sample 62-463-35-1, figures 17-19 cm, figured by Georgescu and Huber (2009, pl. 5, figures 1-2). **E-H** ISMRS: I-2backextended. E-early Turonian, Sample 62-463-35-1, 17-19 cm, figured by Georgescu (2007a, pl. 3, Figure 7). F-early Turonian, Sample 62-463-35-1, 17-19 cm, figured by Georgescu (2007a, pl. 3, Figure 6). G-early Turonian, Sample 62-463-35-1, 17-19 cm, figured by Georgescu (2007a, pl. 3, Figure 4) and Georgescu, Quinney and Anderson (2011, pl. 1, figures 1-2). H-early Turonian, Sample 62-463-35-1, 17-19 cm, figured by Georgescu (2007a, pl. 3, Figure 5) and Georgescu, Quinney and Anderson (2011, pl. 1, figures 3-4). **I-L** FSMRS: F-2backextended. I-late Turonian, Sample 15-150-10-2, 48-6 cm, figured by Georgescu (2007a, pl. 2, Figure 3). J-late Turonian, Sample 15-150-10-2, 48-6 cm, figured by Georgescu (2007a, pl. 2, Figure 5). K-late Turonian, Sample 15-150-10-2, 48-6 cm, figured by Georgescu (2007a, pl. 2, Figure 6) and Georgescu, Quinney and Anderson (2011, pl. 1, figures 6-7, text-Figure 2B). L-late Turonian, Sample 15-150-10-2, 48-6 cm, figured by Georgescu (2007a, pl. 2, Figure 4). **M-O** SSMRS: S-2backextended. M-Coniacian, Sample 10-95-16-5, 100.5-102 cm, figured by Georgescu, Quinney and Anderson (2011, pl. 1, Figure 14). N-Coniacian, Sample 10-95-16-5, 100.5-102 cm, figured by Georgescu, Quinney and Anderson (2011, pl. 1, figures 12-13). O-Coniacian, Sample 10-95-16-5, 100.5-102 cm, figured by Georgescu, Quinney and Anderson (2011, pl. 1, figures 10-11).

Ultrastructure features show that there are no significant changes in the test porosity in the course of evolution from I-2backextended to both F-2backextended and S-2backextended. Ornamentation features in the three SMRS of this branched lineage indicate that the development of longitudinal solid and discontinuous costae independently happened in the evolution from the ISMRS to F-2backextended and S-2backextended respectively. Branched lineage 2backextended became extinct without leaving descendants.

Earliest Evolution of Multichamber Growth in Leptocostate Taxa
(DL: 2multichamber)
(Figure 29)

The earliest lineage consisting of globular-chambered tests that evolved adult stage with multichamber growth stage initiated in the earliest Coniacian and was defined as directional lineage *Texasina* in evolutionary classification by Georgescu (2010); in the new nomenclature associated with the evolutionary classification is herein named 2multichamber. There are two SMRS included in this directional lineage: ISMRS=*Gümbelina papula* Belford 1960 (I-2multichamber) and FSMRS=*Ventilabrella austinana* Cushman 1938 (F-2multichamber). Directional lineage 2multichamber became extinct in the early Campanian.

Figure 29. Specimens of the directional lineage 2multichamber. **A-D** Precursor SMRS: S5. A-early Maastrichtian, Sample 113-690C-19-1, 119-123 cm. B-early Maastrichtian, Sample 113-690C-20-5, 75-76 cm. C-late Santonian, Sample 174AX, 505.35-505.38 m, figured by Georgescu, Saupe and Huber (2008, pl. 3, Figure 2). D-late Santonian, Sample 174AX, 505.35-505.38 m. **E-F** ISMRS: I-2backextended. E-late Santonian, Sample 122-763B-17-1, 70-71 cm. F-late Santonian, Sample 122-763B-17-1, 70-71 cm. **G** FSMRS: F-2backextended. G-Santonian, Nanaimo Group, British Columbia, McGugan Collection, figured by Georgescu (2010, pl. 1, figures 11, 13).

The representatives of this lineage have the test ornamented with discontinuous longitudinal leptocostae and present completely symmetrical tests in edge view. One of the most significant achievements within this lineage is the evolution of some of the largest tests in the entire Cretaceous history of the planktic foraminifera with alternately added chambers with respect to the test growth axis; tests as long as 0.6000 mm are frequent and occasionally they can reach a length of nearly 0.8000 mm.

The precursor SMRS of the 2multichamber directional lineage is the stalk SMRS S5 and this is supported by the similarities in the general test architecture between the ancestor and juvenile stage of the descendant SMRS, namely I-2multichamber. The periapertural structures consisting of symmetrically developed metaflanges, one on each side of the test consistently occurs in the ISMRS of the 2multichamber directional lineage and they are only occasionally recorded in S5 where they co-occur with orthoflanges. Chamber ornamentation consists of longitudinally oriented leptocostae with a thickness of 0.0018-0.0025 mm in S5 and 0.0030-0.0036 mm in I-2multichamber. Georgescu (2010) described the leptocostae in the SMRS of this lineage as discontinuous and recently collected well preserved material from the Coniacian-lower Campanian sediments of ODP Hole 763B (Exmouth Plateau, offshore Australia) confirm the loss of leptocostae continuity along this lineage. There are no significant modifications in the pore diameter in the evolution from S5 (0.0006-0.0009 mm) to I-2multichamber (0.0003-0.0009 mm).

Multichamber growth in the adult stage in the directional lineage 2multichamber evolved in the SMRS F-2multichamber. Such evolutionary development is associated with an increase in chamber number (from 13-18 in I-2multichamber to 13-21 in F-2multichamber), increase in the discontinuous leptocostae thickness (from 0.0030-0.0036 mm in I-2multichamber to 0.0030-0.0069 mm in F-2multichamber) and increase in pore diameter (from 0.0003-0.0009 mm in I-2multichamber to 0.0012-0.0029 mm in F-2multichamber). In parallel there is a reduction in the test length due to the change in chamber addition pattern in the adult stage. The multichamber growth stage consists of the biaperturate progressive chamber followed by two to three sets of two chambers.

Second Evolution of Multichamber Growth in Leptocostate Taxa (DL: 3multichamber)
(Figure 30)

A distinct lineage that initiated from the stalk directional lineage in the early Santonian and led to the evolution of globular-chambered tests and eventually to a multichamber growth stage in the terminal SMRS was recognized by Georgescu (2013); this directional lineage was named *Ehrenbergites* by Georgescu (2013) and is herein renamed 3multichamber.

The directional lineage 3multichamber presents two SMRS: ISMRS=*Textularia striata* Ehrenberg 1838 (I-3multichamber) and FSMRS =*Ventilabrella riograndensis* Martin 1972 (F-3multichamber). This lineage has a complicated taxonomic history; the ISMRS relative simple test morphology, which is globular-chambered throughout and with chambers alternately added with respect to the test growth axis, together with a poor understanding of the high detail test morphological features including wall ultrastructure, ornamentation and porosity features represented a source of confusion in the past that often led to extensive lumping between SMRS of different lineages that independently achieved such test architecture. In addition the FSMRS was sometimes confused for other SMRS with relatively similar gross test architecture but with completely different test ornamentation. The 3multichamber directional lineage became extinct in the proximity or at the Cretaceous/ Paleogene boundary. No descendants of the 3multichamber lineage are known.

Figure 30. Specimens of the directional lineage 3multichamber. **A-D** Precursor SMRS: S5. A-early Maastrichtian, Sample 113-690C-19-1, 119-123 cm. B-early Maastrichtian, Sample 113-690C-20-5, 75-76 cm. C-late Santonian, Sample 174AX, 505.35-505.38 m, figured by Georgescu, Saupe and Huber (2008, pl. 3, Figure 2). D-late Santonian, Sample 174AX, 505.35-505.38 m. **E-J** ISMRS: I-3multichamber. E-late Campanian, Sample 1596b, Ehrenberg Collection, Mississippi River Basin, figured by Georgescu (2013a, pl. 3, figures 10-11). F-late Campanian-Maastrichtian, Sample 2217c, Ehrenberg Collection, Meudon. G-Campanian-Maastrichtian, Sample 2854, Ehrenberg Collection, Puszkary. H-Campanian-Maastrichtian, Sample 2854, Ehrenberg Collection, Puszkary, figured by Georgescu (2013a, pl. 4, figures 2-3). I-late Campanian, Sample 1596b, Ehrenberg Collection, Mississippi River Basin. J-late Campanian, Sample 1596b, Ehrenberg Collection, Mississippi River Basin. **K-M** FSMRS: F-3multichamber. K-late Maastrichtian, Sample 122-761B-24-1, 65-66 cm. L-late Maastrichtian, Sample 122-761B-24-1, 65-66 cm. M-late Maastrichtian, Sample 122-761B-24-1, 71-72 cm.

Recognizing the morphological features of I-3multichamber and clarify its taxonomical status was a problem that could be solved only after the examination of the original material from the C.G. Ehrenberg Collection deposited in the Museum of Natural History in Berlin.

Ehrenberg (1838) described the first planktic foraminiferal SMRS of the group in his pioneering study on chalky rocks components. Planktic foraminiferal tests were collected from a variety of chalky rocks of Europe, northern Africa and Middle East and studied mounted in Canada balsam; such method of study allows only partly the high detail test architecture and Georgescu (2013a) demonstrated that the Canada balsam, which is a natural resin, may affect the chamber ornamentation appearance over time when examined with the aid of a transmitted light microscope. Additional material collected from the bulk samples led to the conclusion that in the Late Cretaceous material from the chalky samples of the Ehrenberg Collection occurs only one SMRS, which has the chambers ornamented with longitudinal leptocostae and aperture bordered by two symmetrically developed orthoflanges, one on each test side.

Test morphological features and stratigraphical distribution indicate that the 3multichamber directional lineage evolved from S5. Both SMRS are ornamented with longitudinal discontinuous leptocostae, which are 0.0018-0.0025 mm thick in S5 and 0.0014-0.0027 mm in I-3multichamber; notably, in I-3multichamber the ornamentation is slightly but visibly stronger over the earlier portion of the test by contrast to S5 where it is uniformly developed over the test surface. Both SMRS present a periapertural pustulose area consisting of scattered dome-like pustules. Pores are circular and situated in the space between the leptocostae, rarely interrupting them. Apparently there are no changes in pore diameter with the initiation of the 3multichamber directional lineage; pore diameter is of 0.0006-0.0009 mm in S5 and 0.0004-0.0009 mm in I-3multichamber. Periapertural structures in I-3multichamber consist of symmetrically developed orthoflanges, one on each side of the test; this contrasts to the periapertural structures of S5 where they consist of both orthoflanges and metaflanges. The most significant morphological difference between the ancestor S5 and descendant I-3multichamber is the evolution of globular chambers in the latter SMRS that differ from the laterally compressed ones of S5; specimens with chamber compression rate between the two are common and they support the inferred evolutionary relationship between S5 and I-3multichamber in which the former SMRS is the ancestor and the latter its descendant.

All these data indicate that S5 is the precursor SMRS of the directional lineage 3multichamber. Test length presents a distinct reduction in size in the evolution from S5 to I-3multichamber but the magnitude of this reduction requires additional biometrical studies for an accurate assessment. Multichamber growth in the adult stage evolved in the SMRS F-3multichamber; therefore, for the most part of its history the directional lineage 3multichamber is represented only by tests with chambers alternately added with respect to the test growth axis throughout the ontogeny. The adult stage consists of the biaperturate progressive chamber, which is followed by up to four sets of chambers that present a gradual decrease in size; chamber number in the adult stage increases between the successive sets by one (e.g., first set-two chambers, second set-three chambers, third set-four chambers, etc.) but random occurrences of successive sets with identical chamber numbers are known.

Chambers are globular to subglobular throughout the ontogeny and with a more variable width/height ratio in the last-formed chamber sets. Ornamentation in F-3multichamber is thicker over the earlier portion of the test due to the addition of successive layers of calcite and consists of leptocostae, which present a distinct trend towards a loss in the longitudinal orientation over the adult stage with multichamber growth. There is a distinct trend of increasing the leptocostae thickness in the evolution from I-3multichamber (0.0014-0.0027 mm) to F-3multichamber (0.0027-0.0044 mm).

Pores are mostly circular in shape and occasionally elliptical and they present an increase in size from 0.0004-0.0009 mm in I-3multichamber to 0.0006-0.0012 mm in F-3multichamber.

Evolution of Leptocostate Bimodal Ornamentation (DL: 3backextended)
(Figure 31)

A new lineage that evolved chamber double backward extension was initiated in the early Santonian; it was described as directional lineage *Leptobimodalia* by Georgescu (2014b); its representatives are frequent occurrences in lower Santonian-Maastrichtian sediments accumulated in tropical and subtropical conditions. This lineage is herein renamed in the nomenclature associated with evolutionary classification as 3backextended. It evolved bimodal ornamentation, retroflanges and supplementary apertures at the posterior end of the chamber double backward extension for the first time in the history of Cretaceous planktic foraminifera with chambers alternately added with respect to the test growth axis.

According to Georgescu (2014b) the directional lineage 3backextended includes of three SMRS: ISMRS=*Leptobimodalia leptobimodalis* Georgescu 2014b (I-3backextended), FSMRS=*Pseudoguembelina costellifera* Masters 1976 (F-3backextended) and SSMRS =*Pseudoguembelina kempensis* Esker 1968 (S-3backextended).

The ISMRS of the directional lineage 3backextended evolved from the stalk terminal SMRS, S5. The gross test architecture shows significant morphological differences between the ancestor S5 and descendant I-3backextended. Two symmetrically developed bulbous backward extensions, one on each side of the test evolved in the last-formed one or two chambers of the SMRS I-3backextended; these structures will develop supplementary apertures in the descendant SMRS. The periapertural structures in I-3backextended consist of two symmetrically developed retroflanges ornamented with leptocostae, one on each side of the test; retroflanges are not rimmed and this leads to the conclusion that I-3backextended evolved from the tests of S5 in which the periapertural structures consist of orthoflanges. In S5 the ornamentation is unimodal and consists of longitudinal discontinuous leptocostae with a thickness of 0.0018-0.0025 mm, whereas the leptocostate ornamentation is bimodal in I-3backextended and the leptocostae present a thickness of 0.0017-0.0036 mm. The absence of a pustulose periapertural area in I-3backextended apparently indicates its loss with the initiation of the directional lineage 3backextended.

There are no significant changes between the pore diameter in S5 (0.0006-0.0009 mm) and I-3backextended (0.0005-0.0010 mm); pore shape is variable in I-3backextended from circular to elliptical, whereas in S5 they are circular.

The evolution from I-3backextended to F-3backextended resulted in the development of supplementary apertures at the posterior end of the chamber double backward extension; such supplementary apertures occur only in the last-formed one to four chambers. Periapertural structures evolved from the simple retroflanges in I-3backextended to rimmed retroflanges in F-3backextended; notably, the periapertural structures in both SMRS are ornamented with leptocostae. Chamber surface in F-3backextended is ornamented with closely spaced leptocostae that display bimodal arrangement; leptocostae thickness is of 0.0017-0.0036 mm in I-3backextended and 0.0021-0.0036 mm in F-3backextended. No ornamentation thickening over the earlier test chambers could be observed in either I-3backextended or F-3backextended; there is no pustulose periapertural area in F-3backextended.

Pores are circular or elliptical and with a diameter or maximum dimension of 0.0005-0.0010 mm in both SMRS. These data show that the evolution from I-3backextended to F-3backextended is mostly apparent in the gross test architecture.

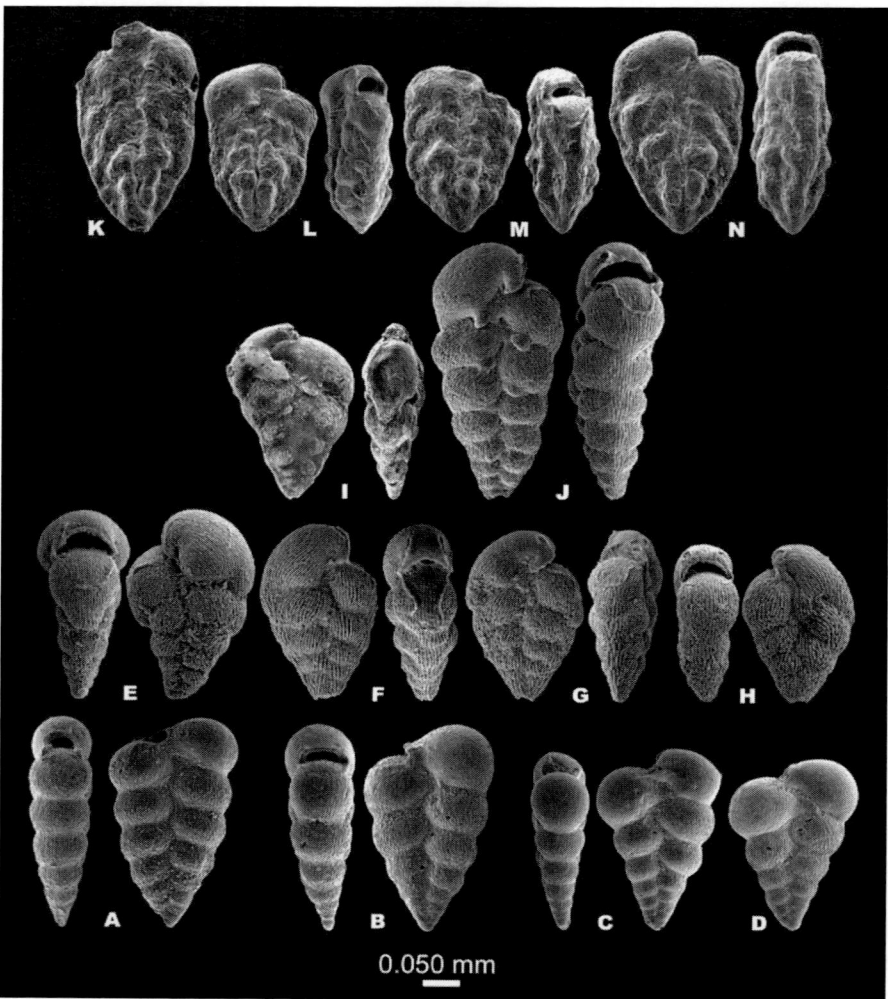

Figure 31. Specimens of the directional lineage 3backextended. **A-D** Precursor SMRS: S5. A-early Maastrichtian, Sample 113-690C-19-1, 119-123 cm. B- early Maastrichtian, Sample 113-690C-20-5, 75-76 cm. C-late Santonian, Sample 174AX, 505.35-505.38 m, figured by Georgescu, Saupe and Huber (2008, pl. 3, Figure 2). D-late Santonian, Sample 174AX, 505.35-505.38 m. **E-H** ISMRS: I-3backextended. E-early Campanian, Sample 32-305-25-1, 100-102 cm, figured by Georgescu (2014b, pl. 3, figures. 9-10). F-early Campanian, Sample 62-463-26-3, 52-54 cm, figured by Georgescu (2014b, pl. 3, Figure 7-8). G-early Campanian, Sample 62-463-26-3, 52-54 cm, figured by Georgescu (2014b, pl. 3, Figure 5-6). H-late Campanian, Sample 32-305-21-2, 100-102 cm, figured by Georgescu (2014b, pl. 3, figures 1-2). **I-J** FSMRS: F-3backextended. I-early Maastrichtian, Prairie Bluff Chalk, figured by Masters (1976, pl. 1, Figure 8) and refigured herein. J-late Campanian, Sample 32-305-18-5, 80-95 cm. **K-N** SSMRS: S-3backextended. K-late Maastrichtian, Sample 171B-1050C-11-cc. L-late Maastrichtian, Sample 171B-1050C-11-cc. M-late Maastrichtian, Sample 171B-1050C-11-cc. N-late Maastrichtian, Sample 171B-1050C-11-cc.

S-3backextended evolved from F-3backextended and the morphological changes are mostly apparent in the gross test architecture. A flexure of the test wall is developed at the test periphery resulting in a rimmed appearance in S-3backextended. Ornamentation over the last-formed chambers consists of discontinuous leptocostae with a thickness of 0.0027-0.0043 mm; the ornamentation over the earlier chambers of the test consists of fused leptocostae resulting in irregular structures that are concentrated in the proximity of the central suture. Pores are circular to elliptical in shape and with a diameter of 0.0005-0.0010 mm; there is no apparent difference between the porosity features of S-3backextended and those of its ancestor F-3backextended.

Growth Rate Diversification in Alternate Taxa (BL: 3alternate)
(Figure 32)

The branched lineage *Planulitella* Georgescu 2010 documents a new evolutionary trend in the Cretaceous evolution of the planktic foraminifera with chambers alternately added with respect to the test growth axis, namely the diversification of the growth rate in chamber addition. *Planulitella* was defined as a unit with significance in evolutionary classification (Georgescu 2010) and is herein renamed 3alternate. The three SMRS included in this branched lineage present chambers alternately added with respect to the test growth axis and there is no trend to develop either multichamber adult stage, transversally elongate chambers or early planispirally coiled stage. The branched lineage 3alternate includes the following SMRS: ISMRS=*Heterohelix sphenoides* Masters 1976 (I-3alternate), FSMRS =*Heterohelix stenopos* Masters 1976 (F-3alternate) and SSMRS =*Gümbelina malocaucasica* Aliyulla in Geodakchan and Aliyulla 1959 (S-3alternate). This lineage evolved after the Coniacian/ Santonian boundary and became extinct before the Santonian/Campanian boundary.

Georgescu (2010) demonstrated that the branched lineage 3alternate evolved from S5. I-3alternate is a laterally compressed SMRS ornamented with discontinuous leptocostae and aperture bordered by symmetrically developed metaflanges, one on each side of the test. The initiation of the 3alternate lineage is marked by an increase in leptocostae thickness, from 0.0018-0.0025 mm in S5 to 0.0031-0.0048 mm in I-3alternate; leptocostae thickness is variable over the chamber surface in I-3alternate, stronger at the periphery when compared to those on the chamber lateral sides, which are finer. Rare specimens present thicker ornamentation over the earliest chambers of the test. Pore diameter increased significantly from 0.0006-0.0009 mm in S5 to 0.0019-0.0044 mm in I-3alternate. A completely new feature in the group history evolved in I-3alternate, namely one small accessory aperture in the proximity of the central test suture; such feature occurs occasionally and is not rimmed. The gross test architecture features show significant changes in the chamber thickness and in addition the last-formed chambers are petaloid in shape. Divergent evolution from the I-3alternate led to evolution of tests with two distinct growth stages in F-3alternate in parallel with an increase in leptocostae thickness from 0.0031-0.0048 mm in I-3alternate to 0.0039-0.0063 mm in F-3alternate; pore diameter shows a decrease from 0.0019-0.0044 mm in I-3alternate to 0.0019-0.0036 mm in F-3alternate. There is a relative decrease in test length in the evolution from I-3alternate to F-3alternate. Evolution from I-3alternate to S-3alternate is apparent in a slight increase in the test size and development of last-formed chamber with antero-posterior elongation; this appears the earliest attempt by this group to develop an increase in chamber volume through a chamber elongation in the direction of test growth.

Figure 32. Specimens of the branched lineage 3alternate. **A-D** Precursor SMRS: S5. A-early Maastrichtian, Sample 113-690C-19-1, 119-123 cm. B-early Maastrichtian, Sample 113-690C-20-5, 75-76 cm. C-late Santonian, Sample 174AX, 505.35-505.38 m, figured by Georgescu, Saupe and Huber (2008, pl. 3, Figure 2). D-late Santonian, Sample 174AX, 505.35-505.38 m. **E-G** ISMRS: I-3alternate. E-late Santonian, Eureka 67-128 well (Gulf of Mexico), Sample top 1659.03 mbsf, figured by Georgescu (2010, pl. 2, Figure 5). F-late Santonian, Eureka 67-128 well (Gulf of Mexico), Sample top 1659.03 mbsf, figured by Georgescu (2010, pl. 2, figures 3-4). G-late Santonian, Eureka 67-128 well (Gulf of Mexico), Sample top 1659.03 mbsf, figured by Georgescu (2010, pl. 2, Figure 1). **H-I** FSMRS: F-3alternate. H-late Santonian, Eureka 67-128 well (Gulf of Mexico), Sample top 1659.03 mbsf, figured by Georgescu (2010, pl. 3, figures 1-2). I-late Santonian, Eureka 67-128 well (Gulf of Mexico), Sample top 1659.03 mbsf, figured by Georgescu (2010, pl. 3, figures 3-4). **J-M** SSMRS: S-3alternate. J-late Santonian, Sample 10-95-14-1, 99-112 cm, figured by Georgescu (2010, pl. 3, Figure 12). K-late Santonian, Sample 10-95-14-1, 99-112 cm, figured by Georgescu (2010, pl. 3, Figure 10). L-late Santonian, Sample 10-95-14-1, 99-112 cm, figured by Georgescu (2010, pl. 3, Figure 5). M-late Santonian, Sample 10-95-14-1, 99-112 cm, figured by Georgescu (2010, pl. 3, figures 8-9).

Test ultrastructure features show an increase in leptocostae thickness (0.0031-0.0048 mm in I-3alternate and 0.0045-0.0062 mm in S-3alternate) and a decrease in pore diameter from 0.0019-0.0044 mm in I-3alternate to 0.0011-0.0040 mm in S-3alternate. Rare specimens in both descendants F-3alternate and S-3alternate present slightly thickened ornamentation over the earlier portion of the test.

Second Evolution of Pore Mounded Ornamentation (DL: 4multichamber)
(Figure 33)

A second evolution of the pore-mounded ornamentation happened in the middle part of the Santonian. Largely this lineage includes the representatives of the genus *Laeviheterohelix* Nederbragt 1991 as successively emended by Georgescu (2009a, 2013a, 2013c). At the time when *Laeviheterohelix* was described by Nederbragt (1991) it was assumed that it ranged from Turonian throughout Maastrichtian. Georgescu (2009a) demonstrated that pore-mounded ornamentation evolved iteratively based on high-resolution observations on the test wall ultrastructure, chamber ornamentation and pores size; the main result of this study is that *Laeviheterohelix* was emended and transformed into a unit with significance in evolutionary classification, namely a directional lineage. This general concept is used herein and this directional lineage is renamed 4multichamber; it has two SMRS: ISMRS=*Textilaria euryconus* Ehrenberg 1854 (I-4multichamber) and FSMRS =*Ventilabrella reniformis* Marie 1941 (F-4multichamber).

I-4multichamber presents chambers alternately added with respect to the test growth axis and periapertural structures consisting of symmetrically developed orthoflanges, one on each side of the test. In addition the ornamentation consists of scattered pore mounds, which in general present a diameter of 0.0017-0.0028 mm. Test wall is simple and perforate by circular pores with a diameter of 0.0004-0.0009 mm. These data indicate that the ancestor of I-4multichamber is the smooth I-1multichamber that presents circular pores with a diameter of 0.0004-0.0010 mm; therefore, I-1multichamber is the precursor of the directional lineage 4multichamber. There are evident similarities in the general test appearance between the ancestor I-1multichamber and descendant I-4multichamber; moreover, the two present periapertural structures consisting of symmetrically developed orthoflanges. Evolution from I-1multichamber to I-4multichamber is apparent in the development of ornamentation consisting of pore mounds and development of a pustulose periapertural area, which consists of scattered dome-like pustules with a diameter of 0.0017-0.0022 mm, in I-4multichamber. More detailed studies are necessary to date with precision the evolutionary occurrence of I-4multichamber. Georgescu (2009a) considered this SMRS occurs in the lower part of upper Coniacian-Santonian sediments from DSDP Site 511 based on a single-specimen occurrence. Subsequently Georgescu (2013a, 2013c) reconsidered the age of this bioevent and dated it as late Santonian and this later conclusion is followed herein.

Evolution from I-4multichamber to F-4multichamber is mostly apparent in the evolution of an adult stage with multichamber growth; the adult stage consists of the biaperturate progressive chamber followed by one set of two relapsed chambers. Chamber surface is ornamented with pore mounds with a diameter of 0.0032-0.0060 mm, which are larger than those in the ancestor SMRS (0.0017-0.0028 mm in I-4multichamber); a pustulose periapertural area occurs in the anterior portion of the chambers and consists of scattered dome-like pustules. There is an increase in the pore diameter from 0.0004-0.0009 mm in I-4multichamber to 0.0006-0.0012 mm in F-4multichamber. F-4multichamber is a rare SMRS and is known only from single and few specimen occurrences in sediments of Campanian age.

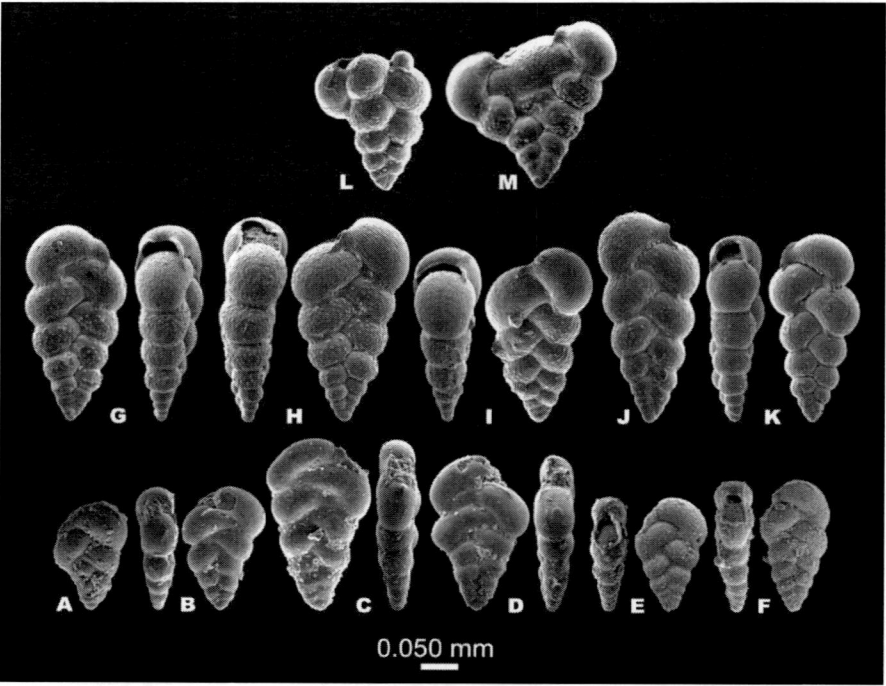

Figure 33. Specimens of the directional lineage 4multichamber. **A-F** Precursor SMRS: I-1multichamber. A-early Maastrichtian, Sample 2404b, Ehrenberg Collection, Rügen Island. **B**-late Santonian, Sample 1595b, Ehrenberg Collection, Missouri River Basin, figured by Georgescu (2013c, Figure 6: 14-15). C-late Santonian, Sample 1595b, Ehrenberg Collection, Missouri River Basin, figured by Georgescu (2013c, Figure 6: 12-13). D-late Santonian, Sample 1595b, Ehrenberg Collection, Missouri River Basin, figured by Georgescu (2013c, Figure 6: 10-11). E-Coniacian to early Santonian, 62-463-26-5, 53-58 cm, figured by Georgescu (2013c, Figure 6: 3-4). F-Coniacian to early Santonian, 62-463-26-5, 53-58 cm, figured by Georgescu (2013c, Figure 6: 5-6). **G-K** ISMRS: I-4multichamber. G-late Campanian, Sample 174AX, 435.59-435.60 m, figured by Georgescu (2009a, Figure 8: 1). H-early Campanian, Sample 79-511-31-5, 33-35 cm, figured by Georgescu (2009a, Figure 8: 3). I-early Campanian, Sample 79-511-32-1, 33-35 cm, figured by Georgescu (2009a, Figure 7: 3). J-early Campanian, Sample 79-511-32-4, 22-25 cm, figured by Georgescu (2009a, Figure 7: 2a). K-early Campanian, Sample 79-511-32-4, 22-25 cm, figured by Georgescu (2009a, Figure 7: 1). **L-M** FSMRS: F-4multichamber. L-early Campanian, Sample 79-511-36-4, 23-27 cm, figured by Georgescu, Saupe and Huber (2008, pl. 2, Figure D-2a). M-early Campanian, Sample 79-511-32-cc, figured by Georgescu, Saupe and Huber (2008, pl. 2, Figure D-1b) and Georgescu (2009a, Figure 8: 4c).

Earliest Evolution of Sutural Ridges (DL: 5multichamber)
(Figure 34)

The earliest descendant from the branched lineage I-3alternate that evolved multichamber growth in the adult stage is the directional lineage *Sigalia*, which is herein defined according to the revision of Georgescu (2010) and renamed 5multichamber in the nomenclature of the evolutionary classification. This lineage shows a gradual development of the adult multichamber stage, which is paralleled by the earliest evolution of sutural ridges among the representatives of the planktic foraminiferal with chambers alternately added with respect to the test growth axis.

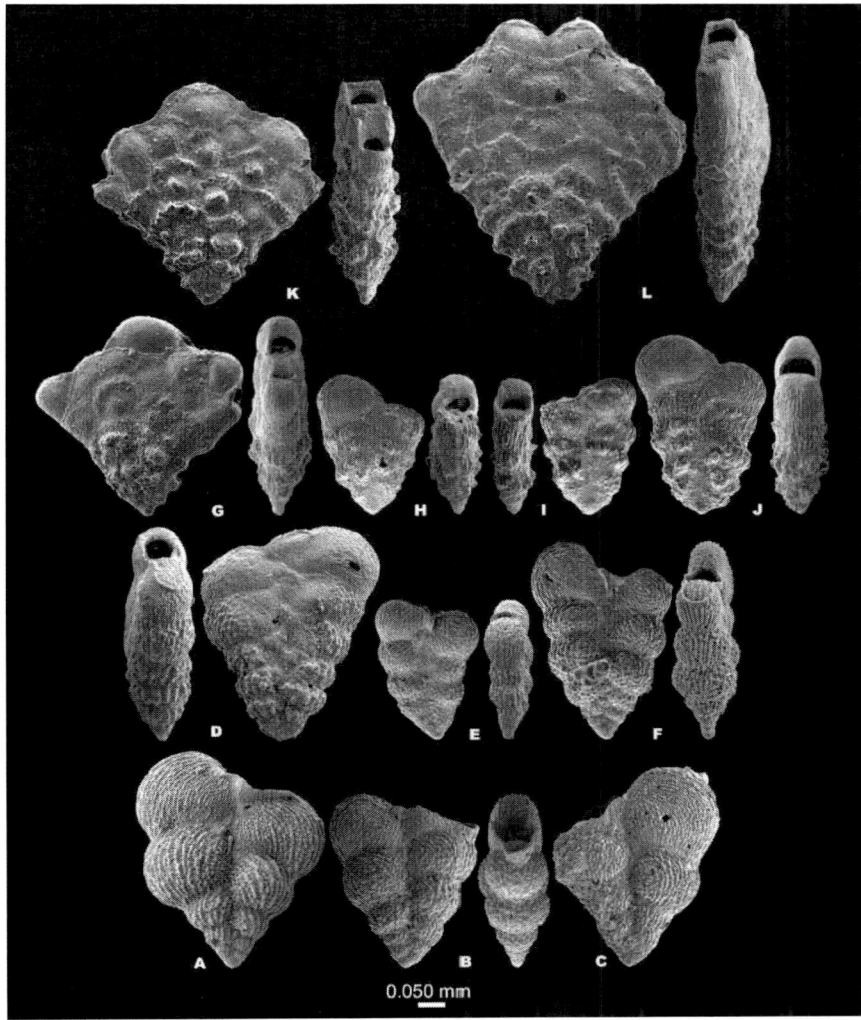

Figure 34. Specimens of the directional lineage 5multichamber. **A-C** Precursor SMRS: I-3alternate. A-late Santonian, Eureka 67-128 well (Gulf of Mexico), Sample top 1659.03 mbsf, figured by Georgescu (2010, pl. 2, Figure 5). B-late Santonian, Eureka 67-128 well (Gulf of Mexico), Sample top 1659.03 mbsf, figured by Georgescu (2010, pl. 2, figures 3-4). C-late Santonian, Eureka 67-128 well (Gulf of Mexico), Sample top 1659.03 mbsf, figured by Georgescu (2010, pl. 2, Figure 1). **D-F** ISMRS: I-5multichamber. D-late Santonian, Eureka 67-128 well (Gulf of Mexico), Sample top 1659.03 mbsf, figured by Georgescu (2010, pl. 4, figures 11-12). E-late Santonian, Eureka 67-128 well (Gulf of Mexico), Sample top 1659.03 mbsf, figured by Georgescu (2010, pl. 4, figures 5, 8). F-late Santonian, Sample 10-95-14-1, 102-105 cm, figured by Georgescu (2010, pl. 4, figures 1, 4). **G-J** FSMRS: F-5multichamber. G-late Santonian, Sample 10-95-13-3, 78-92 cm, figured by Georgescu (2010, pl. 6, figures 6-7). H-late Santonian, Sample 10-95-14-1, 99-112 cm, figured by Georgescu (2010, pl. 5, figures 11-12). I-late Santonian, Sample 10-95-14-1, 99-112 cm, figured by Georgescu (2010, pl. 5, figures 1-2). J-late Santonian, Sample 10-95-14-1, 99-112 cm. figured by Georgescu (2010, pl. 5, figures 6-7). **K-L** SSMRS: S-5multichamber. K-late Santonian, Sample 10-95-14-1, 99-112 cm, figured by Georgescu (2010, pl. 7, figures 5-6). L-late Santonian, Sample 10-95-13-3, 78-92 cm, figured by Georgescu (2010, pl. 7, figures 1-2).

The terminology of Georgescu (2010) that subdivided the sutural ridges function of their degree of development into calyptoridges and phaneroridges is herein used; according to this terminology calyptoridges accommodate the sutural ridges in incipient state of development, whereas phaneroridges include the well-developed ones.

This directional lineage consists of three SMRS: ISMRS=*Sigalia incipiens* Georgescu 2010 (I-5multichamber), FSMRS =*Gümbelina* (*Gümbelina*, *Ventilabrella*) *deflaensis* Sigal 1952 (F-5multichamber) and SSMRS =*Sigalia proliferans* Georgescu 2010 (S-5multichamber). An earlier interpretation of the evolutionary relationships between various SMRS that eventually led to the redefinition of *Sigalia* into an evolutionary classification compatible unit was given by Nederbragt (1991).

Directional lineage 5multichamber evolved from the SMRS I-3alternate, which is this lineage precursor SMRS (Nederbragt 1991; Georgescu 2010). ISMRS of 5multichamber presents a number of morphological similarities with the precursor SMRS such as, general test appearance, ornamentation consisting of leptocostae, periapertural structures consisting of symmetrically developed metaflanges, compressed test in edge view and occasional occurrence of one supplementary aperture formed under the periapertural structure in the proximity of the central test suture. By contrast to I-3alternate where the sutures are distinct and depressed between all the chambers of the test, calyptoridges evolved in I-5multichamber; these structures present a significant degree of variability (Georgescu 2010). Ornamentation shows an increase in leptocostae thickness from 0.0031-0.0048 mm in I-3alternate to 0.0034-0.0099 mm in I-5multichamber; in both SMRS the ornamentation is more prominent over the test earlier portion, but this feature is more prominent in the SMRS I-5multichamber. A pustulose periapertural area consisting of scattered dome-like pustules occurs in both SMRS. There was observed a decrease in the circular pore diameter from 0.0019-0.0044 mm in I-3alternate to 0.0012-0.0038 mm in I-5multichamber.

Evolution from I-5multichamber to F-5multichamber is apparent in the gross test architecture especially in a gradual accentuation of the sutural ridges and their transformation from calyptoridges to phaneroridges and occasional occurrence of a multichamber growth in the adult stage. Sigal (1952) in the original report of this SMRS noted that some specimens present incipient multichamber growth and the occurrence of such specimens was confirmed by Reiss (1957); this fact was rarely taken in consideration (Dowsett 1989; Georgescu 2010). According to Georgescu (2010) the adult stage with multichamber growth occurs in less than 15% of the total number of specimens of the SMRS F-5multichamber; the multichamber growth stage consists of the biaperturate progressive chamber followed by up to four chamber sets, each consisting of two or more rarely three chambers. Ornamentation in F-5multichamber is more prominent over the chambers of the earlier portion of the test. The longitudinally arranged leptocostae are mostly visible over the last-formed chambers; there is a decrease in leptocostae thickness from 0.0034-0.0099 mm in I-5multichamber to 0.0032-0.0068 mm in F-5multichamber. A similar decreasing trend is recorded in the pore diameter from 0.0012-0.0038 mm in I-5multichamber to 0.0010-0.0031 mm in F-5multichamber. Test length remains at relatively similar values in the evolution from I-5multichamber to F-5multichamber; the largest specimens of F-5multichamber are those with multichamber growth in the adult stage.

SMRS S-5multichamber, which is the terminal SMRS of the directional lineage 5multichamber, presents the largest tests, which are around two times longer than those of its ancestor, F-5multichamber.

Phaneroridges are especially prominent between the chambers of the earlier portion of the test and transverse keels over the periphery result from the fusion between the phaneroridges situated on the two sides of the test. The decrease in leptocostae thickness from 0.0032-0.0068 mm in F-5multichamber to 0.0021-0.0060 mm in S-5multichamber represents a continuation of the trend in the earlier portion of the lineage. Pore diameter exhibits a wide range of variability: 0.0010-0.0055 mm; this represents an increase from the pore diameter of 0.0010-0.0031 mm in F-5multichamber.

Second Evolution of Sutural Ridges (DL: 6multichamber)
(Figure 35)

A new directional lineage evolved in the late Santonian and from I-3alternate and shortly after 5multichamber. This lineage was prefigured by Nederbragt (1991) and refined and formalized as directional lineage *Proliferania* by Georgescu (2010).

It is herein renamed 6multichamber in the nomenclature associated with the evolutionary classification. Directional lineage 6multichamber shows a parallel evolution with 5multichamber, which resulted in the evolution of calyptoridges, then phaneroridges and multichamber growth in the adult portion of the test; this lineage became extinct before the Santonian/Campanian boundary. According to Georgescu the directional lineage includes three SMRS: ISMRS=*Proliferania initialis* Georgescu 2010 (I-6multichamber), FSMRS= *Sigalia carpathica* Salaj and Samuel 1963 (F-5multichamber) and SSMRS=*Ventilabrella decoratissima* de Klasz 1953 (S-6multichamber).

The gross test architecture and test ultrastructure features indicate that the directional lineage 6multichamber evolved from the SMRS I-3alternate; therefore, I-3alternate is the precursor SMRS of 6multichamber. Evolution from I-3alternate to I-6multichamber is apparent especially in the evolution of calyptoridges in the latter SMRS; calyptoridges can be best observed under the scanning electron microscope. Notably both SMRS present the chambers following the proloculus alternately added with respect to the test growth axis and the occasional occurrence of one supplementary aperture under the periapertural structures of the last-formed chamber in the proximity of the central test suture. There are significant resemblances between the SMRS I-5multichamber and I-6multichamber, which divergently evolved from I-3alternate; the morphological differences between the two SMRS were presented in detail by Georgescu (2010). Chamber surface shows that both I-3alternate and I-6multichamber are ornamented with longitudinal leptocostae, which have a thickness of 0.0031-0.0048 mm in the ancestor I-3alternate and 0.0048-0.0104 mm in the descendant I-6multichamber; such an increase in leptocostae thickness is associated in many specimens with the formation of an incipient reticulate network of interconnected ridges.

In addition, both SMRS present a pustulose periapertural area consisting of scattered dome-like pustules. Pores are circular and with a diameter of 0.0019-0.0044 mm in I-3alternate and circular to elliptical in shape and with the diameter or maximum dimension of 0.0012-0.0041 mm in I-6multichamber. Georgescu (2010) noted that the openings of the incipient reticulate network of interconnected ridges can be up to 0.0111 mm in maximum dimension.

Figure 35. Specimens of the directional lineage 6multichamber. **A-C** Precursor SMRS: I-3alternate. A-late Santonian, Eureka 67-128 well (Gulf of Mexico), Sample top 1659.03 mbsf, figured by Georgescu (2010, pl. 2, Figure 5). B-late Santonian, Eureka 67-128 well (Gulf of Mexico), Sample top 1659.03 mbsf, figured by Georgescu (2010, pl. 2, figures 3-4). C-late Santonian, Eureka 67-128 well (Gulf of Mexico), Sample top 1659.03 mbsf, figured by Georgescu (2010, pl. 2, Figure 1). **D-G** ISMRS: I-6multichamber. D-late Santonian, Sample 10-95-14-1, 99-112 cm, figured by Georgescu (2010, pl. 8, Figure 9). E-late Santonian, Sample 10-95-13-3, 78-92 cm, figured by Georgescu (2010, pl. 8, Figure 5). F-late Santonian, Sample 10-95-15-4, 99.5-100.5 cm, figured by Georgescu (2010, pl. 8, Figure 11). G-late Santonian, Sample 10-95-15-2, 73-87 cm, figured by Georgescu (2010, pl. 8, figures 1, 4). **H-J** FSMRS: F-6multichamber. H-late Santonian, Sample 10-95-15-6, 104-118 cm, figured by (Georgescu 2010, pl. 10, Figure 8). I-late Santonian, Sample 10-95-15-2, 73-87 cm, figured by (Georgescu 2010, pl. 10, figures 9-10). J-late Santonian, Sample 10-95-15-2, 73-87 cm, figured by (Georgescu 2010, pl. 10, figures 5, 8). **K-L** SSMRS: S-6multichamber. K-late Santonian, Eureka 67-128 well (Gulf of Mexico), Sample top 1662.23 mbsf, figured by Georgescu (2010, pl. 12, figures 11-12). L-late Santonian, Sample 10-95-15-4, 99.5-100.5 cm, figured by Georgescu (2010, pl. 11, figures 1-2).

Evolution from I-6multichamber to F-6multichamber is apparent especially in the development of phaneroridges, which fuse across the periphery resulting in the occurrence of well-developed transverse keels in the descendant SMRS. In addition the test periphery evolved from rounded in I-6multichamber to rounded to subangular in F-6multichamber. A significant change is recorded in the chamber ornamentation, which evolved from leptocostate in I-6multichamber to leptocostate to smooth in F-6multichamber; ornamentation reduction is most apparent over the last-formed chambers, which are often smooth. Vestiges of the ornamentation occur in the phaneroridges where costae with a thickness of 0.0144-0.0239 mm are frequently observed; such thick ornamentation elements are the result of the ornamentation concentration in the phaneroridges and addition of successive layers of calcite during the ontogenetic development. The only ornamentation feature that exhibits little or no variability in the evolution from I-6multichamber to F-6multichamber is the pustulose periapertural area, which consists of scattered dome-like pustules. Ornamentation changes are apparently related to significant changes in pore size, which in F-6multichamber have diameters of 0.0010-0.0026 mm over the last-formed one to four chambers and 0.0024-0.0046 mm over the earlier chambers of the test; the largest pores have a vuggy appearance. The vuggy pores are partly the result of the addition of successive layers of calcite that altered the original porosity features, such as shape, diameter and density. F-6multichamber does not present an adult stage with multichamber growth.

Development of the multichamber growth in the adult portion of the test occurs in S-6multichamber; this stage presents a high variability with respect to the number of chambers of which it consists of. The progressive biaperturate chamber is followed by up to five sets of chambers that increase in number by one (e.g., first set-two chambers, second set-three chambers, third set-four chambers, etc.) but irregularities in the number of added chambers between successive chambers occur frequently. Chamber surface is smooth and the original leptocostate ornamentation appears completely lost; its vestiges can be observed in the phaneroridges where fused leptocostae occur as thick and longitudinally oriented costae with a thickness of 0.0188-0.0327 mm; these structures often present a nodular appearance, which can also be observed in F-6multichamber. A pustulose periapertural area consisting of scattered dome-like pustules is observable in edge and apertural views. Pore diameter presents a significant variability over a test of the SMRS S-6multichamber. Pores are small and with a diameter of 0.0023-0.0032 mm over the last-formed chambers; over the earlier chambers and due to the successive additions of layers of calcite during the ontogeny the pores are irregular in shape (vuggy) and larger, with a maximum dimension of 0.0075-0.0089 mm. One morphological feature related to porosity in S-6multichamber is the high pore density; over the earlier portion of the test the surface occupied by pores is almost equal to that of the solid test wall and additional studies to quantify such development in pores density are necessary. The significance of this evolutionary trend observed throughout the directional lineage 6multichamber is unknown.

There are no planktic descendants from 6multichamber. The report by Küpper (1954) of an early biserial stage in the representatives of the late Cretaceous large-sized benthic foraminifera of the genera *Orbitoides* d'Orbigny 1848 and *Omphalocyclus* Bronn 1853 was interpreted as indicative of the origins of these genera from the planktics with proliferating adult stage. This idea was further developed by van Hinte (1965), Neumann (1972) and Georgescu and Almogi-Labin (2008). Georgescu and Almogi-Labin (2008) showed that S-6multichamber is most likely the ancestor of the large-sized benthic foraminifer *Orbitoides*.

The Earliest Evolution of Scalaropores (BL: 7multichamber)
(Figure 36)

The third lineage that evolved in the late Santonian from I-3alternate resulted in the development of tests with multichamber growth in the adult stage and pycnocostate ornamentation. Ornamentation prominence over the test surface is variable, ranging from almost equally developed on all the chambers to distinctly thicken over the chambers of the earlier portion of the test. This branched lineage largely includes the representatives of the genus *Ventilabrella* Cushman 1928. It was redefined as branched lineage in evolutionary classification by Georgescu (2010) and includes three SMRS: ISMRS=*Ventilabrella eggeri* Cushman 1928 (I-7multichamber), FSMRS=*Ventilabrella alpina* de Klasz 1953 (F-7multichamber) and SSMRS=*Ventilabrella eggeri* var. *glabrata* Cushman 1938 (S-7multichamber). The representatives of the branched lineage 6multichamber can be recognized among the other Santonian taxa with proliferating chambers in the adult stage by the depressed sutures which are not lined by sutural ridges. 6multichamber is the only lineage that evolved in the Santonian and developed multichamber growth in the adult stage that crossed the Santonian/Campanian boundary; it became extinct in the early Campanian.

Evolution from I-3alternate to I-7multichamber is mostly apparent in the gross test architecture in the development of an adult stage with multichamber growth in the descendant SMRS; the juvenile stages in the two SMRS are similar and this is a major feature taken in consideration when considering I-3alternate and I-7multichamber in ancestor-descendant relationship. The adult stage with multichamber growth consists of the biaperturate progressive chamber followed by one or rarely two sets of two chambers. In both SMRS the ornamentation consists of leptocostae, which are thinner in I-3alternate (0.0031-0.0048 mm) and thicker in I-7multichamber (0.0066-0.0129 mm); ornamentation in I-7multichamber shows distinctly thicker leptocostae over the earlier chambers of the test. The progressive chamber ornamentation often presents the leptocostae diverging from its base and this pattern can be occasionally observed in the chamber of the adult stage with multichamber growth. Pores are circular in shape and situated in the space between the leptocostae, rarely interrupting them; pores diameter present close values in I-3alternate (0.0019-0.0044 mm) and I-7multichamber (0.0011-0.0048 mm). Therefore, the SMRS I-3alternate is the precursor of 7multichamber branched lineage.

F-7multichamber presents ornamentation consisting of leptocostae, which are distinctly more prominent over the chambers of the earlier portion of the test (0.0118-0.0193 mm), whereas over the last-formed chambers are thinner (0.0044-0.0081 mm); these show a significant increase in thickness when compared to the leptocostae of the ancestor I-7multichamber where they are 0.0066-0.0129 mm in thickness. Porosity in F-7multichamber shows the evolution of a new type of pores, namely scalaropores, which coexist with the simple ones. Pores are circular or elliptical in shape and with a diameter or maximum dimension of 0.0018-0.0043 mm in F-7multichamber, a range which is largely comparable to that of its ancestor I-7multichamber (0.0011-0.0048 mm); the addition of successive layers of calcite throughout the ontogeny is the possible cause for the slight decrease in pore diameter. The adult stage with multichamber growth in F-7multichamber consists of the biaperturate progressive chamber followed by one to five sets of chambers that present a gradual increase in number (e.g., first set-two chambers, second set-three chambers, third set-four chambers, etc.); its degree of development sharply contrasts to that of I-7multichamber where it is incipiently developed.

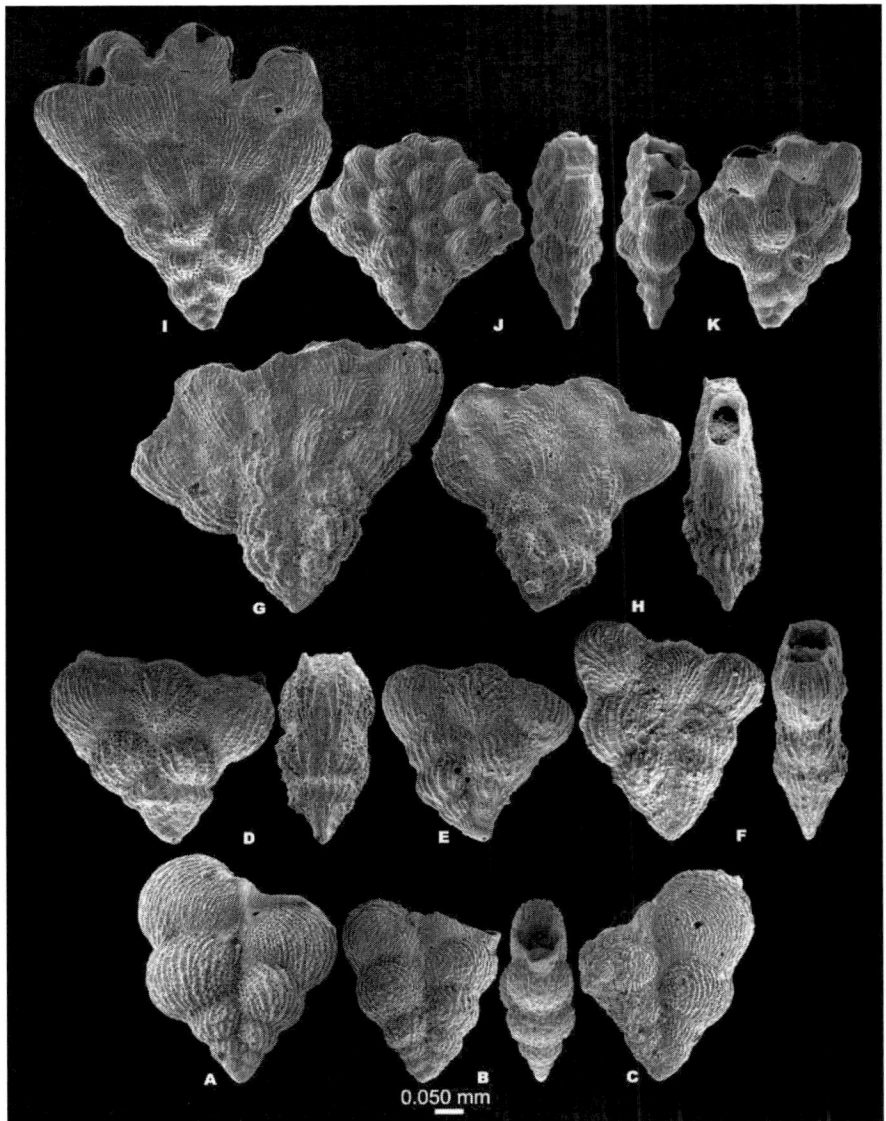

Figure 36. Specimens of the directional lineage 7multichamber. **A-C** Precursor SMRS: I-3alternate. A-late Santonian, Eureka 67-128 well (Gulf of Mexico), Sample top 1659.03 mbsf, figured by Georgescu (2010, pl. 2, Figure 5). B-late Santonian, Eureka 67-128 well (Gulf of Mexico), Sample top 1659.03 mbsf, figured by Georgescu (2010, pl. 2, figures 3-4). C-late Santonian, Eureka 67-128 well (Gulf of Mexico), Sample top 1659.03 mbsf, figured by Georgescu (2010, pl. 2, Figure 1). **D-F** ISMRS: I-7multichamber. D-late Santonian, Sample 10-95-13-3, 78-92 cm, figured by Georgescu (2010, pl. 13, figures 9-10). E-late Santonian, Sample 10-95-15-4, 99.5-100.5 cm, figured by Georgescu (2010, pl. 13, Figure 3). F-late Santonian, Sample 10-95-14-1, 100-102 cm, figured by Georgescu (2010, pl. 13, figures 1-2). **G-H** FSMRS: F-7multichamber. G-late Santonian, Sample 10-95-14-1, 100-102 cm, figured by Georgescu (2010, pl. 15, Figure 1). H-late Santonian, Sample 10-95-14-1, 100-102 cm, figured by Georgescu (2010, pl. 15, figures 2-3). **I-K** SSMRS: S-7multichamber. I-late Santonian, Sample 174AX, 505.35-505.38 m, figured by Georgescu (2010, pl. 16, Figure 4). J-late Santonian, Sample 174AX, 505.35-505.38 m, figured by Georgescu (2010, pl. 16, figures 8-9). K-late Santonian, Sample 174AX, 505.35-505.38 m, figured by Georgescu (2010, pl. 16, figures 1-2).

The SMRS S-7multichamber evolved divergently from I-7multichamber and presents an adult stage with multichamber growth and ornamentation almost equally developed over the entire test surface. The stage with multichamber growth consists of the biaperturate progressive chamber followed by up to six sets of chambers. The first one to three sets consists of two chambers (one relapsed and one biaperturate) and the subsequent ones present the number of chambers increasing by an increment of one; irregularities in chamber number from the last added sets are common. Ornamentation in S-7multichamber is leptocostate in which the leptocostae are mostly longitudinal and occasionally diverging from the chamber base especially in the adult proliferating stage.

The measured leptocostae thickness in S-7multichamber is of 0.0035-0.0057 mm and apparently there is a reduction from a thickness of 0.0066-0.0129 mm recorded in the ancestor SMRS I-7multichamber; the leptocostae continuity and increased thickness occasionally results in the development of pycnocostate appearance. S-7multichamber consistently presents a pustulose periapertural area consisting of scattered dome-like pustules. Pores are simple or scalaropores with a diameter or maximum dimension of 0.0015-0.0037 mm, which are smaller than those of its ancestor I-7multichamber: 0.0011-0.0048 mm.

Earliest Evolution of Early Planispiral Coil (DL: 1planispiral)
(Figure 37)

The lineage that first evolved an early planispiral coil in the evolutionary history of the planktic foraminifera with chambers alternately added with respect to the test growth axis includes the first generic name used in this group; this was *Spiroplecta* who was recognized by Ehrenberg (1844). The occurrence of an earlier planispiral coil in some of the representatives of this group and was extensively used afterwards to track the origins of the planktic foraminifera with chambers alternately added with respect to the test growth axis among the planispiral coiled taxa (Cushman 1927d).

Spiroplecta was often considered a synonym of an earlier alleged genus named by Ehrenberg (1841), namely *Heterohelix*; this taxonomic interpretation was given by Parker and Jones (1872) and reiterated by Jones (1895). A thorough re-study of the works by Ehrenberg (1838, 1841, 1844, 1854 and 1855) led Georgescu (2013a) to the conclusion that *Heterohelix* was a provisional name mentioned by C.G. Ehrenberg in 1841 and published in 1843 and *Spiroplecta* is the formal applied name. The revision of the oldest representatives of the group that were reported and named by C.G. Ehrenberg during the time period between 1838 and 1856 was underwent by Georgescu (2013a) by studying not only the original specimens deposited in the Museum of Natural History in Berlin, but also the bulk samples from the localities investigated by Ehrenberg and deposited in the same institution. Additional historical material was re-discovered in the Jacob Whitman Bailey Collection from the Farlow Herbarium and revised by Georgescu (2013a). These two review studies helped in the transformation of the monospecific genus *Spiroplecta* into a directional lineage, which is herein named 1planispiral and consists of two SMRS: ISMRS=*Textilaria americana* Ehrenberg 1841 (I-1planispiral) and FSMRS= *Spiroplecta clarae* Georgescu 2013, new name for *S. americana* Ehrenberg 1844 (F-1planispiral). Directional lineage 1planispiral presents significant variability in the chamber shape especially in the adult stage and chamber arrangement in the earlier portion of the test.

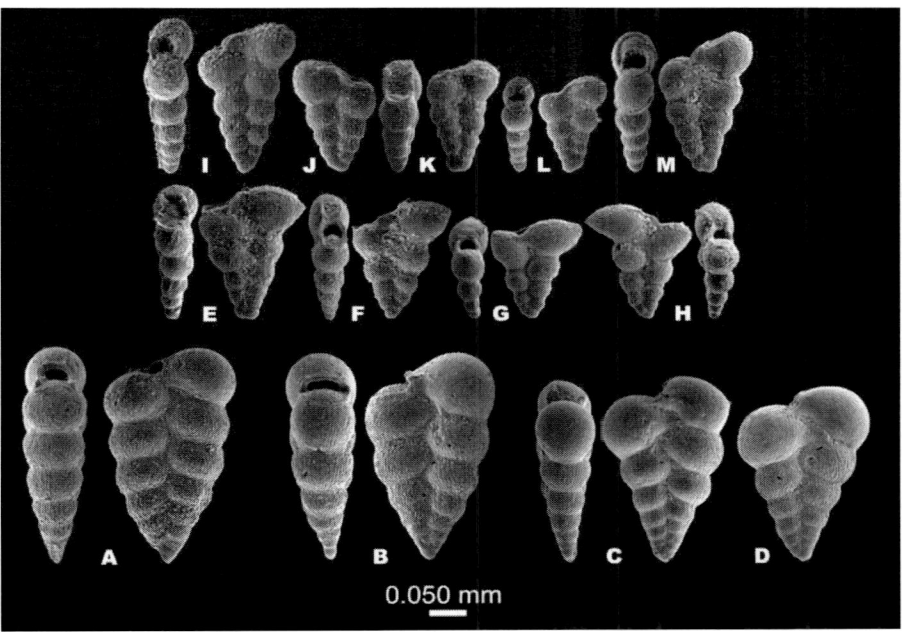

Figure 37. Specimens of the directional lineage 1planispiral. **A-D** Precursor SMRS: S5. A-early Maastrichtian, Sample 113-690C-19-1, 119-123 cm. B-early Maastrichtian, Sample 113-690C-20-5, 75-76 cm. C-late Santonian, Sample 174AX, 505.35-505.38 m, figured by Georgescu, Saupe and Huber (2008, pl. 3, Figure 2). D-late Santonian, Sample 174AX, 505.35-505.38 m. **E-H** ISMRS: I-1planispiral. E-late Santonian, Sample 1595b, Ehrenberg Collection, Missouri River Basin. F-late Santonian, Sample 1595b, Ehrenberg Collection, Missouri River Basin. G-late Santonian, Sample 1595b, Ehrenberg Collection, Missouri River Basin, figured by Georgescu (2013a, pl. 1, figures 4-6). H-late Santonian, Sample 1595b, Ehrenberg Collection, Missouri River Basin. **I-M** FSMRS: F-1planispiral. I-late Santonian, Sample 1595b, Ehrenberg Collection, Missouri River Basin. J-late Santonian, Sample 1595b, Ehrenberg Collection, Missouri River Basin. K-late Santonian, Sample 1595b, Ehrenberg Collection, Missouri River Basin. K-late Santonian, Sample 1595b, Ehrenberg Collection, Missouri River Basin, figured by Georgescu (2013a, pl. 2, figures 10-11). L-late Santonian, Sample 1595b, Ehrenberg Collection, Missouri River Basin, figured by Georgescu (2013a, pl. 2, figures 6-7).

The ISMRS I-1planispiral evolved apparently from S5 through the development of a hysterohelicid test from the ahelicid ones of the ancestor SMRS; the terminology for the test earlier portion is that given by Georgescu (2013). In addition there are significant changes in the chamber shape in the adult stage, which presents laterally pinched last-formed chambers. This feature is uniquely developed within the 1planispiral directional lineage in the entire evolutionary history of the planktic foraminifera with chambers alternately added with respect to the test growth axis. The number of laterally pinched chambers is variable (one to four) and the occurrences of peripheral spines associated with the pinched chambers claimed in the studies by Frerichs and Gaskill (1978) and Masters (1980) were not confirmed by Georgescu (2013) after the re-study of the original material. Test aperture is bordered by two symmetrically developed metaflanges, one on each test side. Such gross test architecture morphological features to which test lateral compression and similarities in the juvenile portion of the test clearly indicate that I-1planispiral evolved from the stalk SMRS S5, which is herein demonstrated the precursor SMRS of 1planispiral.

Chamber ornamentation consists of discontinuous longitudinal leptocostae, which have a thickness of 0.0018-0.0025 mm is S5 and 0.0022-0.0036 mm in I-1planispiral. In addition a pustulose periapertural area consisting of scattered dome-like pustules occurs in both SMRS. Pores are circular in shape in both SMRS and with a diameter of 0.0006-0.0009 mm in S5 and 0.0007-0.0009 mm in I-1planispiral; they are situated in the spaces between the discontinuous leptocostae, rarely interrupting them. Chamber ornamentation and porosity features show that the evolution from S5 to I-1planispiral led to an increase in leptocostae thickness, whereas no distinct trend can be recognized in the pore size and distribution. Notably, chamber ornamentation and porosity of the SMRS I-1planispiral were reassessed on the basis of well-preserved specimens detached from the bulk samples of the Ehrenberg Collection.

Evolution from I-1planispiral to F-1planispiral shows significant morphological changes both in gross test architecture and ultrastructural features. This evolution process led to the formation in F-1planispiral of holohelicid tests, in which the proloculus is completely surrounded by one whorl of small-sized chambers; specimens with intermediate morphological features, namely with the proloculus tangential to the periphery are known in both Ehrenberg Collection and Bailey Collection (Georgescu 2013a, 2013c). Another result of the evolution from I-1planispiral to F-1planispiral is the reduction and eventual loss of the pinched chambers; specimens with holohelicid tests and pinched chambers in the adult stage are known only from below the Santonian/Campanian boundary but only specimens of F-1planispiral without pinched last-formed chambers are the only known in the Campanian-Maastrichtian stratigraphic interval (Pessagno 1967; Smith and Pessagno 1973; Georgescu 2006). Periapertural structures are identical in I-1planispiral and F-1planispiral and consist of symmetrically developed metaflanges, one on each test side. Chamber surface is ornamented in both SMRS with longitudinal discontinuous leptocostae, which have a thickness of 0.0022-0.0036 mm in I-1planispiral and 0.0025-0.0031 mm in F-1planispiral; both SMRS present a pustulose periapertural area consisting of scattered dome-like pustules. Pores are circular in shape, situated in the space between the discontinuous leptocostae, rarely interrupting them and present a slight increase in diameter from 0.0007-0.0009 mm in I-1planispiral to 0.0009-0.0013 mm in F-1planispiral. There are not known descendants from F-1planispiral.

There are contradictory data in the fluctuation of the test length with the initiation of the directional lineage 1planispiral and along its evolution. The detached material from the bulk samples of the Ehrenberg Collection show a significant decrease in test length to nearly one half in the transition from S5 to I-1planispiral and a slight increase from I-1planispiral to F-1planispiral. Frerichs and Gaskill (1978) illustrated from South Dakota specimens assignable to I-1planispiral and which present a test length at least double when compared to the length of the tests of this SMRS from the Ehrenberg Collection and Bailey Collection.

The stratigraphic distribution of the SMRS of the directional lineage 1planispiral was briefly presented by Georgescu (2013a), but with the complete study of the collection material can be re-evaluated. Both SMRS I-1planispiral and F-1planispiral occur consistently in the four slides of the Bailey Collection; these samples were collected from the Smoky Hill Shale Member of the Niobrara Chalk Formation of late Santonian in age (Georgescu 2013a). There are three bulk samples that can be related to the material from which the two SMRS of 1planispiral were described in the Ehrenberg Collection and they are labelled 1595b, 1595e and 1595f (Georgescu 2013). One of them, namely 1595e is represented from a lithological point of view by laminated clayey sandstone and was probably collected from the Pierre Shale; the foraminiferal assemblage yielded by these rocks indicates an early Campanian age

(Georgescu 2013a). The other two samples 1595b (green soft marl) and 1595e (yellow chalky marl) yielded late Santonian foraminiferal assemblages. According to Frerichs and Dring (1981) the two lithologic types represent the unaltered/altered rocks that occur within the Smoky Hill Member of the Niobrara Formation. Therefore, a late Santonian age for these sample is assumed, which is consistent with the interpretation given by Georgescu (2013a, tables 1-2 and text-Figure 4); "Fort Hays Limestone Member" which was mentioned by Georgescu (2013a, p. 9, 20) as possible lithostratigraphic unit of provenance for I-1planispiral and F-1planispiral of the Ehrenberg Collection (Sample 1595b) is at a much lower stratigraphic level to include the representatives of the 1planispiral directional lineage.

Evolution of Longitudinally Zoned Ornamentation (DL: 8multichamber)
(Figure 38)

Another lineage that evolved in the late Santonian from the stalk SMRS S5 and led to the development to tests with multichamber growth in the adult stage is *Magellanina* Georgescu 2014b; it is renamed herein 8multichmaber. The morphological development that characterizes this directional lineage among the other Cretaceous planktic foraminiferal with chambers alternately added with respect to the test growth axis is the longitudinally displayed zones of ornamentation; this ornamentation pattern was reported for the first time by Georgescu, Saupe and Huber (2008). Directional lineage 8multichmaber includes two SMRS: ISMRS=*Magellanina magellani* Georgescu 2014b (I-8multichmaber) and FSMRS=*Gublerina acuta* de Klasz 1953 (F-8multichmaber).

SMRS I-8multichmaber evolved from S5 and this is demonstrated by the general similarities in test appearance, periapertural structures, chamber ornamentation and wall porosity features. Periapertural structures in I-8multichmaber consist of two symmetrically developed metaflanges, one on each test side indicating that this SMRS evolved from the test variety of S5 with similar structures. Another interesting development in test morphology is the occasional occurrence of subangular periphery in I-8multichmaber, which contrasts to the rounded periphery in S5 and most of the specimens of I-8multichmaber. In the ancestor SMRS S5 the chamber ornamentation consists of discontinuous longitudinal leptocostae with a thickness of 0.0018-0.0025 mm; leptocostae are closely spaced and increase in thickness in the descendant SMRS I-8multichmaber to 0.0020-0.0027 mm. Pores and situated in the space between the discontinuous leptocostae, rarely interrupting them. They are circular in shape in S5 and with a diameter of 0.0006-0.0009 mm; in I-8multichmaber pores are circular to elliptical in shape, with a diameter or maximum dimension of 0.0006-0.0014 mm. Notably in I-8multichmaber the pores are larger in the median portion of the test resulting in a characteristic test appearance. These data indicate that S5 is the precursor SMRS of the directional lineage 8multichmaber. The earliest occurrence of I-8multichmaber, which corresponds to that of the directional lineage 8multichmaber, is recorded in the upper Santonian sediments of the Pacific Ocean (DSDP Site 463); the known occurrences show that this SMRS begun expanding its distribution areal above the Santonian/Campanian boundary. Test length shows a significant reduction in the evolution from S5 to I-8multichmaber.

Figure 38. Specimens of the directional lineage 8multichamber. **A-D** Precursor SMRS: S5. A-early Maastrichtian, Sample 113-690C-19-1, 119-123 cm. B-early Maastrichtian, Sample 113-690C-20-5, 75-76 cm. C-late Santonian, Sample 174AX, 505.35-505.38 m, figured by Georgescu, Saupe and Huber (2008, pl. 3, Figure 2). D-late Santonian, Sample 174AX, 505.35-505.38 m. **E-J** ISMRS: I-8multichamber. E-middle Campanian, Sample 62-463-24-1, 50-52 cm, figured by Georgescu (2014b, pl. 1, Figure 11). F-middle Campanian, Sample 62-463-24-1, 50-52 cm, figured by Georgescu (2014b, pl. 1, figures 9-10). G-middle Campanian, Sample 62-463-24-1, 50-52 cm, figured by Georgescu (2014b, pl. 1, figures 7-8). H-late early Campanian, Sample 62-463-25-1, 51-53 cm, figured by Georgescu (2014b, pl. 1, figures 5-6). I-late early Campanian, Sample 62-463-25-1, 51-53 cm, figured by Georgescu (2014b, pl. 1, figures 3-4). J-late early Campanian, Sample 62-463-25-1, 51-53 cm, figured by Georgescu (2014b, pl. 1, figures 1-2). **K-M** FSMRS: F-8multichamber. K-late Maastrichtian, Sample 198-1212B-13-7, 50-52 cm, figured by Georgescu, Saupe and Huber (2008, pl. 4, figures 4a). L-late Maastrichtian, Sample 12-111A-11-2, 5-19 cm, figured by Georgescu, Saupe and Huber (2008, pl. 4, Figure 2). M-late Maastrichtian, Sample 171B-1050C-11-1, 127-130 cm, figured by Georgescu, Saupe and Huber (2008, pl. 4, Figure 3).

Multichamber growth stage in the adult portion of the test evolved in F-8multichmaber, which has the earliest occurrence in the late Campanian; this stage consists of the biaperturate progressive chamber followed by up to two sets of two chambers. The chambers of the adult stage present long metaflanges, which makes the chambers to appear well-distanced from each other rather than adjacent. Tests are distinctly compressed in the adult stage, with the two sides parallel in edge view. Five longitudinal ornamentation zones occur over the test surface. The periphery and central suture region are ornamented with longitudinal closely spaced leptocostae with a thickness of 0.0030-0.0034 mm; pores in these regions are simple and circular in shape, with a diameter of 0.0006-0.0008 mm. Two zones with reticulate ornamentation occur between the central leptocostate zone and each of the two peripheral leptocostate ones; the reticulate network consists of low ridges that are relatively difficult to separate from the test wall. Pores of the two reticulate zones are the largest known in the directional lineage 8multichmaber and they present a diameter of 0.0020-0.0049 mm. Evolution of the adult stage with multichamber growth resulted in a significant increase in test length, the tests of F-8multichmaber being frequently three to four times longer than those of the ancestor I-8multichmaber.

Earliest Evolution of Transversally Elongate Chambers (DL: 4alternate)
(Figure 39)

A new trend in the evolution of Cretaceous planktic foraminifera with chambers alternately added with respect to the test axis is the development of transversally elongate chambers; in this case the chamber elongation is along an approximate plane that includes the successive apertures. Chamber transversal elongation was evaluated in the past function of the chamber width/thickness ratio; according to this system such chambers were referred to as 'thicker than wide' or 'deeper than wide'.

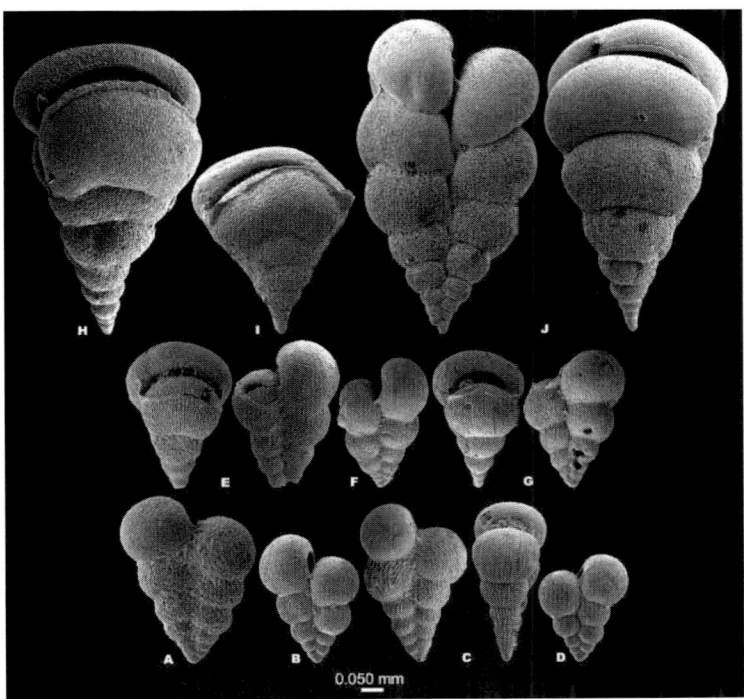

Figure 39. Specimens of the directional lineage 4alternate. **A-D** Precursor SMRS: S-1alternate. A-Coniacian to early Santonian, 122-763B-18-1, 72-73 cm, figured by Georgescu in Georgescu and others (2013, pl. 7, Figure 1 only). B-early Campanian, Sample 71-511-38-6, 13-17 cm, figured by Georgescu in Georgescu and others (2013, pl. 7, Figure 5 only). C-late Santonian, Sample 1595b, Ehrenberg Collection, Missouri River Basin. D-Santonian, Toolonga Calcilutite, western Australia, Sample Huber, figured by Georgescu in Georgescu and others (2013, pl. 7, Figure 9 only). **E-G** ISMRS: I-4alternate. E-late Santonian, Sample 10-95-15-4, 99.5-100.5 cm, figured by Georgescu (2014c, pl. 1, figures 1-2). F-Santonian, Toolonga Calcilutite, western Australia, Sample Huber. G-Santonian, Toolonga Calcilutite, western Australia, Sample Huber, figured by Georgescu (2014c, pl. 1, Figure 8). **H-J** FSMRS: F-4alternate. H-late Maastrichtian, 43-384-14-2, 50-52 cm. I-late Maastrichtian, 171B-1050C-11-1, 75-78 cm. J-late Maastrichtian, 43-384-14-2, 50-52 cm, figured by Georgescu (2014c, pl. 2, figures 9-10).

The general trend in typological classification was to group all the SMRS with transversally elongate chambers into the genus *Pseudotextularia* Rzehak 1891; with only one exception this method was used by all taxonomists. The exception is the study by Montanaro Gallitelli (1956) who defined the monospecific genus *Bronnimannella* having *Gümbelina plummerae* Loetterle 1937 type SMRS, which is the oldest SMRS presenting this test architecture; subsequently Montanaro Gallitelli (1957) considered *Bronnimannella* as one junior synonym of *Pseudotextularia* and this taxonomic decision was thoroughly followed. The development of the evolutionary classification for the Cretaceous planktics with chambers alternately added with respect to the test growth axis shows that transversally elongate chambers independently evolved in distinct lineages (Georgescu 2014c). The earliest of them was initiated in the early Santonian, is herein named 4alternate and has a directional character; it includes two SMRS: ISMRS= *Ventilabrella plummerae* Sandidge 1932 (I-4alternate) and FSMRS= *Gümbelina nuttalli* Voorwijk 1937 (F-4alternate).

Similarities between the test juvenile stages and periapertural structures consisting of symmetrically developed metaflanges, one on each test side indicate that the directional lineage 4alternate evolved from the SMRS S-1alternate, which is therefore the lineage precursor SMRS. Both SMRS are ornamented with continuous longitudinally oriented leptocostae, with a thickness of 0.0032-0.0046 mm in the ancestor S-1alternate and 0.0033-0.0049 mm in the I-4alternate descendant. In addition the ornamentation presents another evolutionary trend, which is the increase in surface of the pustulose periapertural area consisting of scattered dome-like pustules. Wall porosity shows an increase in pore diameter from 0.0006-0.0012 mm to 0.0005-0.0020 mm in the evolution from S-1alternate to I-4alternate. The most evident in the evolutionary changes pertaining of the gross test architecture is the evolution of transversally elongate chambers, which occur in the adult portion of the test, especially in the last one to three, rarely four pairs of chambers.

Evolution from I-4alternate to F-4alternate is especially apparent in the increase of the transversal chamber elongation; as a result the periapertural metaflanges of the last-formed one to eight chambers are often hardly visible in mature and gerontic specimens. High detail features indicate slight increases in leptocostae thickness and pore diameter from 0.0033-0.0049 mm to 0.0032-0.0053 mm and 0.0005-0.0020 mm 0.0008-0.0025 mm respectively. The pustulose periapertural region in F-4alternate is spread over a significant portion of the chamber anterior portion and this is significantly wider than the same ornamentation features known either in the lineage precursor S-1alternate and SMRS I-4alternate. A well-marked increase in test size is recorded in the evolution from I-4alternate to F-4alternate; specimens of F-4alternate with a test length of nearly 0.8000 mm occur especially in the uppermost Campanian and Maastrichtian sediments.

Earliest Evolution of Rimmed Periphery (DL: 2planispiral)
(Figure 40)

A new directional lineage evolved from the stalk in the late Santonian and it will lead to the development of tests that consistently present a peripheral wall flexure resulting in a rimmed appearance at the periphery; at the beginning of its evolution this lineage consisted only of ahelicid tests but holohelicid tests were developed in its late stage. This directional lineage is herein named 2planispiral in the nomenclature associated with the evolutionary classification. Excepting for the changes in the chamber arrangement in the test early portion the representatives of this directional lineage have only chambers alternately added with respect to the test growth axis. The morphology within this lineage was relatively poorly understood in the past and no evolutionary relationships with other taxa could be inferred based only on the gross test architecture features (Masters 1977; Nederbragt 1991). The ahelicid portion of this directional lineage was recognized by Georgescu and Abramovich (2008a) and named *Hendersonia*; the name was changed into *Hendersonites* by Georgescu and Abramovich (2009b) under the rules of the ICZN to avoid the name confusion with the homonym name of one gastropod genus. A distinct genus name was given by Georgescu and Abramovich (2008) to accommodate the holohelicid tests of this directional lineage: *Paraspiroplecta*. The directional lineage 2planispiral includes three SMRS: ISMRS= *Hendersonia hendersoni* Georgescu and Abramovich 2008a (I-2planispiral), FSMRS =*Gümbelina carinata* Cushman 1938 (F-2planispiral) and SDSMRS =*Heterohelix navarroensis* Loeblich 1951 (S-2planispiral).

The general test appearance together with the high resolution morphological features of the test indicate that the directional lineage 2planispiral evolved from the stalk directional lineage and the SMRS S5 and I-2planispiral are in direct ancestor-descendant relationship. Evolution from S5 to I-2planispiral is mostly apparent in the beginning of the formation of the test wall peripheral flexure, which is weak and developed only over the earlier chambers of the test. Periapertural structures in I-2planispiral consist of a mixture of orthoflanges and metaflanges, which is similar to those in the ancestor SMRS. The general test appearance in S5 and I-2planispiral presents numerous similarities and Georgescu and Abramovich (2008a) recognized in I-2planispiral two test varieties function of the degree of resemblance between the two SMRS. Both S5 and I-2planispiral are ornamented with longitudinal leptocostae and there is an increase in leptocostae thickness with the initiation of the 2planispiral directional lineage from 0.0018-0.0025 mm in S5 to 0.0023-0.0031 mm in I-2planispiral. A periapertural pustulose area occurs consistently in the representatives of I-2planispiral. Pores are circular in shape and situated in the space between the leptocostae, rarely interrupting them; pores significantly increase in diameter from 0.0006-0.0009 mm in S5 to 0.0007-0.0024 mm in I-2planispiral.

Evolution from I-2planispiral to F-2planispiral is marked in the general test architecture by the complete development of the test wall flexure at the periphery, which surrounds the test giving it a rimmed appearance. In edge view the peripheral wall flexure confers the test a carinate appearance. Chamber surface is ornamented with leptocostae and there is a distinct increase in leptocostae thickness from I-2planispiral (0.0023-0.0031 mm) to F-2planispiral (0.0029-0.0053 mm); ornamentation appears slightly more prominent in the portion of the central suture. A pustulose periapertural area consisting of scattered dome-like pustules occurs both in I-2planispiral and F-2planispiral. In F-2planispiral pores are simple or scalaropores and with a diameter of 0.0011-0.0027 mm. The occurrence of scalaropores was considered by Georgescu and Abramovich (2008a) a characteristic feature for the SMRS *Hendersonia jerseyensis* Georgescu and Abramovich 2008, new material collected from the uppermost Santonian-lowermost Campanian sediments of DSDP Site 95 (Yucatan Outer Shelf) shows that the two kinds of pores can coexist in F-2planispiral.

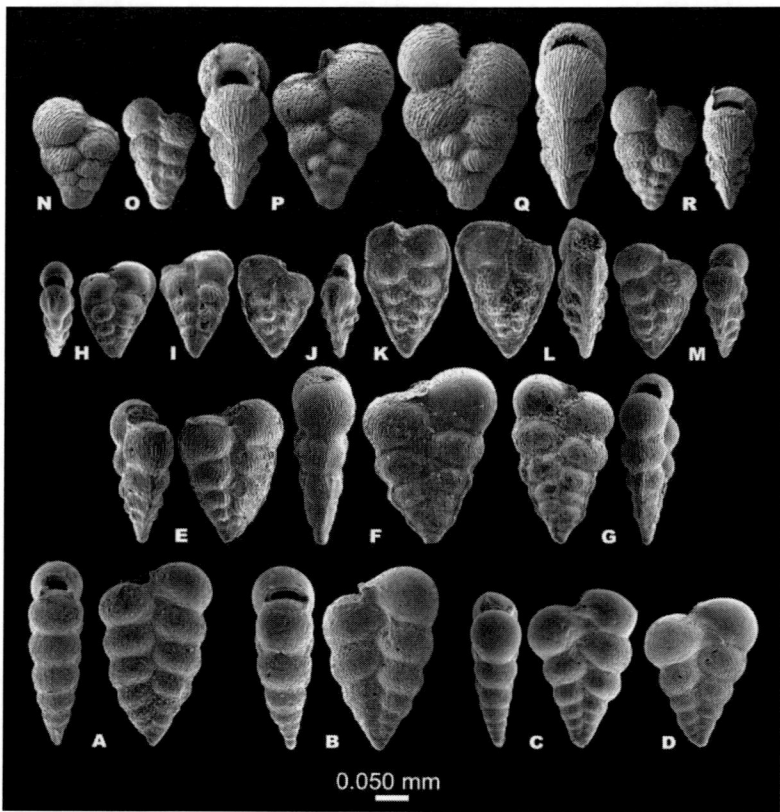

Figure 40. Specimens of the directional lineage 2planispiral. **A-D** Precursor SMRS: S5. A-early Maastrichtian, Sample 113-690C-19-1, 119-123 cm. B-early Maastrichtian, Sample 113-690C-20-5, 75-76 cm. C-late Santonian, Sample 174AX, 505.35-505.38 m, figured by Georgescu, Saupe and Huber (2008, pl. 3, Figure 2). D-late Santonian, Sample 174AX, 505.35-505.38 m. **E-G** ISMRS: I-2planispiral. E-late Campanian, Sample 79-511-24-6, 20-23 cm. F-late Campanian, Sample 79-511-24-6, 20-23 cm, figured by Georgescu and Abramovich (2008a, pl. 1, figures 8-9). G-late Campanian, Sample 79-511-24-6, 20-23 cm, figured by Georgescu and Abramovich (2008a, pl. 1, figures 1, 4). **H-M** FSMRS: F-2planispiral. H-early Campanian, Sample 174AX, 495.30-495.33 cm, figured by Georgescu and Abramovich (2008a, pl. 3, figures 5-6). I-early Campanian, Sample 174Ax, 495.30-495.33 cm, figured by Georgescu and Abramovich (2008a, pl. 3, Figure 10). J-late Campanian, Gulf of Mexico, van Morkhoven Collection, figured by Georgescu and Abramovich (2008a, pl. 3, figures 1-2). K-late Campanian, Gulf of Mexico, van Morkhoven Collection, figured by Georgescu and Abramovich (2008a, pl. 2, Figure 12). L-late Campanian, Gulf of Mexico, van Morkhoven Collection, figured by Georgescu and Abramovich (2008a, pl. 2, figures 8-9). M-late Santonian, Sample 174AX, 505.35-505.38 m, figured by Georgescu and Abramovich (2008a, pl. 2, figures 1, 4). **N-R** SSMRS: S-2planispiral. N-Maastrichtian, Sample Mullinax-3, 77-6-2, 9.79 m, figured by Georgescu and Abramovich (2008a, pl. 4, Figure 9). O-Maastrichtian, Sample Mullinax-3, 77-6-2, 9.79 m, figured by Georgescu and Abramovich (2008a, pl. 4, Figure 8). P-Maastrichtian, Sample Mullinax-3, 77-6-2, 9.79 m, figured by Georgescu and Abramovich (2008a, pl. 4, figures 6-7). Q-Maastrichtian, Sample Mullinax-3, 77-6-2, 9.79 m, figured by Georgescu and Abramovich (2008a, pl. 4, figures 4-5). R-Maastrichtian, Sample Mullinax-3, 77-6-2, 9.79 m, figured by Georgescu and Abramovich (2008a, pl. 4, figures 1-2).

The gross test architecture shows a major development in the evolution from F-2planispiral to S-2planispiral; this is given by the evolution of an early planispirally coiled stage in the descendant SMRS. Therefore, the F-2planispiral to S-2planispiral evolution records the transition from ahelicid to holohelicid tests; seemingly the evolution process was a rapid one and no specimens with intermediate morphological features are known despite the long study history of the two SMRS. Excellent transmitted light illustrations of the early planispiral coil in S-2planispiral were given by Loeblich (1951) and Brown (1969).

Apparently the evolution of holohelicid tests determined a reduction in prominence of the peripheral wall flexure. There is an increase in leptocostae thickness from F-2planispiral (0.0029-0.0053 mm) to S-2planispiral (0.0033-0.0067 mm): the leptocostate ornamentation is thicker in the proximity of the test central suture where the leptocostae may fuse to form more prominent ornamentation elements. A periapertural pustulose area consisting of scattered dome-like pustules occurs in S-2planispiral but over a narrower portion of the chambers when compared to F-2planispiral. An increase in pores diameter is also recorded, from 0.0011-0.0027 mm in F-2planispiral to 0.0015-0.0032 mm in S-2planispiral; scalaropores are not known in S-2planispiral.

Evolution of Simple-Ridged Wall (CL: 5alternate)
(Figure 41)

The first condensed lineage in the history of the planktic foraminifera with chambers alternately added with respect to the test growth axis was initiated in the latest Santonian. It evolved simple-ridged test wall and is the only lineage in the group's history that evolved such a kind of test wall. The SMRS herein included in the condensed lineage 5alternate was described as *Hendersonites pacificus* by Georgescu (2011a). It was included within *Hendersonites* based on the general test appearance, but a more elaborate solution in typological nomenclature is to have this condensed lineage distinctly named at supraspecific level and the name *Neohendersonites* is herein given to it. The condensed lineage 5alternate is the last know lineage of the group that evolved in the Santonian.

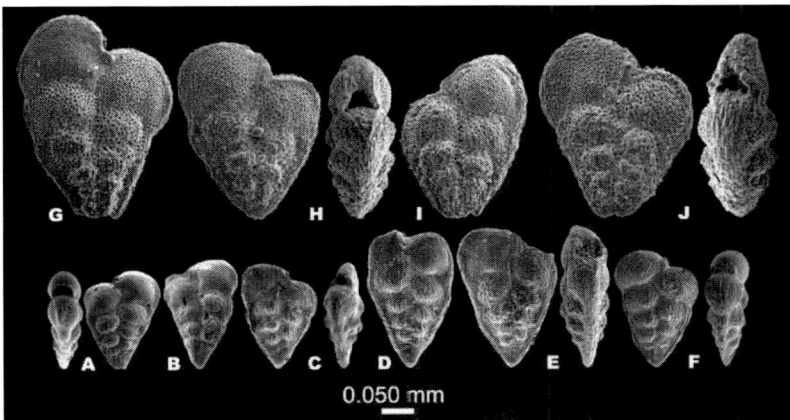

Figure 41. Specimens of the directional lineage 5alternate. **A-F** Precursor SMRS: F-2planispiral. A-early Campanian, Sample 174Ax, 495.30-495.33 cm, figured by Georgescu and Abramovich (2008a, pl. 3, figures 5-6). B-early Campanian, Sample 174Ax, 495.30-495.33 cm, figured by Georgescu and Abramovich (2008a, pl. 3, figures 1-2). C-late Campanian, Gulf of Mexico, van Morkhoven Collection, figured by Georgescu and Abramovich (2008a, pl. 3, Figure 10). D-late Campanian, Gulf of Mexico, van Morkhoven Collection, figured by Georgescu and Abramovich (2008a, pl. 2, Figure 12). E-late Campanian, Gulf of Mexico, van Morkhoven Collection, figured by Georgescu and Abramovich (2008a, pl. 2, figures 8-9). F-late Santonian, Sample 174AX, 505.35-505.38 m, figured by Georgescu and Abramovich (2008a, pl. 2, figures 1, 4). **G-J** ISMRS: I-5alternate. G-latest Santonian-early Campanian, Sample 62-463-26-3, 52-54 cm, figured by Georgescu (2011a, pl. 1, Figure 9). H-latest Santonian-early Campanian, Sample 62-463-26-3, 52-54 cm, figured by Georgescu (2011, pl. 1, Figure 1). I-latest Santonian-early Campanian, Sample 62-463-26-3, 52-54 cm, figured by Georgescu (2011a, pl. 2, figures 3-4). J-latest Santonian-early Campanian, Sample 62-463-26-3, 52-54 cm, figured by Georgescu (2011a, pl. 2, figures 1-2).

Georgescu (2011) showed that I-5alternate evolved from F-2planispiral mainly through the development of simple-ridged test wall, which contrasts to the simple wall of the ancestor SMRS; therefore, F-2planispiral is the precursor of 5alternate. The ornamentation in I-5alternate is leptocostate and the leptocostae present variable thickness over the test surface; they are thicker over the earlier chambers of the test (0.0078-0.0085 mm) and thinner over the last-formed ones (0.0020-0.0027 mm), but thicker leptocostae (0.0068-0.0075 mm) also occur in the peripheral region. Notably, in juvenile specimens the leptocostae present relatively constant thickness over the test surface: 0.0041-0.0065 mm. There appears an increase in the leptocostae thickness when compared to the precursor SMRS where the leptocostae have a thickness of 0.0029-0.0053 mm. A pustulose periapertural area occurs consistently in the anterior portion of the chambers. There is an increase in pore diameter in the evolution from I-5alternate to F-2planispiral. Pore diameter in I-5alternate is variable; larger pores (0.0032-0.0042 mm) occur in the portion of the test with simple-ridged wall, whereas in juvenile specimens they present a diameter of 0.0011-0.0019 mm. The smaller pores range largely overlaps that observed in the ancestor SMRS F-2planispiral: 0.0011-0.0027 mm. Gross test architecture shows that the ancestor F-2planispiral and its descendant I-5alternate present similar periphery with rimmed appearance and periapertural structures consisting of symmetrically developed metaflanges, one on each test side. The last-formed chamber in I-5alternate presents occasionally a slight antero-posterior elongation.

Evolution of Early Planispiral Coil in a Globular-Chambered Lineage (DL: 3planispiral)
(Figure 42)

A new directional lineage evolved from the globular-chambered leptocostate S-1alternate above the Santonian/Campanian boundary. It led to the occurrence of an early planispiral coil and the transition from ahelicid to holohelicid tests was a fast one and happened in the late Campanian. This lineage was defined as *Lazarusina* by Georgescu (2013a) and includes two SMRS as follows: ISMRS=*Gümbelina globocarinata* Cushman 1938 (I-3planispiral) and FSMRS=*Lazarusina lazarusi* Georgescu 2013a (F-3planispiral).

Loeblich and Tappan (1987) included the terminal SMRS of this directional lineage within the genus *Spiroplecta* and supported this taxonomic decision with illustrations of globular-chambered specimens with an early planispiral coil figured by Montanaro Gallitelli (1957); this decision was accepted by Georgescu and Abramovich (2009a). Restudy of the material in the Ehrenberg Collection carried out subsequently by Georgescu (2013) and Bailey Collection by Georgescu (2013b) showed that the specimens included in *Spiroplecta* by Ehrenberg (1841, 1844, 1854, 1856) belong in fact to two distinct lineages.

According to this study the specimens figured from the Smoky Hill Member of the Niobrara Formation (Sample1595b) belong to the *Spiroplecta* directional lineage, which is renamed herein as 1planispiral, whereas those from the Ripley Formation (Sample 1596b) is assignable to the directional lineage *Lazarusina* (3 planispiral). Morphological developments within the 3planispiral further demonstrate that evolution of an early planispiral coil is a late achievement in the group evolution as shown by Brown (1969). The directional lineage 3planispiral evolved from the globular-chambered and leptocostate S-1alternate. Both the ancestor S-1alternate and descendant I-3planispiral present ahelicid tests with globular or subglobular chambers alternately added with respect to the test growth axis. The early portion of the test in I-3planispiral is compressed in edge view conferring the test a pseudocarinate appearance. The earliest chambers are added at a high angle with respect to the direction of growth rather than at an acute angle as in S-1alternate. Aperture is a wide medium-high arch and is bordered by two symmetrically developed metaflanges, one on each side of the test. The longitudinal leptocostae are 0.0032-0.0046 mm thick in the ancestor S-1alternate and 0.0029-0.0053 mm in I-3planispiral. Two other features that occur both in S-1alternate and I-3planispiral further demonstrate the ancestor-descendant relationship between the two SMRS: ornamentation does not present an increased prominence over the earlier chambers of the test and a well-developed pustulose periapertural area consisting of scattered dome-like pustules consistently occurs in the chamber anterior portion. Pores are circular and show an increase in size from a diameter of 0.0006-0.0012 mm in S-1alternate to 0.0010-0.0017 mm in the descendant I-3planispiral. There is a reduction in size with the initiation of the directional lineage 3planispiral.

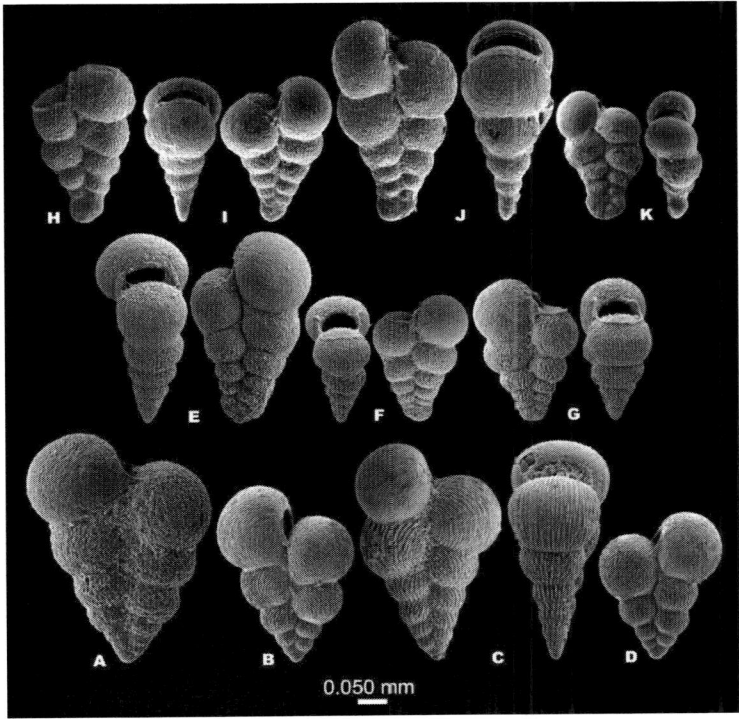

Figure 42. Specimens of the directional lineage 3planispiral. **A-D** Precursor SMRS: S-1alternate. A-Coniacian to early Santonian, 122-763B-18-1, 72-73 cm, figured by Georgescu in Georgescu and others (2013, pl. 7, Figure 1 only). B-early Campanian, Sample 71-511-38-6, 13-17 cm, figured by Georgescu in Georgescu and others (2013, pl. 7, Figure 5 only). C-late Santonian, Sample 1595b, Ehrenberg Collection, Missouri River Basin. D-Santonian, Toolonga Calcilutite, western Australia, Sample Huber, figured by Georgescu in Georgescu and others (2013, pl. 7, Figure 9 only). **E-G** ISMRS: I-3planispiral. E-early Campanian, Sample 32-305-27-2, 106-107 cm. F-early Campanian, Sample 32-305-27-2, 106-107 cm. G-early Campanian, Sample 32-305-27-2, 106-107 cm, figured by Georgescu (2013, pl. 5, figures 1-2). **H-K** FSMRS: F-3planispiral. H-early Maastrichtian, Sample 62-463-8-1, 50-52 cm, figured by Georgescu (2013a, pl. 6, Figure 3). I-late Maastrichtian, Sample 171B-1050C-11-2, 146-149 cm, figured by Georgescu (2013a, pl. 5, figures 9-10). J-early Maastrichtian, Sample 43-384-14-4, 50-52 cm, figured by Georgescu (2013a, pl. 5, figures 7-8). K-late Maastrichtian, Mullinax-1 well, Texas.

Ahelicid tests in the directional lineage 3planispiral evolved in F-3planispiral where the proloculus is surrounded by four to six small-sized subglobular chambers. F-3planispiral has a general test appearance that resembles that of I-3planispiral; one of the most characteristic features of this lineage, namely the early test compression in edge view is further developed in F-3planispiral when compared to its ancestor I-3planispiral where it is in incipient stage. Test periphery is broadly rounded over the last-formed chambers in edge view throughout the directional lineage 3planispiral. There are no apparent changes in the aperture morphology and periapertural structures when compared to those of the ancestor I-3planispiral; the aperture in F-3planispiral is a wide medium high arch bordered by two symmetrically developed metaflanges, one on each side to the test.

In F-3planispiral chamber surface is ornamented with continuous longitudinal leptocostae with a thickness of 0.0035-0.0054 mm; this thickness range is comparable to that observed in the ancestor I-3planispiral, where the leptocostae are 0.0032-0.0046 mm thick. There is a discrete trend to increase the ornamentation prominence over the earlier chambers of the test due to the addition of successive layers of calcite; this feature occurs occasionally in F-3planispiral and is mostly apparent in the less distinct sutures between the chambers of the early planispiral coil. A periapertural pustulose area consisting of dome-like pustules occurs in the chamber anterior part. Wall is calcitic, hyaline and perforate; pores are circular in shape and are situated in the space between the leptocostae, rarely interrupting them. The increase in pore diameter observed in the transition from S-1alternate to I-3planispiral continues in the F-3planispiral where the pore diameters are of 0.0013-0.0023 mm. There are no apparent changes in the test length along the 3planispiral directional lineage.

Evolution of Bimodal Pycnocostate Ornamentation (DL: 4backextended)
(Figure 43)

Leptocostae thickening in the branched lineage 7multichamber led to the development of ornamentation with pycnocostate appearance in S-7multichamber of the late Santonian-early Campanian. Ornamentation consisting completely of leptocostae was developed in a directional lineage that evolved above the Santonian/Campanian boundary and shows in parallel the gradual development of double backward chamber extension.

The representatives of this directional lineage were traditionally included within the genus *Pseudoguembelina* Brönnimann and Brown 1953; *Pseudoguembelina* was revised and redefined in evolutionary classification as directional lineage by Georgescu (2014b) and is renamed herein 4backextended. Three SMRS are included within the directional lineage 4backextended: ISMRS=*Pseudoguembelina praecostulata* Georgescu 2014b (I-4backextended), FSMRS= *Gümbelina costulata* Cushman 1938 (F-4backextended) and SSMRS=*Gümbelina excolata* Cushman 1938 (S-4backextended). Morphological variability and the multitude of specimens with intermediary morphological features between the three SMRS document the occurrence of an evolutionary continuum along the directional lineage 4backextended.

The initiation of the directional lineage 4backextended happened in the earliest Campanian and the transition from the precursor to the ISMRS is recorded in the DSDP Site 463 of the Central Equatorial Pacific Ocean. This lineage evolved from 3backextended, a directional lineage that evolved bimodal ornamentation; the incipient state of development of the double backward chamber extension, one on each side of the test, together with the absence of supplementary apertures in I-4backextended indicate that the lineage 4backextended evolved from I-3backextended.

Test shape in edge view in I-4backextended presents in general a shape similar to that of the ancestral SMRS, but tests with lanceolate shape occur occasionally. The aperture in both SMRS has the shape of a low to medium high arch and is bordered by retroflanges which are not rimmed in both I-3backextended and I-4backextended; notably, in the descendant I-4backextended the periapertural structures are ornamented with pycnocostae and often define a false apertural structure in the anterior portion of the test central suture. Leptocostae in I-4backextended have a thickness of 0.0028-0.0052 mm, which show a significant increase when compared to the leptocostae thickness observed in I-3backextended (0.0017-0.0036 mm); the pycnocostae are parallel to the periphery in the peripheral region and oblique to the growth axis in the central portion of the test. Pores in I-4backextended are circular, elliptical and occasionally irregular in shape and have a diameter or maximum dimension of 0.0009-0.0013 mm; there is an increase in pore size from the ancestor I-3backextended where the pores are circular or elliptical in shape and with a diameter of 0.0005-0.0010 mm. These data indicate that I-3backextended is the precursor of the directional lineage 4backextended.

Evolution from the I-4backextended to F-4backextended is associated with an increase in test size and the descendant F-4backextended frequently yields tests with a double length when compared to that of the ancestral SMRS. Morphological features show that this evolution process resulted in the development of supplementary apertures under the double chamber backward extensions of the last-formed one to four chambers; occasionally the double chamber backward extension appear rimmed towards the posterior part of the test. The periapertural structures in the F-4backextended consist of retroflanges, which can be rimmed or not.

The observed increase in pycnocostae is more than double, namely from 0.0028-0.0052 mm in I-4backextended to 0.0054-0.0118 mm in F-4backextended; in the descendant SMRS the pycnocostae present a distinct trend of reducing the space between them. Pores are situated in the space between the pycnocostae; they have circular, elliptical or irregular shape and a diameter or maximum dimension of 0.0009-0.0013 mm.

The morphological differences in gross test architecture between F-4backextended and S-4backextended are apparent in a higher width/length ratio and periapertural structures that consist only of rimmed retroflanges in S-4backextended, whereas a mixture of rimmed and non-rimmed retroflanges is documented in the ancestor F-4backextended. There is a significant increase in the pycnocostae thickness from 0.0054-0.0118 mm in F-4backextended to 0.0055-0.0187 mm in S-4backextended; the pycnocostae are closely-spaced and their bimodal arrangement is at least partly obscured by the increased thickness; Georgescu (2014b) considered that the pycnocostae thickening in S-4backextended marks the return to unimodal ornamentation. Pores present a circular, elliptical or irregular shape with a diameter or maximum dimension of 0.0009-0.0013 mm. Therefore, no changes in pore size are apparent throughout the directional lineage 4backextended.

0.050 mm

Figure 43. Specimens of the directional lineage 4backextended. **A-D** Precursor SMRS: I-3backextended. A-early Campanian, Sample 32-305-25-1, 100-102 cm, figured by Georgescu (2014b, pl. 3, figures. 9-10). B-early Campanian, Sample 62-463-26-3, 52-54 cm, figured by Georgescu (2014b, pl. 3, Figure 7-8). C-early Campanian, Sample 62-463-26-3, 52-54 cm, figured by Georgescu (2014b, pl. 3, Figure 5-6). D-late Campanian, Sample 32-305-21-2, 100-102 cm, figured by Georgescu (2014b, pl. 3, figures 1-2). **E-I** ISMRS: I-4backextended. E-middle Campanian, Sample 62-463-25-1, 51-53 cm, figured by Georgescu (2014b, pl. 4, figures 10-11). F-middle Campanian, Sample 62-463-25-1, 51-53 cm, figured by Georgescu (2014b, pl. 4, figures 7-8). H-early Campanian, Sample 32-305-27-2, 106-107 cm, figured by Georgescu (2014b, pl. 4, figures 5-6). I-early Campanian, Sample 32-305-27-2, 106-107 cm, figured by Georgescu (2014b, pl. 4, figures 3-4). **J-L** FSMRS: F-4backextended. J-early Maastrichtian, Sample 32-305-17-4, 80-95 cm. K-middle Maastrichtian, Sample 32-305-16-5, 60-76 cm. L-late Campanian, Sample 32-305-18-5, 80-95 cm. **M-O** SSMRS: S-4backextended. M-middle Maastrichtian, Sample 32-305-16-5, 60-76 cm. N-late Campanian, Sample 32-305-18-5, 80-95 cm. O-middle Maastrichtian, Sample 32-305-16-5, 60-76 cm.

There are two additional ornamentation features that occur throughout the directional lineage 4backextended. First, no pustulose periapertural area occurs in the specimens of this lineage and this is consistent with the loss of this feature from the precursor SMRS I-3backextended; this feature was lost with the initiation of the directional lineage 3backextended from its stem precursor S5. Secondly, the ornamentation in the representatives of the directional lineage 4backextended do not present ornamentation thickenings over the earlier portion of the test due to the addition of successive layers of calcite during the ontogenetic development.

Evolution of Multiple Ornamentation Patterns (DL: 9multichamber)
(Figure 44)

Evolution at the level of ornamentation during the Campanian-Maastrichtian stratigraphic interval is probably best illustrated by a directional lineage that was initiated in the late Campanian and led to the evolution of tests with elaborated ornamentation, which show three patterns over the chamber surface. Ornamentation increase in complexity in this lineage happened in parallel with the evolution of a complex adult stage with multichamber growth. This directional lineage includes partly the representatives of the genus *Praegublerina* Georgescu, Saupe and Huber 2008. New data especially from its terminal portion leads to a redefinition of this directional lineage, which is herein named in evolutionary classification nomenclature 9multichamber. The directional lineage 9multichamber includes the following SMRS: ISMRS=*Gümbelina pseudotessera* Cushman 1938 (I-9multichamber), FSMRS= *Gublerina acuta robusta* de Klasz 1953 (F-9multichamber), SSMRS=*Planoglobulina meyerhoffi* Seiglie 1960 (S-9multichamber) and TSMRS=*Ventilabrella multicamerata* de Klasz 1953 (T-9multichamber). The directional lineage 9multichamber evolved directly from the group's stalk lineage.

Gross architecture and high resolution test features of I-9multichamber together with the stratigraphic range indicate that the 9multichamber evolved from the terminal SMRS of the stalk directional lineage, namely S5; therefore, S5 is the precursor of the directional lineage 9multichamber. The general test appearance in the ancestor S5 and descendant I-9multichamber presents evident similarities (e.g., chamber shape, compressed tests in edge view, suture orientation, aperture shape, etc.); notably, the tests of I-9multichamber consist only of chambers alternately added with respect to the test growth axis without an adult stage with multichamber growth. The aperture in I-9multichamber is bordered by symmetrically developed metaflanges, one on each side of the test; by contrast to the periapertural structures in the ancestral S5 where they consist of a mixture of orthoflanges and metaflanges, the metaflanges of I-9multichamber are much longer filling the space at the centre of the test between the two rows of divergent chambers at the periphery. Both S5 and I-9multichamber have the chamber surface ornamented with longitudinal leptocostae and there is a significant increase in the leptocostae thickness from the ancestor (0.0018-0.0025 mm) to descendant (0.0025-0.0040 mm). A major change in the ornamentation occurs with the initiation of the directional lineage 9multichamber: the leptocostate ornamentation is occasionally more prominent over the earlier chambers of the test, which contrasts to the ornamentation in S5 where is equally developed over all the chambers of the test. Pores are circular in shape and in both SMRS they are situated between the leptocostae rarely interrupting them; pore diameter is of 0.0006-0.0009 mm in S5 and 0.0004-0.0008 mm in I-9multichamber.

Figure 44. Specimens of the directional lineage 9multichamber. **A-D** Precursor SMRS: S5. A-early Maastrichtian, Sample 113-690C-19-1, 119-123 cm. B-early Maastrichtian, Sample 113-690C-20-5, 75-76 cm. C-late Santonian, Sample 174AX, 505.35-505.38 m, figured by Georgescu, Saupe and Huber (2008, pl. 3, Figure 2). D-late Santonian, Sample 174AX, 505.35-505.38 m. **E-I** ISMRS: I-9multichamber. E-late Campanian, Gulf of Mexico, van Morkhoven Collection, figured by Georgescu, Saupe and Huber (2008, pl. 3, Figure 8). F-late Campanian, Upper Taylor Marl, Onion Creek, Texas, Loeblich and Tappan Topotype Collection, figured by Georgescu, Saupe and Huber (2008, pl. 3, Figure 11). G-late Campanian, Upper Taylor Marl, Onion Creek, Texas, Loeblich and Tappan Topotype Collection, figured by Georgescu, Saupe and Huber (2008, pl. 3, Figure 10). H-late Campanian, Upper Taylor Marl, Onion Creek, Texas, Loeblich and Tappan Topotype Collection, figured by Georgescu, Saupe and Huber (2008, pl. 3, Figure 9). I-late Maastrichtian, Sample 12-111A-11-6, 71-85 cm, figured by Georgescu, Saupe and Huber (2008, pl. 3, Figure 6). **J-L** FSMRS: F-9multichamber. J-middle Maastrichtian, Sample 32-305-16-5, 60-76 cm, figured by Georgescu, Saupe and Huber (2008, pl. 4, Figure 10a). K-late Maastrichtian, Sample 12-111A-11-2, 5-19 cm, figured by Georgescu, Saupe and Huber (2008, pl. 4, Figure 6). L-late Maastrichtian, Sample 12-111A-11-4, 124-128 cm, figured by Georgescu, Saupe and Huber (2008, pl. 4, Figure 5a). **M-N** SSMRS: S-9multichamber. M-late Campanian, Sample 171B-1050C-18-1, 44-45 cm. N-late Campanian, Sample 171B-1050C-18-1, 44-45 cm. **O-P** TSMRS: T-9multichamber. O-late Maastrichtian, Sample 171B-1050C-15-1, 82-84 cm. P-late Maastrichtian, Sample 171B-1050C-11-1, 75-78 cm.

A significant leap in the evolution of the directional lineage 9multichamber happened with the evolutionary occurrence of F-9multichamber. The general chamber arrangement shows the evolution of an adult stage with multichamber growth; this stage initiates with the progressive biaperturate chamber, which is followed by maximum three sets of chambers that present an increase in number at an increment of one (e.g., first set-two chambers, second set-three chambers, third set-four chambers, etc.). The central portion of the test consists of well-developed metaflanges and this is apparent both in the terminal part of the stage with chambers alternately added with respect to the test growth axis and adult flaring stage. Georgescu, Saupe and Huber (2008) reported and figured transverse biaperturate walls that occur in well-preserved specimens and probably had the role to reinforce the test wall in the region with expanded metaflanges. A significant change in the gross test architecture is apparent in the test shape in edge view, which beginning with F-9multichamber is consistently lanceolate. F-9multichamber is a SMRS with wide variability in ornamentation and Georgescu, Saupe and Huber (2008) reported and illustrated specimens with costate to reticulate ornamentation; therefore, it is evident the parallel and convergent evolution between the directional lineages 8multichamber and 9multichamber. The dominant ornamentation in F-9multichamber is with longitudinal leptocostae, which is best developed over the adult portion of the test, especially in the region of the last added chamber sets. The leptocostae present a thickness of 0.0029-0.0040 mm, which is comparable if not identical to that observed in the directional lineage ISMRS (0.0025-0.0040 mm). In this portion of the test the pores are circular, situated in the space between the leptocostae, rarely interrupting them and with a diameter of 0.0006-0.0009 mm. Reticulate ornamentation occurs over the earlier portion of the test and in this case the pore diameter can be 0.0019 mm in diameter. The successive addition of layers of calcite throughout ontogeny leads occasionally to the development of irregular ornamentation structures over the earliest chambers of the final whorl. The evolutionary occurrence of S-9multichamber documents the beginning of a new stage in the history of the directional lineage 9multichamber. There are significant changes in the gross test architecture, which occur mostly in the morphology of the adult stage with multichamber growth.

There are two changes when compared to the similar stage of the ancestral F-9multichamber: reduction in size of the periapertural structures and increase in the number of chamber sets following the biaperturate progressive chamber. The reduction in length of the metaflanges of the S-9multichamber results in an adult stage in which most of the volume is occupied by chambers rather than periapertural structures as in F-9multichamber. When compared to the ancestor F-9multichamber where there where observed up to four chamber sets in the adult stage following the progressive chamber, in the descendant S-9multichamber there are up to eight chamber sets. The leptocostate ornamentation in S-9multichamber can be best observed on the chambers of the last-added chamber sets of the adult stage; leptocostae present a thickness of 0.0024-0.0049 mm. Ornamentation changes posteriorly to reticulate and eventually becomes irregular over the earliest chambers of the final whorl as the result of the successive addition of layers of calcite throughout the ontogenetic development. Vermicular ornamentation in S-9multichamber was documented by Martin (1972, pl. 2, Figure 5). The pores in the leptocostae ornamented region are circular, elliptical or irregular in shape and with a diameter or maximum dimension of 0.0007-0.00017 mm; in the reticulately ornamented portion of the test the pores are either circular or elliptical and with a diameter of 0.0020-0.0034 mm. There is an increase of the pore dimensions when compared to those observed in the ancestral F-9multichamber where they present dimensions of 0.0006-0.0009 mm over the leptocostate ornamented portion of the test and up to 0.0019 mm over the reticulate one.

Evolution from S-9multichamber to T-9multichamber is marked especially in an increase in the observed number of chamber sets following the biaperturate progressive chamber from eight to eleven. In addition the test shape in the adult stage is highly variable, from triangularly shaped to wide fan-shaped. The three ornamentation zones known from S-9multichamber (from the anterior margin towards proloculus: leptocostate, reticulate and irregular) also occur in T-9multichamber. Vermicular ornamentation in T-9multichamber was occasionally observed in specimens collected from the upper Maastrichtian of the DSDP Hole 111A. Leptocostae thickness in T-9multichamber is of 0.0029-0.0050 mm. Pore diameter or maximum dimension is of 0.0008-0.0017 mm in the leptocostate ornamented portion of the test and of 0.0022-0.0035 mm over the reticulately ornamented one. There is a gradual increase in test length along the directional lineage 9multichamber. T-9multichamber frequently presents tests longer than 0.8000 mm and occasionally tests wider than 1.0000 mm occur.

Evolution of Pycnocostae in a Globular-Chambered Lineage
(DL: 10multichamber)
(Figure 45)

The evolution of pycnocostate ornamentation in a directional lineage consisting of tests with globular chambers happened in the middle Campanian. The lineage documents a new development of an adult stage with multichamber growth stage in the history of the Cretaceous planktic foraminifera with chambers alternately added with respect to the test growth axis; wall ultrastructure presents the evolution of a new kind of porosity, which due to its peculiar appearance can be successfully used in recognizing the ancestor-descendant relationships within this lineage. The representatives of this directional lineage largely include the genus *Planoglobulina* Cushman 1927c; the lineage is named herein 10multichamber.

The directional lineage 10multichamber includes two SMRS: ISMRS=*Planoglobulina sphaeralis* Georgescu 2014d (I-10multichamber) and FSMRS =*Gümbelina acervulinoides* Egger1899 (F-10multichamber). The ISMRS is the first taxonomic unit of this rank which receives a name directly in evolutionary classification and no Latin name is given to it.

The evolutionary occurrence of I-10multichamber is known from the terminal part of the middle Campanian and is herein reported from the upper part of the Demopolis Chalk of Alabama; it also occurs in the late Campanian Ripley Formation and early Maastrichtian Prairie Bluff Chalk and no occurrences outside of the US are known. I-10multichamber consists of globular chambers added alternately with respect to the test growth axis throughout the ontogenetic development; the periapertural structures consist of metaflanges. These morphological features together with the general test appearance indicate that the ancestor of I-10multichamber is S-1alternate. I-10multichamber is ornamented with longitudinal pycnocostae which resemble the continuous leptocostae of S-1alternate but are thicker: 0.0023-0.0084 mm rather than 0.0032-0.0046. There is a complete loss of the pustulose periapertural area with the evolution of the 10multichamber directional lineage.

Pores in I-10multichamber are irregularly distributed in the space between the pycnocostae and present diameters of 0.0004-0.0017 mm; in S-1alternate the pore diameter is of 0.0006-0.0012 mm and they display a longitudinal arrangement in the space between the leptocostae. These data indicate that S-1alternate is the precursor of the directional lineage 10multichamber.

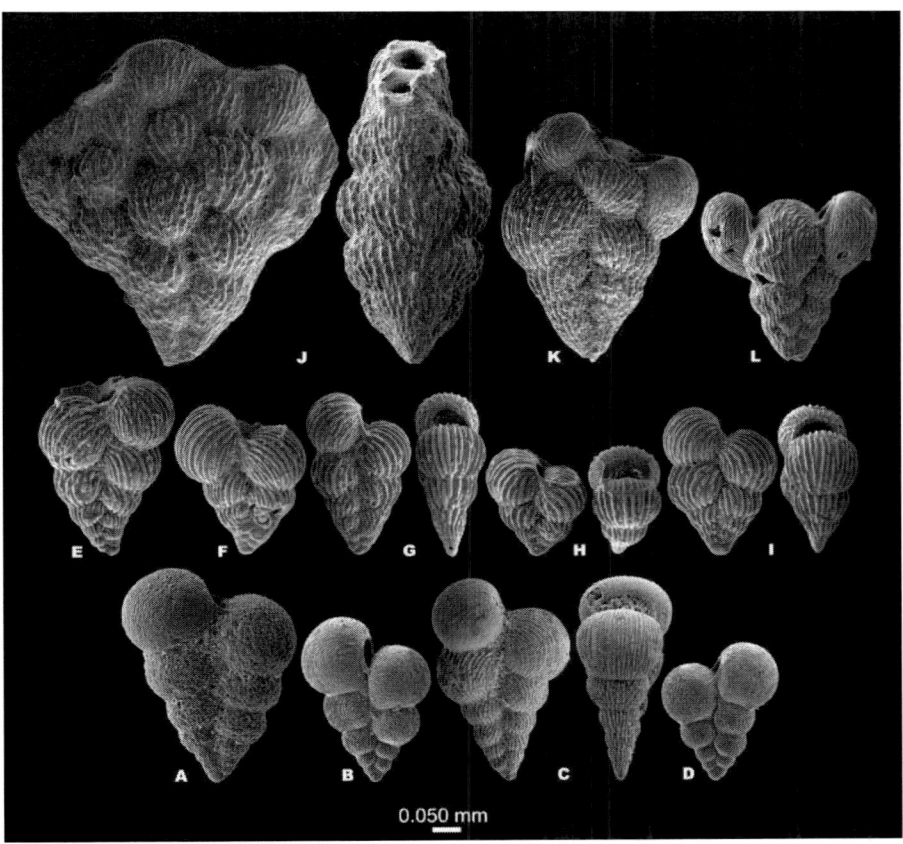

Figure 45. Specimens of the directional lineage 10multichamber. **A-D** Precursor SMRS: S-1alternate. A-Coniacian to early Santonian, 122-763B-18-1, 72-73 cm, figured by Georgescu in Georgescu and others (2013, pl. 7, Figure 1 only). B-early Campanian, Sample 71-511-38-6, 13-17 cm, figured by Georgescu in Georgescu and others (2013, pl. 7, Figure 5 only). C-late Santonian, Sample 1595b, Ehrenberg Collection, Missouri River Basin. D-Santonian, Toolonga Calcilutite, western Australia, Sample Huber, figured by Georgescu in Georgescu and others (2013, pl. 7, Figure 9 only). **E-I** ISMRS: I-10multichamber. E-late middle Campanian, Demopolis Chalk, Alabama, Scott (1968, Stop 7), McGugan Collection, figured by Georgescu (2014d, pl. 1, Figure 8). F-late middle Campanian, Demopolis Chalk, Alabama, Scott (1968, Stop 7), McGugan Collection, figured by Georgescu (2014d, pl. 1, Figure 6). G-late middle Campanian, Demopolis Chalk, Alabama, Scott (1968, Stop 7), McGugan Collection, figured by Georgescu (2014d, pl. 2, figures 1-2). H-late middle Campanian, Demopolis Chalk, Alabama, Scott (1968, Stop 7), McGugan Collection, figured by Georgescu (2014d, pl. 1, figures 4-5). I-late middle Campanian, Demopolis Chalk, Alabama, Scott (1968, Stop 7), McGugan Collection, figured by Georgescu (2014d, pl. 1, fig, 1-2). **J-L** FSMRS: I-10multichamber. J-late Maastrichtian, Sample 12-111A-11-2, 5-19 cm, figured by Georgescu (2014d, pl. 5, figures 4-5). K-late Maastrichtian, Sample 12-111A-11-2, 123-127 cm, figured by Georgescu (2014d, pl. 3, Figure 7). L-early Maastrichtian, Prairie Bluff Chalk, Alabama, Scott (1968, Stop 4), McGugan Collection, figured by Georgescu (2014d, pl. 3, Figure 2).

The development of an adult stage with multichamber growth in the directional lineage 10multichamber occurs in F-10multichamber; in the past by using the typological concept of species the various stages of development of the adult stage were recognized as distinct species, but this interpretation is not supported by the new dataset. Evolution of the adult stage with multichamber growth happened in the late Campanian but tests with well-developed adult stage are recorded earlier in the evolution of F-10multichamber in the late middle Campanian. The adult stage begins with the biaperturate progressive chamber and is followed by up to six sets of chambers that increase in number with an increment of one; the first set of chambers consists of two relapsed chambers. There is a significant increase in test length in the directional lineage 10multichamber with the development of the adult stage with multichamber growth. In F-10multichamber the test surface is ornamented with longitudinal pycnocostae with a thickness of 0.0070-0.0149 mm, which appears straight or undulated due to the addition of successive layers of calcite through the ontogenetic development. Pores have a circular or elliptical outline and present a diameter or maximum dimension of 0.0017-0.0034 mm. Therefore, an increase in pycnocostae thickness and pore size in the evolution from I-10multichamber to F-10multichamber is herein documented.

Ornamentation Loss over the Central Suture in a Leptocostate Lineage (DL: 11multichamber)
(Figure 46)

A distinct evolutionary trend in the group history was initiated in the late Campanian and resulted in occurrence of tests in which the ornamentation is completely lost in the central portion of the test on both sides of the central suture. This lineage was recognized by Georgescu, Saupe and Huber 2008 with the revision of the genus *Gublerina* Kikoïne 1948. It was one of the first lineages of planktic foraminifera with chambers alternately added with respect to the test growth axis in which the ornamentation change through time was studied extensively. Although at the time of transformation of the genus *Gublerina* into a genus that accommodates a lineage by Georgescu, Saupe and Huber (2008) the various kinds of lineages as units with significance in evolutionary classification were not defined yet, it was evident that *Gublerina* is a directional lineage. This revision is followed herein and the directional lineage is renamed 11multichamber. It includes two SMRS: ISMRS=*Gublerina rajagopalani* Govindan 1972 (I-11multichamber) and FSMRS =*Gublerina cuvillieri* Kikoïne 1948 (F-11multichamber).

Georgescu, Saupe and Huber (2008) demonstrated that the directional lineage 11multichamber evolved from I-9multichamber. The general test architecture has many similarities between the two SMRS, including the divergent chamber rows towards the anterior part, aperture in the shape of a low to medium high arch and periapertural structures consisting of symmetrically developed metaflanges, one on each side of the test. I-9multichamber is ornamented with longitudinal leptocostae with a thickness of 0.0025-0.0040 mm; the ornamentation of I-11multichamber is derived from the fused leptocostae and successive addition of layers of calcite during the ontogenetic development, where the fused leptocostae result in the formation of irregular ornamented zones over the chamber surface; ornamentation concentration over the chamber surface is associated with the reduction of the ornamentation prominence over the sutures, which remain smooth. The region along the test central suture remains smooth but vestiges of the leptocostate ornamentation are occasionally observed over this portion of the test surface. The unornamented central region has a subtriangular shape; this zone largely corresponds to the non-septate portion between the two rows of divergent chambers. Strong costae are developed along the periphery, which confer the tests a longitudinally keeled appearance. The porosity features over the last-formed chambers in I-11multichamber are similar to those known in the ancestral I-9multichamber: pores are simple and circular in shape, with a diameter of 0.0005-0.0009 mm and situated in the space between leptocostae, rarely interrupting them. Pores are larger and often irregular in shape over the chambers of the earlier portion of the test.

0.050 mm

Figure 46. Specimens of the directional lineage 11multichamber. **A-E** Precursor SMRS: I-9multichamber. A-late Campanian, Gulf of Mexico, van Morkhoven Collection, figured by Georgescu, Saupe and Huber (2008, pl. 3, Figure 8). B-late Campanian, Upper Taylor Marl, Onion Creek, Texas, Loeblich and Tappan Topotype Collection, figured by Georgescu, Saupe and Huber (2008, pl. 3, Figure 11). C-late Campanian, Upper Taylor Marl, Onion Creek, Texas, Loeblich and Tappan Topotype Collection, figured by Georgescu, Saupe and Huber (2008, pl. 3, Figure 10). D-late Campanian, Upper Taylor Marl, Onion Creek, Texas, Loeblich and Tappan Topotype Collection, figured by Georgescu, Saupe and Huber (2008, pl. 3, Figure 9). E-late Maastrichtian, Sample 12-111A-11-6, 71-85 cm, figured by Georgescu, Saupe and Huber (2008, pl. 3, Figure 6). **F-I** ISMRS: I-11multichamber. F-late Maastrichtian, 122-761B-24-1, 76-77 cm, figured by Georgescu, Saupe and Huber (2008, pl. 5, Figure 5). G-late Maastrichtian, 122-761B-24-1, 76-77 cm, figured by Georgescu, Saupe and Huber (2008, pl. 5, Figure 4). H-late Maastrichtian, 39-356-29-6, 11-25 cm, figured by Georgescu, Saupe and Huber (2008, pl. 5, Figure 3). I-late Maastrichtian, 122-761B-24-1, 89-90 cm, figured by Georgescu, Saupe and Huber (2008, pl. 5, Figure 1). **J-L** FSMRS: F-11multichamber. J-late Maastrichtian, Sample 122-761B-24-1, 111-112 cm, figured by Georgescu, Saupe and Huber (2008, pl. 5, Figure 10). K-late Maastrichtian, Sample 122-761B-22-4, 75-76 cm, figured by Georgescu, Saupe and Huber (2008, pl. 5, Figure 9). L-late Maastrichtian, Sample 122-761B-24-3, 28-29 cm, figured by Georgescu, Saupe and Huber (2008, pl. 5, Figure 6).

An adult stage with multichamber growth evolved in F-11multichamber. The multiserial stage consists of the biaperturate progressive chambers followed by up to three sets of chambers; the first set consists of four chambers with variable shape. Most of the adult stage consists of the well-developed metaflanges. Test is ornamented with coarse irregular areas, which are situated over the chambers; the coarse irregular structures on one chamber are fused across the periphery through thick transverse keels. Sutures and the subtriangular zone over the test central suture are smooth; the subtriangular zone developed on both sides of the central suture largely corresponds to the central non-septate portion of the test between the two rows of divergent chambers. Vestigial leptocostate ornamentation occurs occasionally over the adult stage with multichamber growth; where the occur the leptocostae have a thickness of 0.0028-0.0036 mm. Pores are simple, circular in shape and with a diameter of 0.0007-0.0009 mm. Specimens with intermediate morphological features between I-9multichamber and F-11multichamber occur frequently.

Evolution of Reticulate Ornamentation (DL: 6alternate)
(Figure 47)

A lineage consisting of tests with globular chambers evolved in the late Campanian and led to the occurrence of reticulate ornamentation and for the second time in the history of the planktic foraminifera with chambers alternately added with respect to the test growth axis to the development of leptoflanges. The representatives of this lineage were previously included within *Braunella*, a genus named by Georgescu (2007b) to accommodate a directional lineage. Notably, *Braunella* was one of the first supraspecific units with significance in evolutionary classification defined in this group. The genus *Braunella* is herein renamed 6alternate in the nomenclature associated with the evolutionary classification. Two SMRS are included within the 6alternate directional lineage: ISMRS=*Gümbelina punctulata* Cushman 1938 (I-6alternate) and FSMRS =*Braunella brauni* Georgescu 2007b (F-6alternate).

Georgescu (2007) considered that the directional lineage 6alternate evolved from the SMRS *Heterohelix striata*, but this interpretation was given accepting the taxonomic framework by Masters (1977) and before the revision of the foraminifera of the Ehrenberg Collection (Georgescu 2013a). The illustration of the ancestral SMRS of the directional lineage 6alternate was given by Georgescu (2007b, Figure 2) and clearly belongs to the SMRS S-1alternate. Therefore, S-1alternate is the precursor SMRS of 6alternate. There are significant resemblances in the general test appearance between the ancestral S-1alternate and descendant I-6alternate and these probably led in the past to misidentifications that impeded in recognizing the evolutionary occurrence of this lineage. Chambers are globular in S-1alternate and subspherical and with a more prominent chamber overlapping especially in the adult stage in I-6alternate. There are two growth stages in most of the specimens of I-6alternate: the juvenile stage with faster chamber size increase and the adult one with slower rate of chamber size increase; a rare development of incipient multichamber growth in the adult stage was illustrated by Georgescu (2007) but no taxonomic significance is conferred to this feature that appears rather an occasional occurrence.

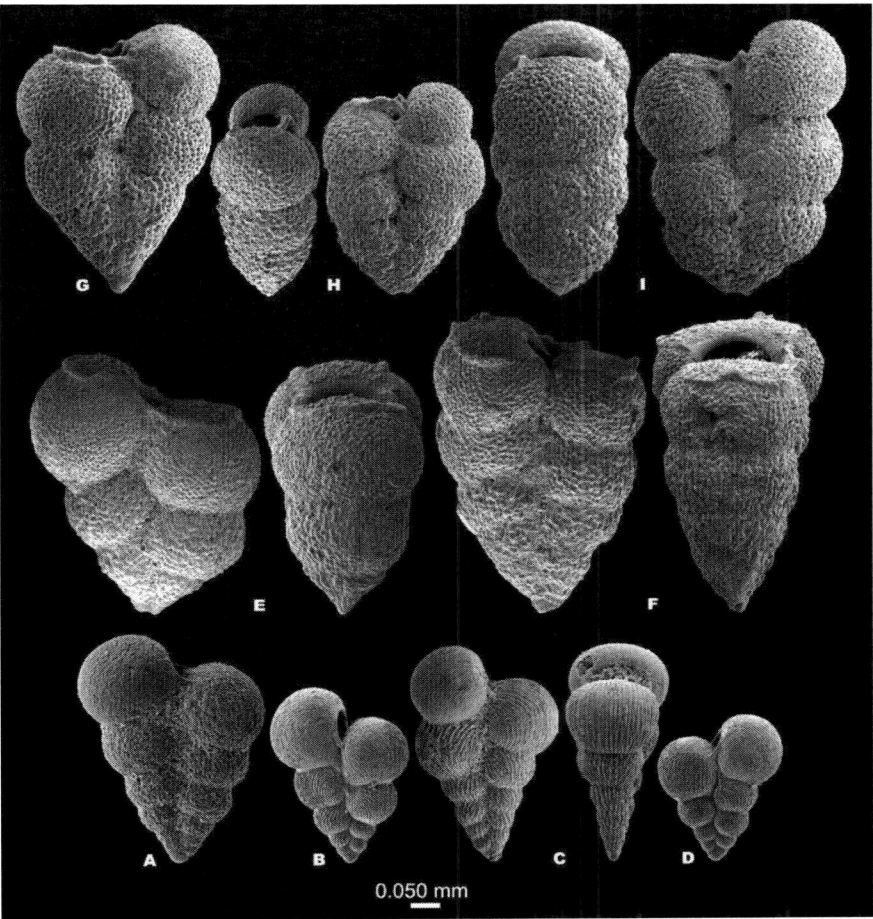

Figure 47. Specimens of the directional lineage 6alternate. **A-D** Precursor SMRS: S-1alternate. A-Coniacian to early Santonian, 122-763B-18-1, 72-73 cm, figured by Georgescu in Georgescu and others (2013, pl. 7, Figure 1 only). B-early Campanian, Sample 71-511-38-6, 13-17 cm, figured by Georgescu in Georgescu and others (2013, pl. 7, Figure 5 only). C-late Santonian, Sample 1595b, Ehrenberg Collection, Missouri River Basin. D-Santonian, Toolonga Calcilutite, western Australia, Sample Huber, figured by Georgescu in Georgescu and others (2013, pl. 7, Figure 9 only). **E-F** ISMRS I-6alternate. E-late Campanian, Sample 10-152-17-1, 65-79 cm, figured by Georgescu (2007b, pl. 1, Figure 3). F-late Campanian, Sample 10-152-17-1, 65-79 cm, figured by Georgescu (2007b, pl. 1, Figure 1). **G-I** FSMRS: F-6alternate. G-late Campanian, Sample 43-384-15-1, 80-86 cm, figured by Georgescu (2007b, pl. 3, Figure 1a). H-late Campanian, Sample 62-463-19-5, 75-76 cm, figured by Georgescu (2007b, pl. 2, Figure 5). I-late Campanian, Sample 62-463-19-5, 75-76 cm, figured by Georgescu (2007b, pl. 2, Figure 1).

In addition the periapertural structures consist of symmetrically developed metaflanges, one on each test side. Georgescu (2007b) mentioned a wide imperforate band bordering the aperture; this stretches between the two metaflanges. Ornamentation features show an increase of the leptocostae thickness with the initiation of 6alternate from 0.0032-0.0046 mm in S-1alternate to 0.0030-0.0067 mm in I-6alternate. There is a distinct tendency to develop an irregular reticulate ornamentation through the leptocostae fusion and possibly addition of successive layers of calcite throughout the ontogenetic development. Pores are circular in shape and their diameter nearly doubles in the evolution of I-6alternate from 0.0006-0.0012 mm in S-1alternate to 0.0013-0.0024 mm. Masters (1977) and Nederbragt (1991) considered that I-6alternate begun its evolution in the late Santonian; Georgescu (2007b) re-evaluated the stratigraphic range of I-6alternate and showed that it evolved in the late Campanian and newly collected specimens with intermediate morphological features between S-1alternate and I-6alternate from the latest middle Campanian of DSDP Site 463 seemingly confirm this reassessment.

F-6alternate also evolved in the late Campanian. There is a major morphological development when compared to the ancestor I-6alternate, namely the evolution of symmetrically developed leptoflanges, one on each side of the test. Leptoflanges define false supplementary apertures along the test central suture in the terminal part of the adult stage, in the region of the last-formed one to four chambers. Leptocostate ornamentation occurs as vestige over the last-formed chambers and most of the test surface is ornamented with a network of ridges defining an irregular reticulate pattern, which was first noted by Pessagno (1967). Pores are circular in shape with a diameter of 0.0019-0.0025 mm, which is comparable to the range known in I-6alternate (0.0013-0.0024 mm).

Ornamentation Loss over the Central Suture in a Pycnocostate Lineage (CL: 7alternate)
(Figure 48)

A condensed lineage is recognized for the tests with pycnocostate ornamentation in the two peripheral regions and a smooth surface on the both sides of the test central suture. This condensed lineage is herein named 7alternate in the nomenclature associated with the evolutionary classification and includes the ISMRS=*Gümbelina semicostata* Cushman 1938 (I-7alternate); Georgescu (2014a) assigned for it the name *Eicheriella* in the Latin-based nomenclature. Condensed lineage 7alternate evolved in the late Campanian and became extinct just above the Campanian/Maastrichtian boundary.

The high resolution morphology of this SMRS was poorly understood in the various revisions of the group (Masters 1977; Weiss 1983; Nederbragt 1991). In the past it was consistently included within the genus *Heterohelix* based on the general test architecture showing chambers alternately added with respect to the test growth axis throughout the ontogenetic development, but no evolutionary relationships between this SMRS and any other of the group could be inferred.

I-7alternate presents pycnocostate ornamentation with the pycnocostae parallel to the peripheral margins. The pycnocostae have a thickness of 0.0060-0.0144 mm but thinner ornamentation structures occur occasionally over the last-formed one or two chambers; notably, the pycnocostae thickness in F-4backextended is of 0.0054-0.0118 mm. Pycnocostae thickness and orientation indicate that F-4backextended is the ancestor of I-7alternate and therefore the precursor of the condensed lineage 7alternate. The evolution from F-4backextended to I-7alternate led to the divergence between the two rows of chambers and a simplification of the periapertural structures, which are significantly reduced in size and without a rim. The characteristic structure, which is unique to this condensed lineage among the Cretaceous planktic foraminifera with the chambers alternately added with respect to the test growth axis, is represented by a calcitic structure without ornamentation that is symmetrically developed over the test central suture on both sides; this morphological structure is named perforate central plate.

Figure 48. Specimens of the directional lineage 7alternate. **A-C** Precursor SMRS: F-4backextended. A-early Maastrichtian, Sample 32-305-17-4, 80-95 cm. B-middle Maastrichtian, Sample 32-305-16-5, 60-76 cm. C-late Campanian, Sample 32-305-18-5, 80-95 cm. **D-F** ISMRS: I-7alternate. D-late Campanian, Sample 122-761B-25-2, 70-71 cm, figured by Georgescu (2014a, pl. 1, figures 10-11). E-late Campanian, Sample 122-761B-25-2, 70-71 cm, figured by Georgescu (2014a, pl. 1, figures 8-9). F-late Campanian, Gulf of Mexico, Eureka 67-128 well, van Morkhoven Collection, figured by Georgescu (2014a, pl. 1, figures 4-5).

Pores over the perforate central plate have diameters of 0.0002-0.0005 mm. Morphological changes in the proximity of the test central suture can be probably correlated with the significant reduction of the periapertural structures and additional studies are necessary to understand such processes that are herein reported for the first time. The primary wall porosity in I-7alternate is observable in the peripheral regions. Pores have a circular, elliptical or irregular outline, present diameters of 0.0006-0.0012 mm and are situated between the pycnocostae; these features are also observed in the inferred ancestor F-4backextended, where they present diameters of 0.0009-0.0013 mm.

First Evolution of Adult Stage with Multi-Plane Chamber Proliferation (DL: 12multichamber)
(Figure 49)

A new pattern in the chamber proliferation in the adult stage was developed in a directional lineage that evolved in the late Campanian. For most part of its lifespan the lineage was represented only by tests with chambers alternately added with respect to the test growth axis, but in the late Maastrichtian it evolved an adult stage with chamber proliferation. Throughout its history this lineage consists of tests with pycnocostate ornamentation. This directional lineage is herein named 12multichamber and includes partly the representatives of the genus *Pseudotextularia* Rzehak 1891, namely the latter type SMRS. The directional lineage 12multichamber includes two SMRS: ISMRS=*Cuneolina elegans* Rzehak 1891 (I-12multichamber) and FSMRS =*Pseudotextularia varians* Rzehak 1895 (F-12multichamber). This directional lineage was reviewed in evolutionary classification by Georgescu (2014e).

General test appearance together with the gross test architecture and high resolution features of the test indicate that the directional lineage 12multichamber evolved from S-1alternate; there are frequent specimens with intermediate morphological features between S-1alternate and I-12multichamber and they were encountered throughout the stratigraphic range of I-12multichamber. The main morphological change that happened with the initiation of 12multichamber is the transversal chamber elongation; this is the second evolution of such a feature from S-1alternate after the evolution of the directional lineage 4alternate in the Santonian. The aperture in S-1alternate and I-12multichamber is wide and in the form of a low to medium high arch; two symmetrical metaflanges border the aperture, one on each side of the test. The initiation of the directional lineage 12multichamber marks the second evolution of the pycnocostate ornamentation in globular-chambered taxa from leptocostate ornamentation; the leptocostae in S-1alternate are 0.0032-0.0046 mm thick, whereas the pycnocostae in I-12multichamber are of 0.0068-0.0148 mm. By contrast to the ancestral SMRS where the ornamentation is equally developed over the test surface in I-12multichamber is more prominent over the earlier chambers of the test. A pustulose periapertural area over the anterior portion of the chambers occurs both in S-1alternate and its descendant I-12multichamber. Pores in S-1alternate have a circular outline and diameters of 0.0006-0.0012 mm and in the descendant SMRS I-12multichamber the pores are circular to elliptical in shape present a diameter or maximum dimension of 0.0010-0.0029 mm. There are no apparent changes in the test size with the initiation of the directional lineage 12multichamber.

The evolution from I-12multichamber to F-12multichamber is marked by a significant increase in test length, which is frequently more than two times longer in the descendant when compared to the ancestor; tests with a length over 0.7000 mm are frequent in F-12multichamber. By contrast to the ancestral SMRS I-12multichamber, which has the chambers alternately added with respect to the test growth axis throughout the ontogenetic development, a short adult stage occurs in F-12multichamber. The adult stage is formed of a chamber set consisting of two chambers followed by a second set of three chambers; Vassilenko (1961) illustrated one specimen of F-12multichamber with adult stage with multichamber growth stage. The aperture is low and bordered by an imperforate band and two symmetrically developed leptoflanges, one on each part of the test. The leptoflanges are free resulting in the occurrence of false supplementary apertures along the test central suture in the region of the last-formed chambers; leptoflanges are bordered by one imperforate rim towards the anterior margin or at both anterior and posterior margins. There are no apparent changes in the ornamentation between I-12multichamber and F-12multichamber; both are ornamented with longitudinal pycnocostae with a thickness of 0.0068-0.0148 mm in I-12multichamber and 0.0063-0.0135 mm in F-12multichamber. The pustulose periapertural area occurs occasionally and over a narrow area. Pores are circular to elliptical in shape in both SMRS and there is an increase in diameter or maximum dimension from I-12multichamber (0.0010-

0.0029 mm) to F-12multichamber (0.0022-0.0043 mm). F-12multichamber is herein reviewed to include the elongate tests with and without the adult stage with multichamber growth.

Figure 49. Specimens of the directional lineage 12multichamber. **A-D** Precursor SMRS: S-1alternate. A-Coniacian to early Santonian, 122-763B-18-1, 72-73 cm, figured by Georgescu in Georgescu and others (2013, pl. 7, Figure 1 only). B-early Campanian, Sample 71-511-38-6, 13-17 cm, figured by Georgescu in Georgescu and others (2013, pl. 7, Figure 5 only). C-late Santonian, Sample 1595b, Ehrenberg Collection, Missouri River Basin. D-Santonian, Toolonga Calcilutite, western Australia, Sample Huber, figured by Georgescu in Georgescu and others (2013, pl. 7, Figure 9 only). **E-G** ISMRS: I-12multichamber. E-late Maastrichtian, Sample 122-761B-24-2, 72-73 cm. F-late Maastrichtian, surroundings of Limestone City, Texas, Sample Huber, figured by Georgescu (2014e, pl. 1, figures 3-4). G-Maastrichtian, Brudendorf, Germany, Loeblich and Tappan Topotype Collection, figured by Georgescu (2014e, pl. 1, figures 1-2). **H-L** FSMRS: F-12multichamber. H-late Maastrichtian, Sample 122-761B-24-1, 25-26 cm, figured by Georgescu (2014e, pl. 1, figures 3-4). I-late Maastrichtian, Sample 122-761B-24-1, 89-90 cm, figured by Georgescu (2014e, pl. 2, figures 5-6). J-late Maastrichtian, Sample 122-761B-24-1, 111-112 cm, figured by Georgescu (2014e, pl. 3, Figure 11). K-late Maastrichtian, Sample 122-761B-24-1, 25-26 cm, figured by Georgescu (2014e, pl. 3, figures 7-8). L-late Maastrichtian, Sample 122-761B-24-1, 89-90 cm.

First Evolution of Multiple Test Architectures in One Lineage
(DL: 1mixed)
(Figure 50)

The earliest lineage that evolved multiple test architectures was initiated in the late Campanian and became extinct at the Cretaceous/Paleogene boundary. The ISMRS presents gross test architecture similar to that of the precursor: the successive chambers are alternately added with respect to the test axis throughout the ontogenetic development. FSMRS evolved double backward chamber extensions and the SSMRS evolved multichamber growth in the adult stage. This directional lineage was named *Nederbragtina* by Georgescu (2014b) and is herein renamed 1mixed. It consists of three SMRS: ISMRS=*Nederbragtina prima* Georgescu 2014b (I-1mixed), FSMRS=*Pseudoguembelina palpebra* Brönnimann and Brown 1953 (F-1mixed) and SSMRS=*Pseudoguembelina hariaensis* Nederbragt 1991 (S-1mixed). The directional lineage 1mixed evolved from S-1alternate as demonstrated by Georgescu (2014b) based on both gross test architecture and high resolution features between S-1alternate and I-1mixed; therefore S-1alternate is the precursor of the directional lineage 1mixed.

There is a distinct test compression in the adult stage in I-1mixed conferring the test a lanceolate shape; due to this feature the aperture in I-1mixed appears less wide than in the ancestor S-1alternate; in both SMRS the aperture is bordered by an imperforate rim. The symmetrically developed metaflanges of S-1alternate, which border the aperture on each lateral side, evolved in I-1mixed in rimmed retroflanges, one on each side of the test. Small-sized false supplementary apertures along the test central suture occur at the junction between the retroflanges and the previous chambers; such structures are known only in the last-formed chambers.The ornamentation consisting of longitudinal leptocostae in S-1alternate evolved in I-1mixed where the leptocostae present incipiently bimodal arrangement, are slightly more prominent over the earlier chambers of the test and thicker (0.0033-0.0068 mm when compared to that in S-1alternate where they are 0.0032-0.0046 mm thick). In addition to these changes in ornamentation with the initiation of the directional lineage 1mixed there is a complete loss of the pustulose periapertural area consisting of dome-like pustules. Pores in I-1mixed have a circular to elliptical outline and a diameter or maximum dimension of 0.0010-0.0021 mm; there is a significant increase in pore dimensions when compared to those in the ancestral SMRS S-1alternate (0.0006-0.0012 mm).

Evolution from I-1mixed to F-1mixed is mostly apparent in the development of double backward chamber extensions, supplementary apertures along the test central suture and an increase in size of the periapertural structures. The ornamentation in F-1mixed consists of leptocostae with bimodal arrangement, which are thicker (0.0061-0.0130 mm) when compared to those in the ancestral I-1mixed where they are 0.0033-0.0068 mm thick; in F-1mixed the ornamentation is distinctly more prominent over the earlier portion of the test. Pores in F-1mixed are circular to elliptical and with a diameter or maximum dimension of 0.0010-0.0025 mm but occasionally larger vuggy pores occur along the central test suture; this range shows a slight increase when compared to that observed in the ancestor SMRS: 0.0010-0.0021 mm.

Figure 50. Specimens of the directional lineage 1mixed. **A-D** Precursor SMRS: S-1alternate. A-Coniacian to early Santonian, 122-763B-18-1, 72-73 cm, figured by Georgescu in Georgescu and others (2013, pl. 7, Figure 1 only). B-early Campanian, Sample 71-511-38-6, 13-17 cm, figured by Georgescu in Georgescu and others (2013, pl. 7, Figure 5 only). C-late Santonian, Sample 1595b, Ehrenberg Collection, Missouri River Basin. D-early Santonian, Toolonga Calcilutite, western Australia, Sample Huber, figured by Georgescu in Georgescu and others (2013, pl. 7, Figure 9 only). **E-G** ISMRS: I-1mixed. E-late Campanian, Sample 32-305-18-2, 100-102 cm, figured by Georgescu (2014b, pl. 5, figures 4-5). F-late Campanian, Sample 32-305-18-3, 100-102 cm, figured by Georgescu (2014b, pl. 5, figures 7-8). G-late Campanian, Sample 32-305-18-2, 100-102 cm, figured by Georgescu (2014b, pl. 5, figures 2-3). **H-J** FSMRS: F-1mixed. H-middle Maastrichtian, Sample 32-305-16-5, 60-76 cm. I-middle Maastrichtian, Sample 32-305-16-5, 60-76 cm. J-middle Maastrichtian, Sample 32-305-16-5, 60-76 cm. **K-N** SSMRS: S-1mixed. K-late Maastrichtian, Sample 171B-1050C-11-1, 127-130 cm. L-late Maastrichtian, Sample 171B-1050C-11-1, 127-130 cm. M-late Maastrichtian, Sample 39-356-30-4, 42-56 cm. N-late Maastrichtian, Sample 39-357-31-3, 80-94 cm.

S-1mixed evolved through the loss of the adult stage with double backward chamber extensions and supplementary apertures along the test central suture and development of an adult stage with multichamber growth. The adult stage begins with the biaperturate progressive chamber, followed by a first set consisting of two relapsed chambers; frequently the adult stage with chamber proliferation exhibit irregular chamber addition. Ornamentation features show an increase in leptocostae thickness from 0.0061-0.0130 mm in F-1mixed to 0.0071-0.0147 mm in S-1mixed. One slight increase in size is also recorded in the pore dimension, from 0.0010-0.0025 mm in F-1mixed to 0.0012-0.0028 mm in S-1mixed. Specimens with intermediate morphological features between F-1mixed and S-1mixed are not known.

Second Evolution of Adult Stage with Multi-Plane Chamber Proliferation (DL: 13multichamber)
(Figure 51)

A new lineage that led to development of multi-plane chamber proliferation in the adult stage evolved in the earliest Maastrichtian; it consists of tests in which the degree of chamber proliferation increases gradually throughout the lineage stratigraphic range. This lineage was recognized by Nederbragt (1991) and its SMRS were traditionally included into two genera: *Pseudotextularia* Rzehak 1891 and *Racemiguembelina* Montanaro Gallitelli 1957 function of the increased degree of chamber proliferation. This interpretation is followed herein with the mention that an evolutionary continuum exists throughout this directional lineage since its emergence from ancestors with chambers alternately added with respect to the test growth axis throughout the ontogenetic development. The directional lineage is herein named 13multichamber and consists of the following three SMRS: ISMRS=*Pseudotextularia intermedia* de Klasz 1953 (I-13multichamber), FSMRS =*Racemiguembelina powelli* Smith and Pessagno 1973 (F-13multichamber) and SSMRS =*Gümbelina fructicosa* Egger 1899 (S-13multichamber).

The directional lineage 13multichamber evolved from I-12multichamber. The initiation of 13multichamber is apparent in the evolution of an adult stage with chamber proliferation. There are one or two chamber sets in I-13multichamber; the first set consists of two chambers and the second set of three chambers; the chamber addition pattern in the adult stage resembles that of F-12multichamber. The ornamentation consists of longitudinal pycnocostae both in the ancestor and descendant, and they have a thickness of 0.0068-0.0148 mm in I-12multichamber and 0.0066-0.0133 mm in I-13multichamber. Pores have a circular o elliptical outline and diameters of 0.0010-0.0029 mm in I-12multichamber and 0.0012-0.0053 mm in I-13multichamber.

Figure 51. Specimens of the directional lineage 13multichamber. **A-C** Precursor SMRS: I-12multichamber. A-late Maastrichtian, Sample 122-761B-24-2, 72-73 cm. B-late Maastrichtian, surroundings of Limestone City, Texas, Sample Huber, figured by Georgescu (2014e, pl. 1, figures 3-4). C-Maastrichtian, Brudendorf, Germany, Loeblich and Tappan Topotype Collection, figured by Georgescu (2014e, pl. 1, figures 1-2). **D-G** ISMRS: I-13multichamber. D-late Maastrichtian, Sample 171B-1050C-10-2, 39-41 cm. E-late Maastrichtian, Sample 171B-1050C-11-cc, figured by Georgescu (2014e, pl. 4, figures 4-5). F-late Maastrichtian, Sample 171B-1052E-18-3, 17-20 cm. G-late Maastrichtian, Sample 171B-1050C-11-cc, figured by Georgescu (2014e, pl. 4, figures 1-2). **H-I** FSMRS: F-13multichamber. H-late Maastrichtian, Sample 39-357-31-5, 80-94 cm, figured by Georgescu (2014e, pl. 5, figures 3-4). I-late Maastrichtian, Sample 122-761B-21-4, 145-146 cm, figured by Georgescu (2014e, pl. 5, figures 1-2). **J-L** SSMRS: S-13multichamber. J-late Maastrichtian, Sample 43-384-13-6, 70-72 cm, figured by Georgescu (2014e, pl. 5, Figure 8). K-late Maastrichtian, Sample 43-384-13-6, 70-72 cm. L-late Maastrichtian, Sample 43-384-14-4, 50-52 cm.

Evolution in F-13multichamber and S-13multichamber is only apparent in the gradual increase of the chambers in the adult stage with multi-plane chamber proliferation. Longitudinal pycnocostae observed thickness is of 0.0050-0.0149 mm in F-13multichamber and 0.0054-0.0206 mm in S-13multichamber; vestigial leptocostate ornamentation is occasionally observed over the last-formed chambers of S-13multichamber. Pores have a circular, elliptical or irregular outline and diameters of 0.0018-0.0053 mm in F-13multichamber to 0.0019-0.0057 mm in S-13multichamber. Bridges connecting the chambers across the umbilicus occur throughout the directional lineage 13multichamber. Test length presents a gradual size increase in the evolution from I-13multichamber to S-13multichamber.

Evolution of Pustulose Leptocostae in Tests with Early Planispiral Coil (DL: 4planispiral)
(Figure 52)

A new lineage evolved in the proximity of the Campanian/Maastrichtian boundary from the long-ranging smooth SMRS I-1multichamber. The lineage led to the occurrence of ornamentation consisting of pustulose leptocostae and tests with early planispiral coil; notably, this is the fourth occurrence of the early planispiral coil in one lineage of Cretaceous planktic foraminifera with chambers alternately added with respect to the test growth axis. In typological classification the lineage is subdivided into two genera. The genus *Fleisherites* Georgescu 2009a includes the tests with chambers alternately added with respect to the test growth axis throughout the ontogeny. Genus *Hartella* Georgescu and Abramovich (2009a) was defined for the tests with early planispiral coil and ornamentation consisting of pustulose leptocostae. The two genera are valid units in typological classification due to the significant morphological differences between them. In the evolutionary classification framework they are grouped into one directional lineage, herein named 4planispiral. The directional lineage 4planispiral includes two SMRS: ISMRS=*Gümbelina glabrans* Cushman 1938 (I-4planispiral) and FSMRS =*Hartella harti* Georgescu and Abramovich 2009a (F-4planispiral).

I-4planispiral consists of tests that have significant resemblances with the ancestor I-1multichamber; the latter is the precursor of the directional lineage 4planispiral. The morphological differences between I-1multichamber and I-4planispiral are apparent in the high resolution morphological features. The periapertural structures consist of orthoflanges in the ancestor I-1multichamber whereas in the descendant I-4planispiral there is a mixture of orthoflanges and metaflanges (Georgescu 2009).

I-4planispiral was considered since its description a SMRS with smooth chambers but Georgescu (2009a) described and illustrated specimens with incipient costate ornamentation and mentioned the consistent occurrence of pustulose periapertural area consisting of small dome-like pustules in the anterior portion of the chambers. Pore diameter is of 0.0004-0.0010 mm in I-1multichamber and 0.0004-0.0008 mm in I-4planispiral. The wide morphological variability in I-4planispiral is further demonstrated by specimens from the Maastrichtian of Texas, which occasionally have a peripheral imperforate band; this feature is herein reported for the first time in the representatives of the Cretaceous planktics with chambers alternately added with respect to the test growth axis.

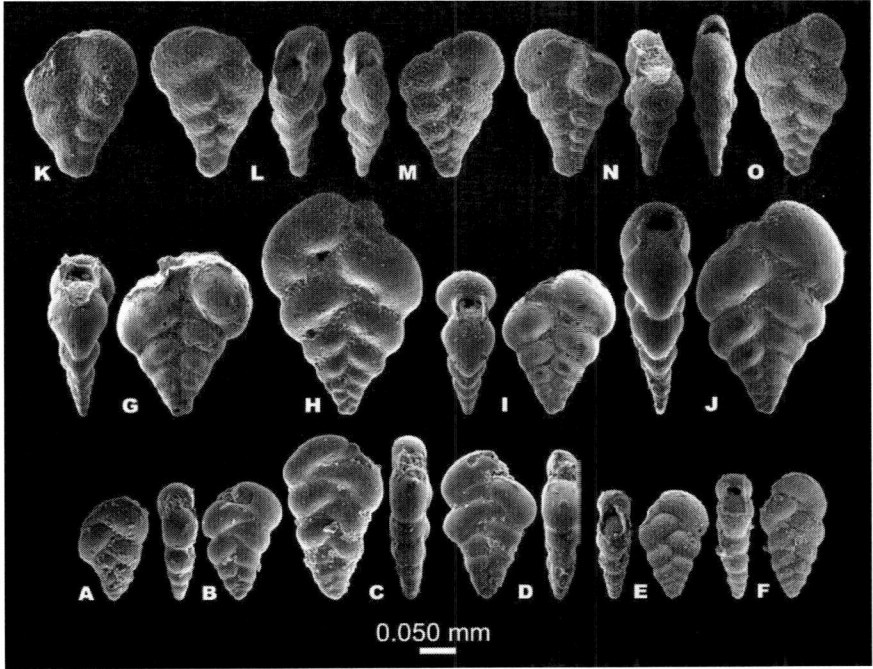

Figure 52. Specimens of the directional lineage 4planispiral. **A-F** Precursor SMRS: I-1multichamber. A-early Maastrichtian, Sample 2404b, Ehrenberg Collection, Rügen Island. B-late Santonian, Sample 1595b, Ehrenberg Collection, Missouri River Basin, figured by Georgescu (2013c, Figure 6: 14-15). C-late Santonian, Sample 1595b, Ehrenberg Collection, Missouri River Basin, figured by Georgescu (2013c, Figure 6: 12-13). D-late Santonian, Sample 1595b, Ehrenberg Collection, Missouri River Basin, figured by Georgescu (2013c, Figure 6: 10-11). E-Coniacian to early Santonian, 62-463-26-5, 53-58 cm, figured by Georgescu (2013c, Figure 6: 3-4). F-Coniacian to early Santonian, 62-463-26-5, 53-58 cm, figured by Georgescu (2013c, Figure 6: 5-6). **G-J** ISMRS: I-4planispiral. G-late Maastrichtian, surroundings of Limestone City, Texas, Sample Huber, figured by Georgescu (2009a, Figure 12: 2). H-late Maastrichtian, surroundings of Limestone City, Texas, Sample Huber, figured by Georgescu (2009a, Figure 12: 5). I-late Maastrichtian, surroundings of Limestone City, Texas, Sample Huber, figured by Georgescu (2009a, Figure 12: 4). J-late Maastrichtian, surroundings of Limestone City, Texas, Sample Huber, figured by Georgescu (2009a, Figure 12: 1). **K-O** FSMRS: F-4planispiral. K-late Maastrichtian, Sample 171B-1050C-11-2, 146-149 cm, figured by Georgescu and Abramovich (2009a, Figure 7: 8). L-late Maastrichtian, Sample 171B-1050C-11-2, 146-149 cm, figured by Georgescu and Abramovich (2009a, Figure 7: 11-12). M-late Maastrichtian, Sample 171B-1050C-13-1, 65-67 cm, figured by Georgescu and Abramovich (2009a, Figure 7: 9-10). N-late Maastrichtian, Sample 171B-1050C-11-2, 146-149 cm, figured by Georgescu and Abramovich (2009a, Figure 7: 5-6). O-late Maastrichtian, Sample 171B-1050C-11-1, 127-130 cm, figured by Georgescu and Abramovich (2009a, Figure 6: 1-2).

Evolution from I-4planispiral to F-4planispiral involves the development of an early planispiral coil, periapertural structures consisting of metaflanges and ornamentation consisting of longitudinal pustulose leptocostae over the entire test surface. The pustulose leptocostae consist of aligned pustules with diameters of 0.0019-0.0035 mm. Pore diameter is of 0.0004-0.0008 mm throughout the directional lineage 4planispiral. Specimens with intermediate morphological features between I-4planispiral and F-4planispiral were described from the Maastrichtian of Poland by Gawor-Biedowa (1992), which further demonstrates the evolutionary continuum within this lineage.

Evolution of Sutural Ridges in One Mixed Lineage (DL: 2mixed)
(Figure 53)

One of the most evident cases of iterative evolution in the group of Cretaceous planktics with chambers alternately with respect to the test growth axis is represented by the development of phaneroridges. This feature occurred for the first time in the group history in two directional lineages of the Santonian that became extinct before the Santonian/ Campanian boundary: 5multichamber and 6multichamber. No lineage developed phaneroridges for the most part of the Campanian; a new directional lineage that evolved in the latest Campanian will develop this feature and is herein named 2mixed; by contrast to the two lineages of the Santonian in 2mixed there is no early stage consisting of calyptoridges before the development of phaneroridges. The terminal SMRS was used by Georgescu and Abramovich (2008b) to describe the monophyletic genus *Lipsonia*. This lineage includes two SMRS: ISMRS=*Lipsonia shatskyensis* Georgescu 2014b (I-2mixed) and FSMRS=*Lipsonia lipsonae* Georgescu and Abramovich 2008b (F-2mixed).

The morphological features and stratigraphic distribution indicate that I-2mixed evolved from I-11multichamber in the latest Campanian; therefore I-11multichamber is the precursor SMRS of the directional lineage 2mixed. The ancestor I-11multichamber has the test with prominent ornamentation, which consists of irregularly ornamented structures over the chamber surface and longitudinal costae at the peripheral margins. The initiation of the directional lineage 2mixed marks the migration of the irregularly ornamented structures towards the suture regions resulting in the formation of sutural ridges; in parallel the peripheral ornamentation is lost but the corresponding phaneroridges on the two sides of the test fuse across the periphery resulting in a transverse-keeled appearance. As a result the chamber surface in I-2mixed remains smooth but vestigial leptocostate ornamentation occasionally occurs in the sutural ridge. One small supplementary aperture protected by a short backward lid-like wall extension occurs in the posterior part of the last-formed chamber in I-2mixed. The pores in I-11multichamber present diameters of 0.0005-0.0009 mm, whereas in I-2mixed the observed range of the pores diameter is of 0.0008-0.0014 mm.

Evolution of an adult stage with multichamber growth is documented in F-2mixed; there are significant morphological changes in this directional lineage during the evolution from I-2mixed to F-2mixed. The supplementary aperture on the last-formed chamber in I-2mixed is lost and no longer encountered in F-2mixed. Ornamentation over the earlier portion of the test in F-2mixed becomes more prominent and irregular due to the addition of successive layers of calcite during the ontogenetic development. There is an increase in pore size along the directional lineage 2mixed from 0.0008-0.0014 mm in I-2mixed to 0.0008-0.0017 mm over the last-formed chambers in F-2mixed. Larger vuggy pores with a maximum dimension of 0.0036-0.0081 mm occur over the earlier portion of the test in which the ornamentation is irregular and more prominent.

Figure 53. Specimens of the directional lineage 2mixed. **A-D** ISMRS: I-11multichamber. A-late Maastrichtian, 122-761B-24-1, 76-77 cm, figured by Georgescu, Saupe and Huber (2008, pl. 5, Figure 5). B-late Maastrichtian, 122-761B-24-1, 76-77 cm, figured by Georgescu, Saupe and Huber (2008, pl. 5, Figure 4). C-late Maastrichtian, 39-356-29-6, 11-25 cm, figured by Georgescu, Saupe and Huber (2008, pl. 5, Figure 3). D-late Maastrichtian, 122-761B-24-1, 89-90 cm, figured by Georgescu, Saupe and Huber (2008, pl. 5, Figure 1). **E-H** ISMRS: I-2mixed. E-early Maastrichtian, Sample 32-305-17-6, 100-102 cm, figured by Georgescu (2014b, pl. 2, Figure 10). F-early Maastrichtian, Sample 32-305-17-6, 100-102 cm, figured by Georgescu (2014b, pl. 2, figures 7-8). G-early Maastrichtian, Sample 32-305-17-4, 100-102 cm, figured by Georgescu (2014b, pl. 2, figures 4-5). H-early Maastrichtian, Sample 32-305-17-6, 100-102 cm, figured by Georgescu (2014b, pl. 2, figures 1-2). **I-J** FSMRS: F-2mixed. I-late Maastrichtian, Sample 198-1212B-13-3, 148-150 cm, figured by Georgescu and Abramovich (2008b, pl. 1, Figure 1a). J-late Maastrichtian, Sample 198-1212B-13-1, 40-42 cm, figured by Georgescu and Abramovich (2008b, pl. 2, Figure 1a).

The general test appearance indicates the occurrence of an iterative evolution between the terminal species of the directional lineages 6multichamber of the late Santonian and 2mixed of the late Maastrichtian. With the S-6multichamber recognized as the most probable ancestor of the large-sized benthic *Orbitoides* and in the absence of a demonstrated ancestor-descendant relationship between *Orbitoides* and *Omphalocyclus* of the Maastrichtian a second possible origination of the Latest Cretaceous large-sized benthics can be inferred, namely that *Omphalocyclus* to have evolved in the Maastrichtian from F-2mixed.

CONCLUSION AND DISCUSSION

Continuous developments during the "ultrastructure revolution" period gradually bring their contribution in understanding the evolution of the Cretaceous planktic foraminiferal group and especially those in which at least one ontogenetic stage consists of chambers alternately added with respect to the test growth axis. The evolutionary classification system, which was advocated by Steineck and Fleisher (1978), was developed as a direct and practical result of the "ultrastructure revolution" in order to accommodate the significant influx of new data on the group morphology, stratigraphy and evolution. According to this classification system, the units are grouped as a function of a mixture of resemblances and differences given by the common ancestry and evolutionary divergence respectively.

The lineage is herein defined as the fundamental unit in evolutionary classification, which contrasts to the typological and phylogenetic classifications in which the fundamental unit is the species. The multitude of specimens with intermediate morphological features between what were defined as "species" as well as their occurrences at different stratigraphic levels are considered indicative in showing that the lineages are fundamental units in classification. The concepts of species used in paleontology from morphospecies to composite paleontological species proved insufficient to accommodate the specimens with intermediate morphology between two or more "species" within a lineage and the idea that the paleontological species are arbitrarily defined within a lineage occurs in the works of some authors (e.g., Pearson 1993). The concept of stage of morphological relative stability is herein introduced to define the relative stable morphological appearances within the evolutionary history of a lineage; the flexibility in interpretation of such a concept provides a more accurate background for the occurrences of specimens with morphologies outside one stage of morphological relative stability.

Lineages, which are the new fundamental units in the evolutionary classification, are dynamic units capable accommodating the essence of the evolution process. One of the first outcomes in the change of fundamental unit from species in typological classification to lineage in evolutionary classification is that selection and designation of specimens as types is no longer necessary.

A new system to represent one lineage needs to be developed and this is clearly demonstrated by the new illustration methodology of the representatives of each lineage. Iterative evolution is the dominant evolutionary pattern in the group history and its wide occurrence is used herein to develop a new nomenclature system adapted for the evolutionary classification; this new system is named herein evolutionary nomenclature. Evolutionary nomenclature is not unique and can be subject of experimentation and development as new data become available. The evolutionary nomenclature adopted herein uses six English words: stalk, alternate, backextended, multichamber, planispiral and mixed.

The stalk is the oldest and longest lineage of the group; all the other lineages are direct or indirect descendants from it. All the other lineages are named as a function of the gross test morphological achievements in the terminal portion of each lineage as follows: alternate for those with chambers alternately added with respect to the test growth axis throughout the ontogeny, backextended for those lineages that evolved one or two chamber backward extensions at the periphery or on the lateral side, multichamber for those with adult proliferating stage, planispiral for those that evolved early planispiral coil and mixed for those lineages that evolved two or more of these morphologies throughout their evolutionary history. A cipher is added before each such name to show the order in which evolved a peculiar lineage that achieved the nominal morphology (e.g., 4planispiral). The stages of morphological relative stability are marked with one letter in front of the lineage name as follows: I for the initiating stage, F for the first descendant, S for the second descendant, etc (e.g., F-5multichamber). The compression rate in the name number between typological (Latin-based) nomenclature and evolutionary (English-based) nomenclature is higher than 18:1; there are used six English names rather than 110 Latin names.

Names in the evolutionary classification are not followed by the author's name and year of publication as in the typological system; the new system should reflect the temporary validity of a nomenclatural system and its readiness to be changed if a new and better understanding of one group's evolution is achieved.

The stability in the evolutionary nomenclature is given only by the thorough application of the principles of science; limitations such as conditioning one unit's name validity by the condition of being published in a peer-reviewed journal, or its acceptance by a certain panel of specialists in the field, etc cannot be considered scientific, and therefore, are discarded.

International Codes Botanical and Zoological Nomenclature (ICBN and ICZN) that provide the rules in typological classification and International Code of Phylogenetic Nomenclature (ICPN) that provides the rules of the phylogenetic nomenclature are not compatible with the evolutionary nomenclature; the units are of different nature and their representation follows different principles.

The new advances in evolutionary classification should not be considered purely theoretical. The development of a new and independent nomenclatural system for evolutionary classification is only one step towards a next stage, which is represented by the evolutionary monographs.

ACKNOWLEDGMENTS

During the studies on the Cretaceous planktic foraminiferal taxonomy, classification and evolution we received the help of many specialists in the field or other micropaleontology domains of study; through access to collections, fossil material loan, various editorial processes and constructive criticism they contributed to the coagulation of the ideas presented in this work. I am grateful to all of them. Special thanks are to Dr Charles M. Henderson (University of Calgary) for the permanent support during the last years. The idea of the new illustration of the foraminiferal tests in which one scale is followed throughout the article belongs to him; unused in any of the previous studies in planktic foraminiferal evolution it clarified the general picture of using biometrical data in the evolution of this planktic foraminiferal group and recognizing additional and new evolutionary patterns in the group history. Dr Leonard V. Hills (University of Calgary) is thanked for the continued help during the development of the evolutionary classification. It is a privilege to acknowledge the extraordinary help from Nova Science Publishers in publishing my innovative ideas. I also thank to the thousands of students at the University of Calgary to whom I taught various paleontology courses in the last years; their reaction to some of the new ideas presented in this work was a permanent support in developing the English-based evolutionary nomenclature.

Last, but not least, I thank my family for the enthusiastic support during the long period in which the ideas published in this work were developed.

REFERENCES

Albriton, C. C. Jr., 1937. Upper Jurassic and Lower Cretaceous foraminifera from the Malone Mountains, Trans-Pecos, Texas. *Journal of Paleontology*, 11, 19-23.

Aliyulla, K., 1965. On the state of the knowledge of the family Heterohelicidae and the way of its subsequent study. *Voprosii Mikropaleontologii*, 9, 215-228 [in Russian].

Aliyulla, K., 1977. *Upper Cretaceous and foraminiferal development in the Lesser Caucasus (Azerbaijan)*. Akademiya Nauk Azerbaijanskoy SSR, Institut Geologii im Akad. I. M. Gubkina, Elm-Baku, 232 pp. [in Russian].

Ansary, S. E., Tewfik, N. M., 1968. Planktonic foraminifera and some benthonic species from the subsurface Upper Cretaceous of Ezz El Orban area, Gulf of Suez. *Journal of Geology of the United Arab Republic*, 10, 37-76.

Bailey, J. W., 1844. Notice of a memoir by C. G. Ehrenberg, "On the Extent and Influence of Microscopic Life in North and South America". *American Journal of Science and Arts*, 46, 297-313.

Barr, F. T., 1968. Late Cretaceous planktonic foraminifera from the coastal area east of Susa (Apollonia), northeastern Libya. *Journal of Paleontology*, 42, 308-321.

Belford, D. J., 1960. Upper Cretaceous foraminifera from the Toolonga Calcilutite and Gingin Chalk, Western Australia. *Bureau of Mineral Resources, Geology and Geophysics Bulletin*, 57, 1-198.

Berggren, W. A., 1968. Phylogenetic and taxonomic problems of some Tertiary planktonic foraminiferal lineages. *Tulane Studies in Geology and Paleontology*, 6, 1-12.

Bergstresser, T. J., Frerichs, W. E., 1982. Planktonic foraminifera from the Upper Cretaceous Pierre Shale at Red Bird, Wyoming. *Journal of Foraminiferal Research*, 12, 353-361.

Berthelin, G., 1880. Mémoire sur les foraminifères fossils de l'étage Albien de Montcley (Doubs). *Mémoires de la Societé Géologique de France*, 1(5), 1-84.

Bock, W. J., 1973. Philosophical foundations of classical evolutionary classification. *Systematic Zoology*, 22, 375-392.

Boersma, A., 1981. Cretaceous and early Tertiary foraminifers from Deep Sea Drilling Project Leg 62 Sites in the Central Pacific. In: *Initial Reports of the Deep Sea Drilling Project*, Volume 62 (Stout, L. N., Ed.), Washington, DC: United States Government Printing Office, 377-396.

Brotzen, F., 1936. Foraminiferen aus dem Schwedischen, Untersten Senon von Eriksdal in Schonen. *Sveriges Geologiska Undersökning Årsbok*, 30, 1-69.

Brown, J., 1853. Note on the artesian well at Colchester; and remarks on some of the microscopic fossils from the Colchester Chalk. *The Annals and Magazine of Natural History, including Zoology, Botany, and Geology*, 12, 240-242.

Brown, N. K., Jr., 1969. Heterohelicidae Cushman, 1927, amended, a Cretaceous planktonic foraminiferal family. In: *Proceedings of the First International Conference on Planktonic*

Microfossils, Geneva 1967 (P. Brönnimann, P. and H. H. Renz, Eds.). Leiden: E.J. Brill, 2, 21-67.

Brönnimann, P., Brown, N. K. Jr., 1953. Observations on some planktonic Heterohelicidae from the Upper Cretaceous of Cuba. *Contributions from the Cushman Foundation for Foraminiferal Research,* 4, 150-156.

Caron, M., 1975. Late Cretaceous planktonic foraminifera from the northwestern Pacific: Leg 32 of the Deep Sea Drilling Project. In: *Initial Reports of the Deep Sea Drilling Project,* Volume 32 (Larson, R. L., Moberly, R. and others, Eds.). Washington, DC: United States Government Printing Office, 719-724.

Carsey, D. O., 1926. Foraminifera of the Cretaceous of Central Texas. *University of Texas Bulletin,* 2612, 1-56.

Chapman, F., 1892. Microzoa from the phosphatic chalk of Taplow. *Quarterly Journal of the Geological Society,* 48, 514-518.

Cushman, J. A., 1926. Some foraminifera from the Mendez Shale of eastern Mexico. *Contributions from the Cushman Laboratory for Foraminiferal Research,* 2, 16-26.

Cushman, J. A., 1927a. The American Cretaceous foraminifera figured by Ehrenberg. *Journal of Paleontology,* 1, 213-217.

Cushman, J. A., 1927b. Notes on the foraminifera in the collection of Ehrenberg. *Journal of the Washington Academy of History,* 17, 487-491.

Cushman, J. A., 1927c. Some new genera of foraminifera *Contributions from the Cushman Laboratory for Foraminiferal Research,* 3, 77-81.

Cushman, J. A., 1927d. An outline of a re-classification of the Foraminifera. *Contributions from the Cushman Foundation for Foraminiferal Research,* 3, 1-105.

Cushman, J. A., 1928. Additional genera of foraminifera. *Contributions from the Cushman Laboratory for Foraminiferal Research,* 4, 1-10.

Cushman, J. A., 1931. A preliminary report on the foraminifera of Tennessee. *State of Tennessee Division of Geology Bulletin,* 41, 1-62.

Cushman, J. A., 1932. *Rectogümbelina,* a new genus from the Cretaceous. *Contributions from the Cushman Laboratory for Foraminiferal Research,* 3, 4-7.

Cushman, J. A., 1938. Cretaceous species of *Gümbelina* and related genera. *Contributions from the Cushman Laboratory for Foraminiferal Research,* 14, 2-28.

Darmoian, S. A., 1975. Planktonic foraminifera from the Upper Cretaceous of southeastern Iraq: Biostratigraphy and systematics of the Heterohelicidae. *Micropaleontology,* 21, 185-214.

Darwin, C., 1859. *On the Origin of Species by Means of Natural Selection, or the Preservation of Favoured Races in the Struggle for Life.* London: John Murray, 502 p.

Djafarov, D. I., Agalarova, D. A., Khalilov, D. M., 195_. *Dictionary of microfauna of the Cretaceous deposits of Azerbaijan.* Baku: Gosudarstvenoe Nauchno-technicheskoe Izdatelstvo Neftianoi i Gorno-toplivnoi Lineraturyi Azerbaijanskoe Otdelenie, 128 p. [in Russian].

Douglas, R. G., 1969. Upper Cretaceous Planktonic Foraminifera in Northern California. Part 1-Systematics. *Micropaleontology,* 15, 151-209.

Dowsett, H. J. 1989. Documentation of the Santonian-Campanian and Austinian-Tayloran Stage Boundaries in Mississippi and Alabama Using Calcareous Microfossils. *United States Geological Survey Bulletin,* 1884, 1-19.

Egger, J. G., 1899. Foraminiferen und Ostrakoden aus den Kreidemergeln der Oberbayerischen Alpen. *Abhandlungen der Mathematisch-Physikalischen Klasse der Königlich Bayerischen Akademie der Wissenschaften*, 21, 3-230 [published in 1902].

Ehrenberg, C. G., 1838. Über die Bildung der Kreidefelsen und des Kreidemergels durch unsichtbare Organismen. *Abhandlungen der Königlichen Akademie der Wissenschaften zu Berlin*, 1838, 59-147 [published in 1839].

Ehrenberg, C. G., 1841. Verbreitung und Einflufs des mikroscopischen Lebens in Süd- und Nord- Amerika. *Abhandlungen der Königlichen Akademie der Wissenschaften zu Berlin*, 1841, 291-445 [published in 1843].

Ehrenberg, C. G., 1844. Eine Mittbeilung über 2 neue Lager von Gebirgsmassen aus Infusorien als Meeres–Absatz in Nord–Amerika und eine Vergleichung derselben mit den organischen Kreide–Gebilden in Europa und Afrika. *Bericht über die zur Bekanntmachung geeigneten Verhandlungen der Königlich Preußischen Akademie der Wissenschaften zu Berlin*, 1844, 57-98.

Ehrenberg, C. G., 1854. *Mikrogeologie*. Leipzig: L. Voss, 374 p.

Ehrenberg, C. G., 1855. Über den Grünsand und seine Erläuterung des organischen Lebens. *Abhandlungen der Königlichen Preußischen Akademie der Wissenschaften zu Berlin*, 1855: 85-176 [published in 1856].

Eicher, D. L., Worstell, P., 1970a. *Lunatriella*, a Cretaceous heterohelicid foraminifer from the western interior of the United States. *Micropaleontology*, 16, 117-121.

Eicher, D. L., Worstell, P., 1970b. Cenomanian and Turonian foraminifera from the Great Plains, United States. *Micropaleontology*, 16, 269-324.

Esker, G. C., 1968. A new species of *Pseudoguembelina* from the Upper Cretaceous of Texas. *Contributions from the Cushman Foundation for Foraminiferal Research*, 19, 168-169.

Finlay, H. J., 1939. New Zealand Foraminifera: key species in stratigraphy, No. 2. *Transactions of the Royal Society of New Zealand*, 69, 309-329.

Fleisher, R. S., 1974. Cenozoic planktonic foraminifera and biostratigraphy, Arabian Sea, Deep Sea Drilling Project, Leg 23A. In: *Initial Reports of the Deep Sea Drilling Project*, Volume 23 (Whitmarsh, R. B., Weser, O. E. and others, Eds.). Washington, DC: United States Government Printing Office, 1001-1071.

Frerichs, W. E., 1971. Evolution of planktonic foraminifera and paleotemperatures. *Journal of Paleontology*, 45, 963-968.

Frerichs, W. E., 1979. Planktonic foraminifera from the Sage Breaks Shale, Centennial Valley, Wyoming. *Journal of Foraminiferal Research*, 9, 159-184.

Frerichs, W. E., Dring, N. B., 1981. Planktonic Foraminifera from the Smoky Hill Shale of West Central Kansas. *Journal of Foraminiferal Research*, 11, 47-69.

Frerichs, W. E., Gaskill, C. H., 1978. *Textilaria americana* Ehrenberg: type species of *Heterohelix*. *Journal of Foraminiferal Research*, 8, 143-146.

Gawor-Biedowa, E., 1992. Campanian and Maastrichtian foraminifera from the Lublin Upland, Eastern Poland. *Palaeontologica Polonica*, 52, 1-187.

Geodakchan, A. A., Aliyulla, K., 1959. Representatives of the genus *Gumbelina* in the Upper Cretaceous strata of Azerbaijan. *Elmi Eserler Ucenie Zapisi, Geologo-geograficeskaia seria, Azerbaijanskovo Gosiudarstovo Universiteta*, 4, 51-62 [In Russian].

Georgescu, M. D., 1995. Upper Cretaceous Heterohelicidae in the Romanian Western Black Sea offshore. *Revista Española de Micropaleontología*, 27, 91-106.

Georgescu, M. D., 1997. Upper Jurassic-Cretaceous planktonic biofacies succession and the evolution of the western Black Sea Basin. In: *Regional and petroleum geology of the*

Black Sea and surrounding region (A. G. Robinson, Ed.). *The American Association of Petroleum Geologists Memoir*, 68, 169-182.

Georgescu, M. D., 2000. Late Albian-Turonian planktonic foraminifera in the Romanian western Black Sea offshore. *Revista Española de Micropaleontología*, 32, 157-173.

Georgescu, M. D., 2006. Santonian-Campanian planktonic foraminifera in the New Jersey coastal plain and their distribution related to the relative sea-level changes. *Canadian Journal of Earth Sciences*, 43, 101-120.

Georgescu, M. D., 2007a. A new planktonic heterohelicid foraminiferal genus from the Upper Cretaceous (Turonian). *Micropaleontology*, 53, 212-220.

Georgescu, M. D., 2007b. Taxonomic re-evaluation of the late Cretaceous serial planktonic foraminifer *Gümbelina punctulata* Cushman, 1938 and relates species. *Revista Española de Micropaleontología*, 39, 155-167.

Georgescu, M. D., 2008a. A new planktonic foraminifer (family Hedbergellidae Loeblich and Tappan, 1961) from the lower Campanian sediments of the Falkland Plateau, South Atlantic Ocean (DSDP Site 511). *Journal of Foraminiferal Research*, 38, 157-161.

Georgescu, M. D., 2008b. A new planktonic foraminiferal taxon of the Family Hedbergellidae Loeblich and Tappan 1961 from the Upper Cretaceous (Upper Turonian-Coniacian) of the Caribbean region. *Israel Journal of Earth Sciences*, 57, 55-63.

Georgescu, M. D., 2009a. Taxonomic revision and evolutionary classification of the biserial Cretaceous planktic foraminiferal genus *Laeviheterohelix* Nederbragt, 1991. *Revista Mexicana de Ciencias Geológicas*, 26, 315-334.

Georgescu, M. D., 2009b. On the origins of Superfamily Heterohelicacea Cushman, 1927 and the polyphyletic nature of plantic foraminifera. *Revista Española de Micropaleontología*, 41, 107-144.

Georgescu, M. D., 2009c. Upper Albian-lower Turonian non-schackoinid planktic foraminifera with elongate chambers: morphology reevaluation, taxonomy and evolutionary classification. *Revista Española de Micropaleontología*, 41, 255-293.

Georgescu, M. D., 2010. Origin, taxonomic revision and evolutionary classification of the late Coniacian-early Campanian (Late Cretaceous) planktic foraminifera with multichamber growth in the adult stage. *Revista Española de Micropaleontología*, 42, 59-118.

Georgescu, M. D., 2011a. A new type of test wall in the Late Cretaceous (Late Santonian-Campanian) heterohelicid planktic foraminifera. *Revue de Micropaléontologie*, 54, 105-114.

Georgescu, M. D., 2011b. Iterative evolution, taxonomic revision and evolutionary classification of the praeglobotruncanid planktic foraminifera, Cretaceous (late Albian-Santonian). *Revista Española de Micropaleontología*, 43, 173-207 [published in 2012].

Georgescu, M. D., 2011c. New data on the evolutionary classification of the Late Cretaceous (late Coniacian–Santonian) planktic foraminifera with elongate chambers. *Revista Española de Micropaleontología*, 43, 39-54.

Georgescu, M. D., 2012a. Restudy of the type specimens of *Phanerostomum asperum* Ehrenberg 1854 (Late Cretaceous, Foraminiferida). *Micropaleontology*, 58, 291-303.

Georgescu, M. D., 2012b. Morphology, taxonomy, stratigraphic distribution and evolutionary classification of the schackoinid planktic foraminifera (late Albian-Maastrichtian, Cretaceous). In: *Deep-Sea Marine Biology, Geology, and Human Impact* (Bailey, D. R. and S. E. Howard, Eds.). New York: Nova Publishers, 1-62.

Georgescu, M. D., 2013a. Revised evolutionary systematics of the Cretaceous planktic foraminifera described by C. G. Ehrenberg. *Micropaleontology*, 59, 1-49.

Georgescu, M. D., 2013b. Cretaceous planktic foraminifera from the Jacob Whitman Bailey Collection (Farlow Herbarium, Harvard University). In: *Foraminifera. Aspects of Classification, Stratigraphy, Ecology and Evolution* (Georgescu, M. D., Ed.). New York: Nova Science Publishers, 101-118.

Georgescu, M. D., 2013c. New advances in understanding the heterohelicid planktic foraminifer early evolution. *Studia Universitatis Babeş-Bolyai*, 58, 19-28.

Georgescu, M. D., 2014a. Evolution of central perforate plate in the new condensed lineage *Eicheriella*. In: *Evolutionary Classification and English-based Nomenclature in Cretaceous Planktic Foraminifera* (Georgescu, M. D., Ed.). New York: Nova Science Publishers [this volume].

Georgescu, M. D., 2014b. New Late Cretaceous (Santonian-Maastrichtian) heterohelicid planktic foraminifera from the Pacific and Indian Oceans and their biostratigraphic and evolutionary significance. In: *Evolutionary Classification and English-based Nomenclature in Cretaceous Planktic Foraminifera* (Georgescu, M. D. and Henderson, C.M., Eds). New York: Nova Science Publishers [this volume].

Georgescu, M. D., 2014c. Reinstatement of the Cretaceous planktic foraminifer *Bronnimannella* Montanaro Gallitelli 1956 as directional lineage in evolutionary classification. In: *Evolutionary Classification and English-based Nomenclature in Cretaceous Planktic Foraminifera* (Georgescu, M. D. and Henderson, C.M., Eds). New York: Nova Science Publishers [this volume].

Georgescu, M. D., 2014d. Taxonomic revision of *Planoglobulina* Cushman 1927 as directional lineage in evolutionary classification. In: *Evolutionary Classification and English-based Nomenclature in Cretaceous Planktic Foraminifera* (Georgescu, M. D. and Henderson, C.M., Eds). New York: Nova Science Publishers [this volume].

Georgescu, M. D., 2014e. Evolution and evolutionary classification of the late Campanian-Maastrichtian planktic foraminifera that evolved multiplane chamber proliferation (*Pseudotextularia* and *Racemiguembelina*). In: *Evolutionary Classification and English-based Nomenclature in Cretaceous Planktic Foraminifera* (Georgescu, M. D. and Henderson, C.M., Eds). New York: Nova Science Publishers [this volume].

Georgescu, M. D., Abramovich, S., 2008a. Taxonomic revision and phylogenetic classification of the Late Cretaceous (Upper Santonian-Maastrichtian) serial planktonic foraminifera (Family Heterohelicidae Cushman, 1927) with peripheral test wall flexure. *Revista Española de Micropaleontología*, 40, 97-114.

Georgescu, M. D., Abramovich, S. 2008b. A new serial Cretaceous planktic foraminifer (Family Heterohelicidae Cushman, 1927) from the Upper Maastrichtian of the equatorial Central Pacific. *Journal of Micropaleontology*, 27, 117-123.

Georgescu, M. D., Abramovich, S., 2009a. A new Late Cretaceous (Maastrichtian) serial planktonic foraminifera (Family Heterohelicidae) with early planispiral coil and revision of *Spiroplecta* Ehrenberg, 1844. *Geobios*, 42, 687-698.

Georgescu, M. D., Abramovich, S., 2009b. Short nomenclature note: A new name for the Upper Cretaceous planktic foraminiferal genus *Hendersonia* Georgescu and Abramovich, 2008. *Revista Española de Micropaleontología*, 41, 215.

Georgescu, M. D., Almogi-Labin, A., 2008. New data to support the phylogenetic relationship between the serial planktonic foraminifera (Family Heterohelicidae CUSHMAN, 1927) and some large-sized benthic foraminifera (Family Orbitoididae SCHWAGER, 1876) of the Late Cretaceous. *Revue de Paléobiologie*, 27, 15-24.

Georgescu, M. D., Arz, J. A., Macauley, R. V., Kukulski, R. B., Arenilas, I., and Pérez-Rodriguez, I., 2011. Late Cretaceous (Santonian-Maastrichtian) serial foraminifera with pore mounds or pore mound–based ornamentation structures. *Revista Española de Micropaleontología*, 43, 109-139.

Georgescu, M. D., Burke, R. M., Heikkinen, C. J., 2014. New data and insights on the polyphyletic origin of the Cretaceous planktic foraminifera. In: *Evolutionary Classification and English-based Nomenclature in Cretaceous Planktic Foraminifera* (Georgescu, M. D. and Henderson, C.M., Eds). New York: Nova Science Publishers [this volume].

Georgescu, M. D., Carrigy, C., 2012. Evolutionary classification of the coiled Upper Cretaceous (Turonian-Lower Campanian) planktic foraminifera with simple-ridged test wall. *Revista Española de Micropaleontología*, 44, 79-98.

Georgescu, M. D., Huber, B. T., 2006. *Paracostellagerina* nov.gen., a meridionally costellate planktonic foraminiferal genus of the Middle Cretaceous (latest Albian-earliest Cenomanian). *Journal of Foraminiferal Research*, 36, 368-373.

Georgescu, M. D., Huber, B. T., 2007. Taxonomic revision of the late Campanian-Maastrichtian (Late Cretaceous) planktic foraminiferal genus *Rugotruncana* Brönnimann and Brown, 1956, and a new paleontological species concept for planktic foraminifera. *Journal of Foraminiferal Research*, 37, 150-159.

Georgescu, M. D., Huber, B. T., 2009. Early evolution of the Cretaceous serial planktic foraminifera (late Albian-Cenomanian). *Journal of Foraminiferal Research*, 39, 335-360.

Georgescu, M. D., Quinney, A. E., Anderson, K. D., 2011. New data on the taxonomy, evolution and biostratigraphical significance of the Turonian-Coniacian (Late Cretaceous) planktic foraminifer *Huberella* Georgescu 2007. *Micropaleontology*, 57, 247-254.

Georgescu, M. D., Saupe, E. E., Huber, B. T., 2008. Morphometric and stratophenetic basis for phylogeny and taxonomy in Late Cretaceous gublerinid planktonic foraminifera. *Micropaleontology*, 54, 397-424 [published in 2009].

Georgescu, M. D., Sawyer, M. S., Heikkinen, C. J., Burke, R. M., 2013. New and revised Cretaceous (Albian-Campanian) planktic foraminifera of the Atlantic, Indian and Pacific Oceans. In: *Foraminifera. Aspects of Classification, Stratigraphy, Ecology and Evolution* (Georgescu, M. D., Ed.). New York: Nova Science Publishers, 59-100.

Govindan, A., 1972. Upper Cretaceous planktonic foraminifera from the Ponicherry area, south India. *Micropaleontology*, 19, 160-193.

Gradstein, F. M., Ogg, J. G., Schmitz, M. D., Ogg, G. M. (Eds.), 2012. *The Geologic Time Scale 2012*. Amsterdam and eleven others: Elsevier, 1144 p.

Halkyard, E., 1917. The fossil foraminifera of the Blue Marl of Cote des Basques, Biarritz. *Memoirs and Proceedings of the Manchester Literary and Philosophical Society*, 62/6, 1-145 [published in 1918].

Hattin, D. E., 1962. Stratigraphy of the Carlile Shale (Upper Cretaceous) in Kansas. *Bulletin of the State Geological Survey of Kansas*, 156, 5-155.

Hay, W. H., DeConto, R. M., Wold, C. N., Wilson, K. M., Voigt, S., Schulz, M., Wold, A. R., Dullo, W, Ronov, A. B., Balukhovsky, A. N., Söding, E., 1999. Alternative global Cretaceous paleogeography. In: *Evolution of the Cretaceous Ocean-Climate System* (E. Barrera and C. C. Johnson, Eds.). *The Geological Society of America Special Publication*, 332, 1-47.

Hinte, J. E. van, 1963. Zur Stratigraphie und Mikropaläontologie der Oberkreide und des Eozäns des Krappfeldes (Kränten). *Jahrbuch der Geologischen Bundesanstalt*, 8, 1-147.

Hinte, J. E. van, 1965. An approach to *Orbitoides*. *Proceedings of the Koninklijke Nederlandse Akademie van Wetenschappen*, series B, 68, 57-71.

Huber, B. T., 1990. Maestrichtian planktonic foraminifer biostratigraphy of the Maud Rise (Weddell Sea, Antarctica): ODP Leg 113 Holes 689B and 690C. In: *Proceedings of the Ocean Drilling Program, Scientific Results,* Volume 113 (Barker, P. F., Kennett, J. P. and others, Eds.). College Station: Ocean Drilling Program, 489-513.

Huber, B. T., 1992. Upper Cretaceous planktic foraminiferal biozonation for the Austral Realm. *Marine Micropaleontology*, 20, 107-128.

Huber, B. T., Hodell, D. A., Hamilton, C. P., 1995. Middle-Late Cretaceous climate of the southern high latitudes: Stable isotopic evidence for minimal equator-to-pole thermal gradients. *Geological Society of America Bulletin*, 107, 1164-1191.

Jones, T. R., 1895. The Cretaceous Series in the Upper Missouri; and the Chalk of North America and its foraminifera. *Geological Magazine*, 2, 425-429.

Kavary, E., Frizzell, D. L., 1963. Upper Cretaceous and Lower Cenozoic oraminifera from west central Iran. *University of Missouri School of Mines and Metallurgy Bulletin*, 102, 1-89.

Kikoïne, J., 1948. Les Heterohelicidae du Crétacé supérieur pyrénéen. *Bulletin de la Société Géologique de France*, 18, 15-35.

Klasz, I., de, 1953. Einige neue oder wenig bekannte Foraminiferen aus der helvetischen Oberkreide der bayerischen Alpen südlich Traunstein (Oberbayern). *Geologica Bavarica*, 17, 223-240.

Klasz, I. de, Calvez, Y. Le, Rerat, D., 1969. Noveaux foraminifères du basin sédimenataire du Gabon (Afrique Équatoriale). *Third African Micropaleontological Colloquium,* Cairo, March 4-10 1968, 269-287.

Klasz, I. de, Klasz, S. de, Saint-Marc, P., 1995. Heterohelicids from the Turonian of Senegal (West Africa) with particular emphasis on *Heterohelix americana. Micropaleontology*, 41, 359-368.

Krasheninnikov, V. A., Basov, I. A., 1983. Stratigraphy of Cretaceous sediments of the Falkland Plateau based on planktonic foraminifers, Deep Sea Drilling Project, Leg 71. In: *Initial Reports of the Deep Sea Drilling Project,* Volume 71 (Ludwig, W. J., Krasheninnikov, V. A. and others, Eds.). Washington, DC: United States Government Printing Office, 789-820.

Küpper, K., 1954. Notes of the Upper Cretaceous larger Foraminifera. II. Genera of the Subfamily Orbitoidinae with remarks on the microspheric generation of *Orbitoides* and *Omphalocyclus. Contributions from the Cushman Foundation for Foraminiferal Research*, 5, 179-184.

Lipps, J. H., 1966. Wall structure, systematics and phylogeny of Cenozoic planktonic foraminifera. *Journal of Paleontology*, 40, 1257-1263.

Loeblich, A. R., Jr., 1951. Coiling in the Heterohelicidae. *Contributions from the Cushman Foundation for Foraminiferal Research*, 2, 106-110.

Loeblich, A. R. Jr., Tappan, H., 1987. *Foraminiferal Genera and Their Classification.* New York: Van Nostrand Reinhold Company, 970 p.

Loetterle, G. J., 1937. The micropaleontology of the Niobrara Formation in Kansas, Nebraska, and South Dakota. *Nebraska Geological Survey Bulletin*, 12, 1-73.

Marie, P., 1941. Les foraminifères de la craie a *Belemnitella mucronata* du Bassin de Paris. *Mémoires du Muséum National d'Histoire Naturelle*, 12, 1-296.

Martin, S. E., 1972. Reexamination of the Upper Cretaceous planktonic foraminiferal genera *Planoglobulina* Cushman and *Ventilabrella* Cushman. *Journal of Foraminiferal Research*, 2, 73-92.

Masella, L., 1959. Una nueva specie di *Heterohelix* del Cretaceo della Sicilia. *Rivista Mineraria Siciliana*, 55, 15-17.

Masters, B. A., 1976. Planktic foraminifera from the Upper Cretaceous Selma Group, Alabama. *Journal of Paleontology*, 50, 318-330.

Masters, B. A., 1977. Mesozoic planktonic foraminifera. A world–wide review and analysis. In: *Oceanic Micropaleontology* (A. T. S. Ramsay, Ed.). London-New York-San Francisco: Academic Press, 1, 301-731.

Masters, B. A., 1980. Reevaluation of selected types of Ehrenberg's Cretaceous planktonic foraminifera. *Eclogae Geologicae Helvetiae*, 73, 95-107.

Mayr, E., 1968. Theory of biological classification. *Nature*, 220, 545-548.

Mayr, E., 1974. Cladistic analysis or cladistics classification? *Zeitschrift fur Zoologische Systematik und Evolutionsforschung*, 12, 94-128.

Mayr, E., 1981. Biological classification: toward a synthesis of opposing methodologies. *Science*, 214, 510-516.

Mayr, E., Ashlock, P. D., 1991. *Principles of Systematic Zoology*. New York: McGraw-Hill, 475 p.

McGowran, B., 1968. Reclassification of early Tertiary *Globorotalia*. *Micropaleontology*, 14, 61-80.

McGowran, B., 1971. On foraminiferal taxonomy. *Proceedings of the II Planktonic Conference Roma 1970* (A. Farinacci, Ed.). Roma: Edizioni Tecnoscienza, 813-820.

McNeely, B. W., 1973. Biostratigraphy of the Mesozoic and Paleogene pelagic sediments of the Campeche Embankment area. In: *Initial Reports of the Deep Sea Drilling Project*, Volume 10 (Worzel, J. L., Bryant, W. and others, Eds.). Washington, DC: United States Printing Office, 679-695.

McNulty, C. L., 1979. Smaller Cretaceous foraminifers of Leg 43, Deep Sea Drilling Project. In: *Initial Reports of the Deep Sea Drilling Project*, Volume 43 (Tucholke, B. E., Vogt, P. R. and others, Eds.). Washington, DC: United States Government Printing Office, 487-505.

Miller, K. G. and nineteen others. 1998. Bass River Site. In: *Proceedings of the Ocean Drilling Program, Initial Reports*, Volume 174AX (Miller, K. G., Sugarmann, P. G., Browning, J. V., and others). College Station: Ocean Drilling Program, 5-43.

Montanaro Gallitelli, E., 1956. *Bronnimannella*, *Tappanina* and *Trachelinella*, three new foraminiferal genera from the Upper Cretaceous. *Contributions from the Cushman Foundation for Foraminiferal Research*, 7, 35-39.

Montanaro Gallitelli, E., 1957. A revision of the foraminiferal family Heterohelicidae. In: *Studies in foraminifera* (A. R. Jr. Loeblich, Ed.). Washington, DC: *United States National Museum History Bulletin*, 215, 133-154.

Montanaro Gallitelli, E., 1958. Specie nuove e note di foraminiferi del Cretaceo superior di Serramazzoni (Modena). *Academia di Scienze Lettere e Arti di Modena, Atti e Memorie*, 16, 127-150.

Nakkady, S. E., 1950. A new foraminiferal fauna from the Esna Shales and Upper Cretaceous chalk of Egypt. *Journal of Paleontology*, 24, 675-692.

Nederbragt, A. J., 1989a. Chamber proliferation in the Cretaceous planktonic foraminifera Heterohelicidae. *Journal of Foraminiferal Research*, 19, 105-114.

Nederbragt, A. J., 1989b. Maastrichtian Heterohelicidae (planktic foraminifera) from the West North Atlantic. *Journal of Micropaleontology*, 8, 183-206.

Nederbragt, A. J., 1991. Late Cretaceous biostratigraphy and development of Heterohelicidae (planktic foraminifera). *Micropaleontology*, 37, 329-372.

Nelson, G., 1974. Darwin-Hennig classification: a reply to Ernst Mayr. *Systematic Zoology*, 23, 452-458.

Neumann, M., 1972. A propos des Orbitoïdés du Crétacé supérieur et de leur signification stratigraphique. I-Genre *Orbitoides* d'Orbigny (1847). *Revue de Micropaléontologie*, 14, 197-226.

Oláníví Odébòdé, M. 1982. Senonian Heterohelicidae from the Calabar Flank, southeastern Nigeria. *Revista Española de Micropaleontología*, 14, 231-246.

Olsson, R. K., Hemleben, C., Berggren, W. A., Huber, B. T., 1999. Atlas of Paleocene Foraminifera. *Smithsonian Contributions to Paleobiology*, 85, 1-252.

Padian, K., 1999. Charles Darwin's views of classification in theory and practice. *Systematic Biology*, 48, 352-364.

Pandey, J., 1980. Cretaceous foraminifera of Um Sohryngkew River section, Meghalaya. *Journal of the Palaeontological Society of India*, 25, 53-74 [published in 1981].

Parker, F. L., 1962. Planktonic foraminiferal species in Pacific sediments. *Micropaleontology*, 8, 219-254.

Parker, F. L., 1967. Late Tertiary biostratigraphy (planktonic foraminifera) of tropical Indo-Pacific deep-sea cores. *Bulletins of American Paleontology*, 52(235), 115-187.

Parker, W. K., Jones, T. R., 1872. On the nomenclature of foraminifera. The species figured by Ehrenberg. *The Annals and Magazine of Natural History including Zoology, Botany, and Geology*, 10, 184-200.

Pearson, P. N., 1993. A lineage phylogeny for the Paleogene planktonic foraminifera. *Micropaleontology*, 39, 193-232.

Pessagno, E. A. Jr., 1967. Upper Cretaceous planktonic foraminifera from the Western Gulf coastal plain. *Palaeontographica Americana*, 5(37), 243-445.

Pessagno, E. A. Jr., Longoria, J. F. T., 1973. Mesozoic foraminifera, Leg 15, Deep Sea Drilling Project. In: *Initial Reports of the Deep Sea Drilling Project,* Volume 15 (Worzel, J. L., Bryant, W. and others, Eds.). Washington, DC: United States Government Printing Office, 549-552.

Petters, S. W., 1983. Gulf of Guinea planktonic foraminiferal biochronology and geological history of the South Atlantic. *Journal of Foraminiferal Research*, 13, 32-59.

Petters, S. W., El-Nakhal, H. A., Cifelli, R., 1983. *Costellagerina*, a new Late Cretaceous globigerine foraminiferal genus. *Journal of Foraminiferal Research*, 13, 247-251.

Pflaumann, U., Krasheninnikov, V. A., 1978. Early Cretaceous planktic foraminifers from Eastern North Atlantic, DSDP Leg 41. In: *Initial Reports of the Deep Sea Drilling Project,* Volume 41 (Lancelot, Y., Seibold, E. and others, Eds.). Washington, DC: United States Government Printing Office, 539-564.

Plummer, H. J., 1931. Some Cretaceous foraminifera in Texas. *University of Texas Bureau of Economic Geology and Technology Bulletin*, 3101, 109-203.

Premoli Silva, I., Boersma, A., 1977. Cretaceous planktonic foraminifers-DSDP Leg 39 (South Atlantic). In: *Initial Reports of the Deep Sea Drilling Project,* Volume 39 (Supko,

P. R., Perch-Nielsen, K. and others, Eds.). Washington, DC: United States Government Printing Office, 615-641.

Premoli Silva, I., Bolli, H. M., 1973. Late Cretaceous to Eocene planktonic foraminifera and stratigraphy of Leg 15 sites in the Caribbean Sea. In: *Initial Reports of the Deep Sea Drilling Project,* Volume 15 (Edgar, N. T., Saunders, J. B. and others, Eds.). Washington, DC: United States Government Printing Office, 499-547.

Reiss, Z., 1957. Notes on foraminifera from Israel. *Sigalia* - a new genus of foraminifera. *Bulletin of the Research Council of Israel,* 6b, 239-244.

Reuss, A. E., 1845. *Die Versteinrungen der Böhmischen Kreideformation.* Stuttgart: E. Schweizerbart, 58 p.

Reuss, A. E., 1854. Beiträge zur Charakteristik der Kreideschichten in den Ostalpen, besonders im Gosauthale und am Wolfgangsee. *Denkschriften der Kaiserlichen Akademie der Wissenschaften, Mathematisch-Naturwissenschaftliche Classe,* 7, 1-156.

Reuss, A. E., 1860. Die Foraminiferen der Westphälischen Kreideformation. *Sitzungsberichte der Mathematisch–Naturwissenschaftlichen Classe der Kaiserlichen Akademie der Wissenschaften,* 40, 147-238.

Rzehak, A., 1891. Die Foraminiferenfauna der alttertiären Ablagerungen von Bruderndorf in Nieder-Osterreich, mit Berüchsichtigung des angeblichen Kreidevorkommens von Leitzersdorf. *Annalen des K.K. Naturhistorischen Hofmuseums,* 10, 213-230.

Rzehak, A., 1895. Ueber einige merkwürdige Foraminiferen aus dem österreichischen Tertiär. *Annalen des K.K. Naturhistorischen Hofmuseums,* 6, 1-12.

Salaj, J., 1983. Quelques problèmes taxinomiques concernant les foraminifères planctiques et la zonation du Sénonien supérieur d'El Kef. *Geologický Zborník,* 34, 187-212.

Salaj, J. and Samuel, O. 1963. Mikrobiostratigrafia srednej a vrchnej Kreidy z východnej Časti Bradloveho Pasma. *Geologicke Prace,* 30, 93-112.

Sandidge, J. R., 1932. Significant foraminifera from the Ripley Formation of Alabama. *American Midland Naturalist,* 13, 190-202.

Schreiber, O. S., 1979. Heterohelicidae (Foraminifera) aus der Pemberger-Folge (Oberkreide) von Klein-Sankt Paul am Krappfeld (Kärnten). *Beiträge zur Paläontologie Österreich,* 6, 27-59.

Scott, J. C., Watson, H. M., Copeland, C. W., Cepek, P., Hay, W. W., Masters, B. A., Worsley, T. R., 1968. *Facies changes in the Selma Group in Central and Eastern Alabama. A guidebook for the Sixth Annual Field Trip of the Alabama Geological Society,* December 6-7, 1968. Tuscaloosa: Alabama Geological Society, 69 p.

Seiglie, G. A., 1958. Notas sobre algunas especies de Heterohelicidae del Cretacico superior de Cuba. *Boletín de la Asociatión Mexicana de Geólogos Petroleros,* 11, 51-62.

Seiglie, G. A., 1960. Una nueva especies de Heterohelicidae del Cretacico superior de Cuba. *Memorias de la Sociedad Cubana de Historia Natural,* 24, 121-124.

Shipboard Scientific Party, 1972. Site 111. In: *Initial Reports of the Deep Sea Drilling Project,* Volume 12 (Laughton, A. E. and others, Eds.). Washington, DC: United States Government Printing Office, 33-159.

Shipboard Scientific Party, 1973. Site 95. In: *Initial Reports of the Deep Sea Drilling Project,* Volume 10 (Worzel, J. L., Bryant, W. and others, Eds.). Washington, DC: United States Government Printing Office, 259-295.

Shipboard Scientific Party, 1973. Site 150. In: *Initial Reports of the Deep Sea Drilling Project,* Volume 15 (Edgar, N. T., Saunders, J. B. and others, Eds.). Washington, DC: United States Government Printing Office, 277-299.

Shipboard Scientific Party, 1975. Site 305: Shatsky Rise. In: *Initial Reports of the Deep Sea Drilling Project,* Volume 32 (Larson, R. L., Moberly, R. and others, Eds.). Washington, DC: United States Government Printing Office, 75-158.

Shipboard Scientific Party, 1977a. Site 356: São Paulo Plateau. In: *Initial Reports of the Deep Sea Drilling Project,* Volume 39 (Supko, P. R., Perch-Nielsen, K. and others, Eds.). Washington, DC: United States Government Printing Office, 141-230.

Shipboard Scientific Party, 1977b. Site 357: Rio Grande Rise. In: *Initial Reports of the Deep Sea Drilling Project,* Volume 39 (Supko, P. R., Perch-Nielsen, K. and others, Eds.). Washington, DC: United States Government Printing Office, 231-327.

Shipboard Scientific Party, 1978. Site 370: Deep Basin off Morocco. In: *Initial Reports of the Deep Sea Drilling Project,* Volume 41 (Lancelot, Y., Seibold, E. and others, Eds.). Washington, DC: United States Government Printing Office, 421-491.

Shipboard Scientific Party, 1979a. Site 384: The Cretaceous/Tertiary Boundary, Aptian Reefs, and the J-Anomaly Ridge. In: *Initial Reports of the Deep Sea Drilling Project,* Volume 43 (Tucholke, B. E., Vogt, P. R. and others, Eds.). Washington, DC: United States Government Printing Office, 107-154.

Shipboard Scientific Party, 1979b. Site 398. In: *Initial Reports of the Deep Sea Drilling Project,* Volume 47-2 (Sibuet, J.-C., Ryan, W. B. F. and others, Eds.). Washington, DC: United States Government Printing Office, 25-233.

Shipboard Scientific Party, 1981. Site 463: western Mid-Pacific Mountains. In: *Initial Reports of the Deep Sea Drilling Project,* Volume 62 (Thiede, J., Vallier, T. L. and others, Eds.). Washington, DC: United States Government Printing Office, 33-156.

Shipboard Scientific Party, 1983. Site 511. In: *Initial Reports of the Deep Sea Drilling Project,* Volume 62 (Ludwig, W. J., Krasheninnikov, V. A. and others, Eds.). Washington, DC: United States Government Printing Office, 21-109.

Shipboard Scientific Party, 1988. Site 690. In: *Proceedings of the Ocean Drilling Program, Initial Reports,* Volume 113 (Barker, P. R., Kennett, J. P. and others, Eds.). College Station: Ocean Drilling Program, 183-292.

Shipboard Scientific Party, 1990. Site 761. In: *Proceedings of the Ocean Drilling Program, Initial Reports,* Volume 122 (Haq, B. U., von Rad, U. and others, Eds.). College Station: Ocean Drilling Program, 161-211.

Shipboard Scientific Party, 1990. Site 762. In: *Proceedings of the Ocean Drilling Program, Initial Reports,* Volume 122 (Haq, B. U., von Rad, U. and others, Eds.). College Station: Ocean Drilling Program, 213-288.

Shipboard Scientific Party, 1990. Site 763. In: *Proceedings of the Ocean Drilling Program, Initial Reports,* Volume 122 (Haq, B. U., von Rad, U. and others, Eds.). College Station: Ocean Drilling Program, 289-352.

Shipboard Scientific Party, 1998. Site 1050. In: *Proceedings of the Ocean Drilling Program, Initial Reports,* Volume 171B (Norris, R. D., Kroon, D., Klaus, A., and others, Eds.). College Station: Ocean Drilling Program, 93-169.

Shipboard Scientific Party, 1998. Site 1052. In: *Proceedings of the Ocean Drilling Program, Initial Reports,* Volume 171B (Norris, R. D., Kroon, D., Klaus, A., and others, Eds.). College Station: Ocean Drilling Program, 241-319.

Sigal, J., 1952. Apercu stratigraphique sur la micropaleontologie du Cretace. *Alger, 19th International Geological Congress, Monographies regionales, 1[re] ser.,* Algerie, 26, 1-52.

Sigal, J., 1979. Chronostratigraphy and ecostratigraphy of Cretaceous formations recovered on DSDP Leg 47B, Site 398. In: *Initial Reports of the Deep Sea Drilling Project,* Volume

47 (Laughter, F. H. and Fagerberg, E. M., Eds.). Washington, DC: United States Government Printing Office, 287-326.

Simpson, G. G., 1961. *Principles of Animal Taxonomy*. New York and London: Columbia University Press, 247 p.

Smith, C. C., Pessagno, E. A. Jr., 1973. Planktonic foraminifera and stratigraphy of the Corsicana Formation (Maestrichtian), north-central Texas. *Cushman Foundation for Foraminiferal Research, Special Publications*, 13, 5-68.

Steineck, P. L., 1971. Phylogenetic reclassification of Paleogene planktonic foraminifera. *Texas Journal of Science*, 23, 167-178.

Steineck, P. L., Fleisher, R. L., 1978. Towards the classical evolutionary reclassification of Cenozoic Globigerinacea (Foraminiferida). *Journal of Paleontology*, 52, 618-635.

Tappan, H., 1940. Foraminifera from the Grayson Formation of northern Texas. *Journal of Paleontology*, 14, 93-126.

Van der Sluis, J. P., 1950. Geology of East Ceram. *Geological, petrographical and palaeontological results of explorations, carried out from September 1917 till June 1919 in the Island of Ceram, Third series, Geology*, 3: 7-66

Vassilenko, V. P., 1961. Upper Cretaceous foraminifera of the Mangyshlak Peninsula (descriptions, phylogenetical schemes for some groups and stratigraphic analysis). *Trudy VNIGRI*, 171, 1-487.

Voorwijk, G. H., 1937. Foraminifera from the Upper Cretaceous of Habana, Cuba. *Proceedings of the Koninklijke Akademie van Wetenschappen te Amsterdam*, 40, 190-198.

Wan, X., 1985. Cretaceous strata and foraminifera of Gangba region, Xizang. *Contributions to the Geology of the Qinghai-Xizang (Tibet Plateau)*, 16, 203-228 [in Chinese].

Weiss, W., 1983. Heterohelicidae (seriale planktonische Foraminiferen) der tethyalen Oberkreide (Santon bis Maastricht). *Geologisches Jahrbuch*, A72, 3-93.

White, M. P., 1929. Some index foraminifera of the Tampico Embayment area of Mexico. Part III. *Journal of Paleontology*, 3, 30-57.

Wonders, A. A. H., 1992. Cretaceous planktonic foraminiferal biostratigraphy, Leg 122, Exmouth Plateau, Australia. In: *Proceedings of the Ocean Drilling Program, Scientific Results Leg 122* (Haq, B. U., von Rad, U. and others, Eds.). College Station: Ocean Drilling Program, 587-599.

APPENDIX

Units of uncertain systematic position, which probably evolved through different originations when compared to the study group.

Taxon	Original report	Comments	Origination
Guembelina spinifera	Cushman 1931, p. 43, pl. 7, figs 6-7.	The holotype of this species described from the Selma Chalk of Tenneessee was examinned in the Cushman Collection; chamber surface is smooth and the spines reported by Cusman in the original description are of diagenetic origin. It cannot be assigned at the present state of knowledge to the main plexus of the planktics with chambers alternately added with respect to the test growth axis.	Unknown
Rectogümbelina cretacea	Cushman 1932, p. 6, pl. 1, figs 11-12.	Taxon described originally from the Arkadelphia Marl of Arkansas and reviewed by Huber in Olsson and others (1999); the holotype was examinned in the Cushman Collection. The adult portion of this smooth taxon is uniserial; the early stage presents resemblances with *G. spinifera* Cusman 1931, but additional data are necessary to confirm or reject a possible evolutionary relationship between the two.	Unknown, possibly from *G. spinifera*
Zeauvigerina	Finlay 1939, p. 541.	Genus of late Maastrichtian-Eocene age with variable gross test architecture according to the revision by Huber in Olsson and others (1999). The ornamentation is variable ranging from smooth to finaley pustulose. Huber in Olsson and others (1999) considered the genus derived probably from *Laeviheterohelix*, a hypothesis that cannot be validated.	Unknown
Heterohelix olssoni	Georgescu 2000, p. 162, pl. 1, figs 1-2.	Globular-chambered taxon with pitted chamber surface unlike any other representative of the main plexus of the study group.	Unknown

Editors' Contact Information

Dr. M. Dan Georgescu,
University of Calgary, Department of Geosciences
2500 University Dr. NW, Calgary, Alberta T2N 1N4, Alberta Canada
Email: dgeorge@ucalgary.ca

Dr. Charles M. Henderson,
University of Calgary, Department of Geosciences
2500 University Dr. NW, Calgary, Alberta T2N 1N4, Alberta Canada
Tel: (403) 210-7759
Email: Charles.henderson@ucalgary.ca

INDEX

H

I

J

K

L

Q

R

S